AF595580

Feature Papers in Mathematical and Computational Applications

Feature Papers in Mathematical and Computational Applications

Editors

Gianluigi Rozza
Oliver Schütze
Nicholas Fantuzzi

MDPI • Basel • Beijing • Wuhan • Barcelona • Belgrade • Manchester • Tokyo • Cluj • Tianjin

Editors
Gianluigi Rozza
International School for Advanced Studies
Trieste
Italy

Oliver Schütze
CINVESTAV-IPN
Mexico City
Mexico

Nicholas Fantuzzi
University of Bologna
Bologna
Italy

Editorial Office
MDPI
St. Alban-Anlage 66
4052 Basel, Switzerland

This is a reprint of articles from the Topical Collection published online in the open access journal *Mathematical and Computational Applications* (ISSN 2297-8747) (available at: https://www.mdpi.com/journal/mca/topical_collections/Feature_Pap).

For citation purposes, cite each article independently as indicated on the article page online and as indicated below:

LastName, A.A.; LastName, B.B.; LastName, C.C. Article Title. *Journal Name* **Year**, *Volume Number*, Page Range.

ISBN 978-3-0365-6756-3 (Hbk)
ISBN 978-3-0365-6757-0 (PDF)

Contents

About the Editors

Gianluigi Rozza

Gianluigi Rozza is a Full Professor in Numerical Analysis and Scientific Computing at SISSA MathLab—International School for Advanced Studies, Trieste, Italy. He received his Master of Science in Aerospace Engineering (2002) at Politecnico di Milano and his Ph.D in Applied Mathematics (2005) at Ecole Polytechnique Fédérale de Lausanne, Switzerland. He was a Research Assistant (2002-06) and Researcher and Lecturer (2008-12) at École Polytechnique Fédérale de Lausanne; a Post-Doctoral Associate Researcher (2006-08) at Massachusetts Institute of Technology, Boston MA, USA; and a Researcher (2012-14) and Associate Tenured Professor (2014-17) at the International School for Advanced Studies, Trieste, Italy. His research interests include: Numerical Analysis; Numerical Simulation; Scientific Computing; Reduced Order Modeling and Methods (with a special focus on viscous flows and complex geometrical parametrizations); Efficient Reduced-Basis Methods for Parametrized PDEs and Posteriori Error Estimation; Computational Fluid Dynamics with applications in Aero-Naval-Mechanical Engineering and Environmental Fluid Dynamics; Fluid-Structure Interaction Problems; Parametrized Navier–Stokes Equations for Bifurcations and Stability of Flow; Optimal Control; Flow Control based on PDEs; Optimal Shape Design; Shape Optimization; Shape Reconstruction; Shape Registration; Uncertainty Quantification; Data Assimilation; Parameter Estimation; Machine Learning; Deep Learning; and Neural Networks. He has authored 180 scientific papers, receiving more than 4500 citations (H-index 32). He has received the Bill Morton CFD Prize (2004) from the Institute of Computational Fluid Dynamics, University of Oxford (UK), the ECCOMAS Ph.D Award (2005) from the European Community on Computational Methods in Applied Sciences, the Springer Computational Science and Engineering Prize (2009), the ECCOMAS Jacques Louis Lions Award in Computational Mathematics (2014), the ERC consolidator grant 'Advanced Reduced Order Methods with Applications in Computational Fluid Dynamics' (AROMA-CFD, 2016-2021), and the ERC Proof of Concept Grant 'Advanced Reduced Groupware Online Simulation' (ARGOS, 2022).

Oliver Schütze

Oliver Schütze is a Full Professor at the Cinvestav—IPN in Mexico City, Mexico. His main research interests are numerical and evolutionary optimization. He is the co-author of more than 150 publications, including 2 monographs, 5 school textbooks, and 10 edited books. Two of his papers have received the IEEE Transactions on Evolutionary Computation Outstanding Paper Award (in 2010 and 2012). He is the founder of the Numerical and Evolutionary Optimization (NEO) workshop series. He is Editor-in-Chief of the journal *Mathematical and Computational Applications*, and is a member of the Editorial Board of the journals *Engineering Optimization, Computational Optimization and Applications, IEEE Transactions on Evolutionary Computation, Research in Control and Optimization*, and *Applied Soft Computing*. He is a member of the Mexican Academy of Sciences and the National System of Researchers (SNI III).

Nicholas Fantuzzi

Nicholas Fantuzzi is an Associate Professor at the University of Bologna since 2021, where he teaches Advanced Structural Mechanics and Modeling of Offshore Structures at the Ravenna Campus. Here, he performs research on numerical modeling of advanced composites with innovative numerical methods. He has co-organized 14 International Conferences within the composite materials field and he has been invited to deliver seven keynote lectures at international events. He has received three international awards, and he has been a visiting professor in Croatia, China, and Hong Kong. He is a reviewer for more than 110 international journals and the author of more than 130 publications in international journals, nine books, and four book chapters, with more than 100 abstracts in national and international conferences.

Preface to "Feature Papers in Mathematical and Computational Applications"

This book comprises the first collection of papers submitted by the Editorial Board Members (EBMs) of the journal *Mathematical and Computational Applications* (MCA), as well as outstanding scholars working in the core research fields of MCA. Therefore, this collection typifies the most insightful and influential original articles that discuss key topics in these fields. More precisely, this book contains 11 chapters from 11 research articles published in MCA between January and November 2022. The papers are in the following shortly presented, organized chronologically by their publication times.

In Chapter 1, Monika Stipsitz and Helios Sanchis-Alepuz present a proof-of-concept study of the application of convolutional neural networks to accelerate thermal simulations. Hereby, the focus is on the thermal aspect of electronic systems to provide accurate approximations of full solutions to quickly select promising designs. To this end, a custom network architecture that captures the long-range correlations present in heat conduction problems is proposed and tested.

In Chapter 2, Roy M. Howard utilizes a spline-based integral approximation to define a sequence of approximations to the error function that converge at a significantly faster manner than the default Taylor series. Specifically, two generalizations are investigated, both leading to significantly improved accuracy.

In Chapter 3, Sorena Sarmadi et al. perform automated analysis to quantify the growth dynamics of a population of bacilliform bacteria. To this end, they propose an innovative approach to the frame-sequence tracking of deformable-cell motion by the automated minimization of a new specific cost function. Initial tests using experimental image sequences of E. coli colonies yield convincing results, with a registration accuracy ranging from 90% to 100%.

In Chapter 4, Guzel Khayretdinova et al. propose a method for semi-supervised image segmentation based on geometric active contours. The main novelty of the proposed method is the initialization of the segmentation process, which is performed with a polynomial approximation of a user-defined initialization. The method is compared with other segmentation algorithms, and experimental results are given related to several medical and geophysical applications.

In Chapter 5, Moriz A. Habigt et al. conduct a porcine animal model to parameterize and evaluate a computer simulation model. The results of an animal model on thirteen healthy pigs were used to generate consistent parameterization data for the full heart computer simulation model. Numerical results show that the simulation model used in this study was able to adapt to the high physiological variability in the animal model.

In Chapter 6, Harri Hakula deals with harmonic extension finite elements for the numerical solution of partial differential equations defined on complicated domains. It is shown that, in combination with simple replacement rule-based mesh generation, the performance of the method is equivalent to that of the standard p-version in problems where the boundary layers dominate the solution. The performance over a parameter range is demonstrated in an application of computational asymptotic analysis, where known estimates are recovered via computational means only.

In Chapter 7, Daniele Mortari proposes a least-squares-based numerical approach to estimate the boundary value geodesic trajectory and associated parametric velocity on curved surfaces. Numerical examples are provided for several two-dimensional quadrics for which the estimated geodesic solutions yield residuals at the machine-error level.

In Chapter 8, Lindomar Soares Dos Santos et al. derive analytical solutions of microplastic particles dispersion using a Lotka–Volterra predator–prey model with time-varying intraspecies coefficients. Based on this, they solve analytically particular situations of ecological interest, which are characterized by extreme effects on predatory performance, and propose a second-order differential equation as a possible next step to address this model.

In Chapter 9, Quoc Khanh Nguyen et al. present a spectral analysis of the coefficient matrices associated with the linear systems stemming from the finite-element discretization of a linearly elastic problem for an arbitrary coefficient field in three spatial dimensions. Their analysis is then used to design and study an optimal multigrid method in the sense that the (arithmetic) cost for solving the problem up to a fixed desired accuracy is linear in the corresponding matrix size.

In Chapter 10, Sohail A. Khan and Tasawar Hayat examine the impacts of Dufour and Soret in a radiative Darcy–Forchheimer flow. They investigate physical interpretations of the concentration, entropy rate, velocity, and temperature parameters, and compare their observations with previously published results, leading to an excellent agreement.

Finally, in Chapter 11, Gabriel Thomaz de Aquino Pereira et al. present a new material model for an agglomerated cork based solely on well-known hypotheses of continuum mechanics using fewer parameters than the classical model. Furthermore, a finite-element framework is used to validate the new model against experimental data. This work represents an important step toward the production of materials that are less polluting and harmful to the environment following the UN 2030 agenda for sustainable development.

Gianluigi Rozza, Oliver Schütze, and Nicholas Fantuzzi
Editors

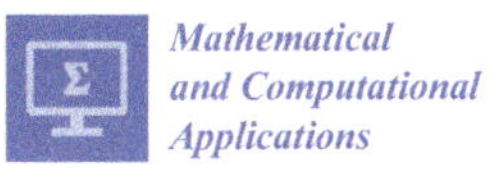

Mathematical and Computational Applications

Article

Approximating the Steady-State Temperature of 3D Electronic Systems with Convolutional Neural Networks

Monika Stipsitz and Hèlios Sanchis-Alepuz *

Silicon Austria Labs GmbH, Inffeldgasse 33, 8010 Graz, Austria; monika.stipsitz@silicon-austria.com
* Correspondence: helios.sanchis-alepuz@silicon-austria.com

Abstract: Thermal simulations are an important part of the design process in many engineering disciplines. In simulation-based design approaches, a considerable amount of time is spent by repeated simulations. An alternative, fast simulation tool would be a welcome addition to any automatized and simulation-based optimisation workflow. In this work, we present a proof-of-concept study of the application of convolutional neural networks to accelerate thermal simulations. We focus on the thermal aspect of electronic systems. The goal of such a tool is to provide accurate approximations of a full solution, in order to quickly select promising designs for more detailed investigations. Based on a training set of randomly generated circuits with corresponding finite element solutions, the full 3D steady-state temperature field is estimated using a fully convolutional neural network. A custom network architecture is proposed which captures the long-range correlations present in heat conduction problems. We test the network on a separate dataset and find that the mean relative error is around 2% and the typical evaluation time is 35 ms per sample (2 ms for evaluation, 33 ms for data transfer). The benefit of this neural-network-based approach is that, once training is completed, the network can be applied to any system within the design space spanned by the randomized training dataset (which includes different components, material properties, different positioning of components on a PCB, etc.).

Keywords: physics simulations; neural networks; electronic design; heat equation

Citation: Stipsitz, M.; Sanchis-Alepuz, H. Approximating the Steady-State Temperature of 3D Electronic Systems with Convolutional Neural Networks. *Math. Comput. Appl.* **2022**, *27*, 7. https://doi.org/10.3390/mca27010007

Academic Editor: David Ryckelynck

Received: 22 October 2021
Accepted: 14 January 2022
Published: 14 January 2022

1. Introduction

Physics simulations are becoming an essential aspect in the design of electronic systems. For an optimally designed electronic system, the interplay of many different physical domains has to be taken into account. For instance, electronic and thermal co-simulations for the design of an efficient power converter have been studied in [1]. Such coupled simulations lead, however, to large computational requirements. Often, the long simulation time render automatic optimizations of designs impossible.

Machine learning (ML) techniques are possible candidates to increase the computational speed. The application of ML techniques to the design of electronic systems focuses mainly on two aspects [2]. The first aims at reducing the number of required design iterations [3,4]. For example, a genetic algorithms required around 1000–10,000 FEM simulations to find the optimal placement of a chip in a system [5], which takes a considerable amount of computation time. Alternatively, ML is used to accelerate the evaluation of individual designs, i.e., replace the time consuming simulations by neural networks [6–9].

In addition to thermal simulations, NNs are successfully applied to perform many types of physics simulations. For instance, NNs have been explored for fluid dynamics problems [10–12], for electro-convection [13], transport problems [14,15], magnetic field estimation [16], classical mechanics [17] and in replicating multi-particle evolutions by learning the pair-wise interaction function [18].

Recent works aim at embedding prior physics knowledge in machine learning: For physics-informed NNs (PINNs) the physics knowledge is used to guide the convergence of the NN towards physics conforming solutions [19]. Typically, the auto-gradient feature of the NN is used to construct a loss term based on the underlying differential equation. PINNs led to promising results in many areas of computational science [20–26]. A review on the application of PINNs to heat transfer problems can be found here [27]. In this work we follow an alternative way to include some physics knowledge: The NN architecture is chosen such that it forces the output to satisfy physical constraints [19]. For instance, purely convolutional networks automatically preserve translational invariance [28].

Although realistic systems are three-dimensional, ML-based physics simulations have mostly focused on lower-dimensional approximations. Especially, the representation of three-dimensional full-field solutions for complex geometries remains a challenge. One technical difficulty when going from 2D to 3D systems is the limited memory of the GPUs [29]. A way to ensure a low memory requirement could be to use a more suitable representation of the geometry instead of a stack of images [30]. Although some alternatives, like point clouds based on CAD geometries [31–33] or octrees were proposed [34], most available NN architectures are still developed for images.

The main contribution of this paper is to explore, for the first time, the approximation of the 3D temperature distribution in an electronic system with complex geometry using NNs. Note that in [35] the thermal problem on a 3D chip was studied but in a much more simplified setting than we consider in this work. In addition to the difficulties discussed above related to memory limitations, another major obstacle when studying 3D systems is the generation of enough high quality training data. We have developed an automatized workflow that creates 3D CAD geometries representing electronic circuits and performs meshing and FEM simulations on them. The workflow generates data in a form suitable for ML requiring no human intervention. In this first proof-of-concept study, we focus on FEM solutions for the steady-state, equilibrium thermal configuration of the systems.

The paper is organized as follows. In Section 2.1, we describe our workflow for the data generation and discuss how the postprocessing of CAD geometries and FEM results into ML training data is performed. The chosen NN architecture and the objective function is discussed in Section 2.2. We present our results and conclusions in Sections 3 and 4, respectively. Details on the generation of the training and evaluation datasets are given in Appendix A. For completeness, the general aspects of NNs most relevant for the understanding of this work are discussed in the Appendix B.

2. Materials and Methods

2.1. Dataset Generation

In this work a fully-convolutional neural network (FCN) was trained to approximate, on 3D electronic systems, the FEM solution of the time-independent heat equation

$$-\vec{\nabla} \cdot \left(k(x)\vec{\nabla}T(x)\right) = \rho(x)h(x)\,, \tag{1}$$

where ρ is the density, k the heat conductivity and h a heat source. We use supervized learning, which requires a large dataset of random systems with corresponding FEM solutions. The dataset needs to be representative of the design space. This entails that all properties that one wants to change during the design process (component type/number/placement, material properties, heat sink specifications, external temperature), were randomized in the dataset. In total, 464 unique systems were generated using an automatized workflow in python (see Figure 1 for a graphical summary) consisting of the following steps:

1. System generation: For each system the number and type of basic components were randomly chosen and they were placed at random locations on a PCB. Material parameters were assigned to the different parts of each component.
2. Generation of FEM solutions: A constant temperature was set at the bottom of the PCB T_{ext}. A heat sink on top of the large IC was mimicked by a heat flux boundary condition. All other outer surfaces were modelled as heat flux boundaries to air. For each of the created systems, an external temperature T_{ext} and heat transfer coefficient

to air α and for the sink α_{sink} were randomly chosen. Heat sources (i.e., electric losses) with random magnitude were assigned to some of the components. The systems were meshed. FEM simulations were performed to obtain the temperature solutions.

3. Voxelization: During postprocessing the systems and the FEM solutions were converted to a set of 3D images per system as input for the NN. Four 3D-images were created per system, one for the distribution of a material property, the external temperature, the heat sources and the heat transfer coefficient.

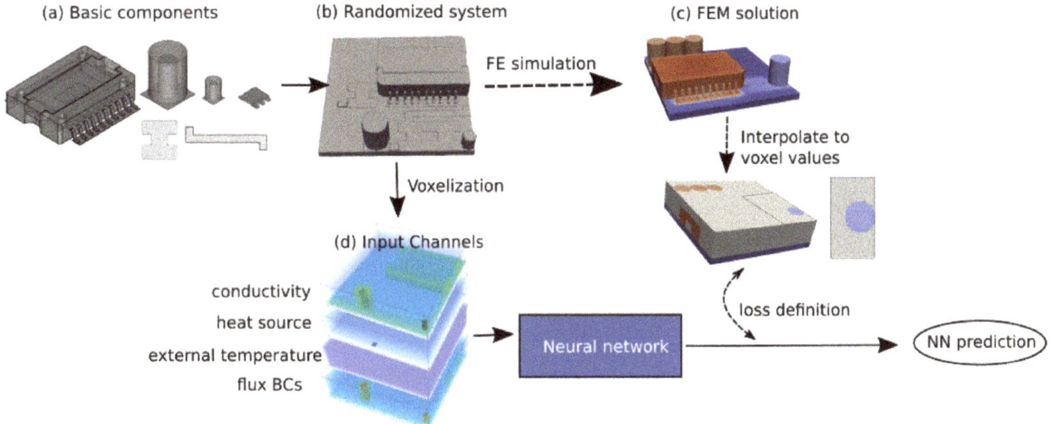

Figure 1. Illustration of the automatized workflow: Randomized systems are generated by randomly choosing and placing basic components. After assigning randomized material properties and BC values, the system is voxelized to create a stack of four 3D-images as input for the NN. Solutions for the supervized training procedure are created using FE simulations.

Detailed information on the individual steps for generating the systems can be found in Appendix A.

2.2. NN Architecture

A 3D fully-convolutional NN is used. The input of the NN consists of 4-channel 3D images and the output is a single-channel 3D temperature image.

As discussed in Section 1, a major constraint for designing an appropriate NN architecture comes from the limited GPU memory. A batch of the input images requires already a significant amount of the available memory. Thus, a network with a relatively small number of parameters is required to be trainable. Therefore, instead of applying a general-purpose network, the NN architecture must be chosen to match the physical problem.

An initial rough design of the architecture was based on physical arguments and some manual exploration, that we discuss next. Once the rough design was fixed, an automatic exploration of the remaining free parameters was performed (e.g., kernel size, type of activation, number of skip connections, etc.). The impact of this fine-tuning was not very large (of the order of a percent in the relative L_1 loss, see below).

2.2.1. Properties of Heat Propagation

During the design of the architecture, we found it instructive to explore exact analytical solutions of the heat equation, even though they apply in very idealized situations only. For example, the 1D heat kernel solution of the (time-dependent) heat equation reads

$$T(x,t) = \int_0^t dt' \int_0^\infty dy \frac{h(y,t')}{\sqrt{4\pi k(t-t')}} \left(\exp\left(-\frac{(x-y)^2}{4k(t-t')} \right) - \exp\left(-\frac{(x+y)^2}{4k(t-t')} \right) \right). \quad (2)$$

Inspecting the heat kernel solution, it is possible to anticipate (1) the presence of exponentials as a general feature of solutions of the heat equation, and (2) the fact that the contributions from different spatial points are aggregated via an integral. These two features are reflected in the NN architecture by means of the choice of activation functions and what we refer to as fusion blocks, as described next.

2.2.2. Long-Range Correlations. Fusion Blocks

One of the most important aspects, that the NN must capture in order to successfully approximate the steady-state temperature distribution is the presence of long-range effects. Most of the convolutional NNs (CNNs) available in the literature have been developed for the classification of images (in the case of 2D CNNs) or of video sequences (in the case of 3D CNNs). The performance of several of the standard low-resource 3D CNNs (e.g., SqueezeNet [36]) on the dataset described in Section 2.1 was analyzed but the temperature fields obtained with such networks were very sharp, not resembling a realistic situation where around a hot spot a temperature gradient is to be expected. Moreover, heat propagation from a heat source to a distant but thermally connected point in the system was absent.

To recover the long-range effects in the system, convolutional layers with large dilations were applied. Dilations have been used previously, for example in [16], to capture long-range correlations without increasing the convolutional kernel size. The maximum dilations in the Fusions were chosen such that the receptive field of the NN spans the whole system size (see Figure 2). This is required since in the steady-state case the temperature at a point can depend on any (arbitrarily) far away point in the system. The kernel sizes were chosen small ($k = 3$ or 4) since the NN parameters and the size of the backpropagation tree, and, thus, the required GPU memory for training, increase quickly with the kernel sizes.

Moreover, as stressed in Section 2.2.1, the solution is an aggregation of long-range effects (via an integration in the case of the heat kernel solution). To enable these aggregations in the network we defined "Fusion blocks": Each fusion block is composed of N 3D convolutional layers with different dilations $[d_1, d_2, \ldots, d_N]$. The result of those convolutions is concatenated along the channel dimension. An illustration of a simple (2D) fusion block is shown in Figure 3.

The fusion block turned out to be the decisive element to obtain realistic and accurate enough predictions from the network.

2.2.3. Choice of Activation Functions

From the heat kernel solution, it can be seen that general solutions contain exponential functions. To approximate that feature, whilst keeping the network small, it is convenient to use non-linearities (activation functions) that contain an exponential. Using too many of them, however, turns out to be damaging for the performance of the network. The reason is that, several consecutive layers containing an exponential function lead to a $\exp(-\exp(-\exp(\ldots)))$-type functions acting on the input. After some exploration, a SELU activation function [37] was chosen for one of the layers only (see Figure 2). For the rest of the layers a LeakyReLU [38] was used.

The impact of using a SELU was not large and it mostly helped in achieving the desired accuracy with less training epochs.

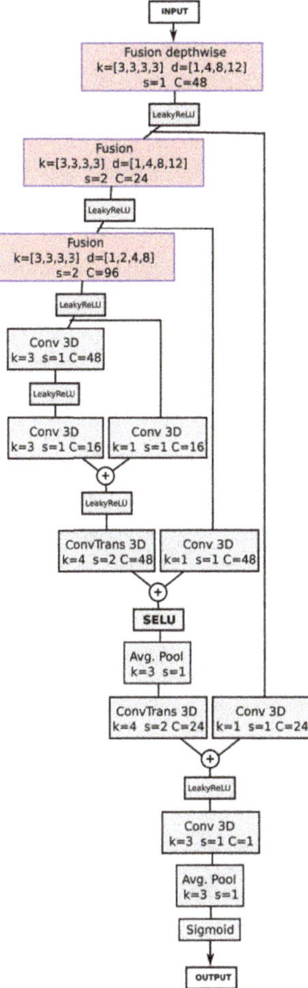

Figure 2. Architecture used in this work. In each block, k refers to the kernels size, s to the stride, d to the dilations and C to the output channels of each layer (the same for all dimensions). See Section 2.2 for further details.

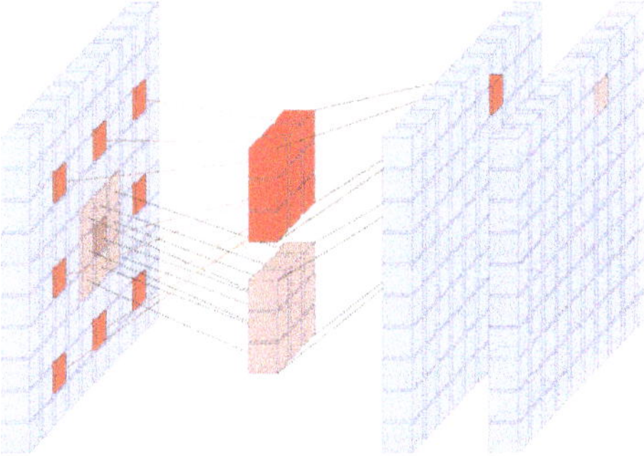

Figure 3. Schematic representation of the action of a 2D fusion block consisting of two convolutional layers with 3×3 kernels and dilations 3 and 1.

2.2.4. Input to the Network

The input of NNs is typically rescaled to be in the $[0, 1]$ range since that makes the training of the networks more stable numerically. In this work, a more physically meaningful rescaling scheme was chosen. Characteristic values were introduced for each of the relevant dimensional quantities: the maximum heat power $P_{\text{max}} = 20\,\text{W}$, the maximum temperature $T_{\text{max}} = 1000\,\text{K}$, and a length scale $l_0 = 0.4\,\text{mm}$. Note, that the maximum values defined here do not represent a hard limit at which the network will fail to work since the network does not strictly require all values to be smaller than one.

The dimensionless channels are then defined as:

$$C_0 = k\frac{T_{\text{max}} l_0}{P_{\text{max}}} \tag{3}$$

$$C_1 = h\rho V_{\text{body}} \frac{1}{P_{\text{max}}} \tag{4}$$

$$C_2 = T_{\text{ext}} / T_{\text{max}} \tag{5}$$

$$C_3 = \frac{1}{\alpha A_{\text{body}}} \frac{P_{\text{max}}}{T_{\text{max}}}, \tag{6}$$

where V_{body} is the total volume of the sub-part where the heat density is applied (e.g., the die volume of an IC), and A_{body} is the total surface area over which the heat transfer is specified (e.g., the surface area of the heat sink). The Dirichlet boundary condition is not fed to the network as an input channel, but softly imposed via the loss (see Section 2.2.6).

2.2.5. Network Architecture

We performed a fairly thorough (manual and automatic) exploration of possible network architectures containing the features discussed so far. We assessed manually the impact of using Fusion layers or not. Furthermore, after several numerical experiments we concluded that:

- Too many downsampling layers had a damaging effect on the accuracy of the output. Downsampling in CNNs is used to extract useful features from images. In our case, the most relevant features are already part of the input, as discussed above. The main reasons for downsampling in our case are to aggregate long-range effects in addition to the dilation in the fusion blocks, and to reduce the memory requirements. Thus, only two downsampling layers were used.
- As is well known in FCNs, skip connections help avoid the usual checkerboard artifacts in the output. In this work we found that using three additive skip connections led to the best results. We had skip connections from the output of the fusion blocks to the input of the two upsampling layers (transpose convolutions) and the final convolutional layer, respectively.
- An initial depthwise fusion block (depthwise means that channels are not mixed) provides the necessary additional preprocessing of the input data.

This fixed the rough architecture. The optimal size of the kernels, dilation values and the type of activations (except for the SELU unit) were found using an automatic optimisation.

As a result, we propose the architecture shown in Figure 2. In the figure, k refers to the kernels size, s to the stride, d to the dilations and C to the output channels of each layer (the same for all dimensions). This network has only ~370 K trainable parameters, which is a tiny number compared to standard architectures. For comparison, the 3D version of SqueezeNet [36,39], which was designed to be a small network, has 2.15 M parameters).

2.2.6. Objective Function and Training Process

For the loss function a mean L_1 relative loss term, $L_1 = \text{mean}(|(T_{\text{NN}} - T_{\text{FEM}})/T_{\text{FEM}}|)$, was combined with a penalty term for the Dirichlet boundary condition at the bottom layer of the PCB (L_{PCB}). Here, T_{NN} and T_{FEM} are the temperature distributions predicted by the NN and obtained from the FEM simulation, respectively. The air region was excluded from the loss evaluation because no FEM solution is available there. A physics-informed loss as proposed by [12] did not improve the solutions. In fact, it drove the system to a uniform temperature distribution since in all parts of the system where no heat source is present, the differential heat equation error can be minimized by minimizing the temperature gradient. This is somewhat unexpected and further exploration on physics-informed losses for the present problem will be the subject of future work.

We trained the network using the RMSProp optimiser [40] with a batch size of 7. The network is trained in several steps. First a warm-up period of $\sim$10 epochs, with a learning rate of 10^{-4}. After that only the Fusion layers are trained (with all other layers frozen) for 10 more epochs at a learning rate of 10^{-5}. Next the Fusions are frozen and everything else is trained for 10 epochs at a learning rate of 10^{-5}. Finally the network is trained for 100–200 epochs, with a decreasing learning rate from 10^{-5} down to 5×10^{-6}. Each epoch takes approximately 10 min.

3. Results

The size of the dataset was increased by four without further computational effort via three 90 degree in-plane rotations of each system, so that the total dataset consisted of 1856 systems. 75% of the generated dataset was used as training set and the remaining 25% as test set. It is the accuracy in reproducing the latter that truly measures the quality of the network solution. Thus, all results reported in this section are taken from the test set.

The mean relative L_1 error per system is at the percent level, with values below 2% in most systems in the test set: This can be seen in Figure 4, where a histogram of all systems of the test set binned according to their corresponding L_1 error is shown. Additionally, the Dirichlet BC at the bottom of the PCB was fulfilled by the network with an average relative error of $L_{\text{PCB}} = 0.1\%$. The present work was intended as a proof-of-principle for an NN-based tool to evaluate the viability of potential system designs from a thermal point of view. The obtained accuracies are judged good enough for that goal.

The mean value of the relative error, however, does not fully represent the (in-)accuracy of the NN results, as large but very localized error values can be smeared out when averaging over the large number of voxels in a system. More insightful is thus to compare FEM and NN solutions directly in their 3D representations. In Figure 4, a representative collection of 3D solutions and 3D maps of the relative temperature differences is shown, picking one sample from most of the bins in the histogram discussed above, as indicated in the figure. For the majority of systems the NN solution reproduces the FEM solution reasonably well. The main discrepancies appear in relation with the small chip and on the surfaces of the components and the PCB.

Further details are shown in Figure 5, where a section of a selected system is presented. There, on the surface of the PCB and also around the small chip, the L_1 error is somewhat larger. Furthermore, in the inner part of the large chip some localized discrepancies can be seen, which are a reminiscence of the checkerboard artefacts of FCNs. However, these are clearly localized and do not affect the overall temperature of the chip.

The second aspect to consider when judging whether the presented NN approach is a valuable tool in the design workflow of electronic systems is the evaluation speed. In Table 1, a comparison of the evaluation speeds between the NN and the FEM solutions is shown. The NN was run on an NVIDIA Titan RTX GPU whilst the FEM solver was run on a single thread of an Intel Xeon W-2145 CPU. This implies that the FEM evaluation time in Table 1 does not correspond to the time a fully optimized and parallelized FEM simulation would require. Nevertheless, we see that the NN provides a solution in $\sim$35 ms (with most of the time, in fact, used in transferring the 3D system to the GPU memory), which is certainly a significant speed-up over any conceivable full-field 3D FEM simulation.

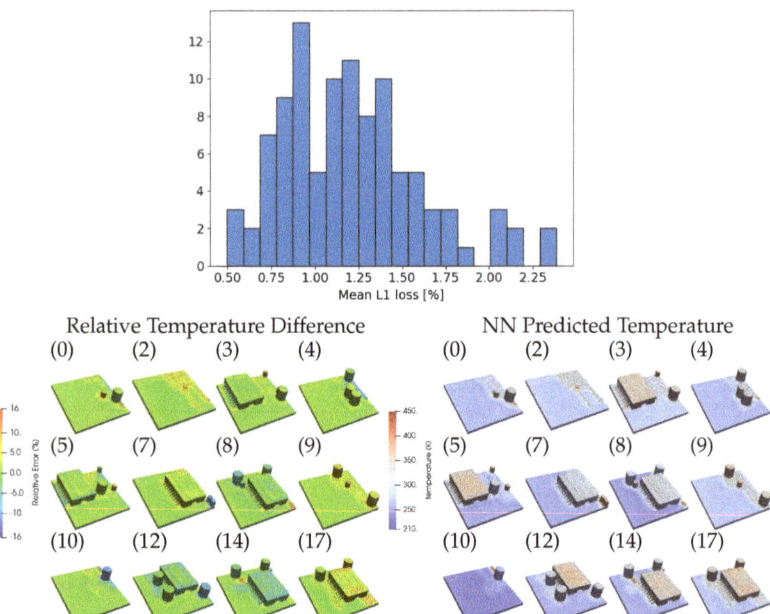

Figure 4. Histogram of the average relative L_1 error per test system (**top**). Below the temperature distributions estimated by the NN (**right**) and the relative temperature difference (**left**) for selected systems of the test dataset (the corresponding error bin of the histogram is indicated in brackets, from 0 indicating the lowest mean error to 19 for the worst mean error). The average relative L_1 error (**top**) is the mean of the absolute values of the relative temperature differences (**bottom left**) per sample.

Figure 5. FE solution (**left**), NN prediction (**center**), and relative L_1 error of the temperature distribution (**right**) on a horizontal cut of a selected system. High predictive errors are mostly found on the surface of the system while the internal temperature distribution is well represented.

Table 1. Average evaluation time for a NN solution and a FEM solution (note that, the time for the FEM simulation is not taken from a performance-optimized solver).

NN, GPU Transfer	NN Inference	Total NN	FEM (Single Core)
0.033 s	0.002 s	0.035 s	160 s

This high evaluation speed comes at the cost of creating a training data set and training the network. The time to create one system of the training/testing data set was about 17 min. The systems were created fully automatized, creating 8 systems in parallel, so that in one day the full data set was created without requiring any manual intervention. Note, that the workflow was not optimized for performance, 90% of the time was used for the system creation in `FreeCAD` and preprocessing the system (meshing, identifying surfaces to apply the BCs, etc.). The voxelization of the system took only fractions of a second. Thus, approximating FEM simulations with NNs is clearly not the right approach if only the result of one specific system is of interest. However, possible application scenarios could lie in design applications, where the impact of design variations could be studied in real time using pre-trained NNs.

It is worth discussing now some of the limitations of the FCN architecture suggested in this work. The first limitation stems from the fact that the architecture sets the maximum effective receptive field on the input images, which is determined by the combination of maximum dilation and number of downsamplings. Thus, even though the architecture is fully convolutional—which implies that it can immediately process larger systems than those it has been trained on—it will not be able to capture any heat transfer effects over larger distances than those covered in the original receptive field. We have not tested what the implications of this limitation are. A second limitation arises from the typical length scales in the system. A minimum length scale is implicitly set by choosing the voxel resolution and, in principle, if a different voxel resolution has to be chosen the network would have to be retrained. The voxelized representation of the systems has, however, the advantage that no tedious and time consuming meshing and preprocessing is required for the evaluation of the NN on a system.

The main advantage of the approach presented here is that, once the network has been trained, it can approximate the temperature field of any system within the design space spanned by the randomized training dataset. This means that the network can be applied to systems with different number of components, different placement of those components, different material properties and different BCs than those it has been trained on. Therefore, the simulation time spent in generating the dataset is compensated by the gain in future simulations, as long as many of them are needed for a certain design task.

Confidence Estimation

Finally, we would like to discuss a possible confidence estimator for the NN solutions. The goal of a confidence estimator in this context is to assess the quality of the NN approximation when no FE result exists to compare with, as would be the case in any realistic use of our tool.

Our suggested confidence estimator is based on an integral form of the steady-state heat equation

$$\int_{\partial V} k\partial_i T n_i dA + \int_V \rho h dV = 0, \tag{7}$$

where Gauss' theorem was used to transform the integral over the system volume V to a surface integral in the first term. Since the NN prediction is available at each voxel, the equation is discretized to obtain an approximation per voxel e

$$\sum_{i=1}^{6} {}^{e}k(\partial_i T) n_i \, {}^{e}\Delta A_i + {}^{e}\rho^e h^e \Delta V \approx 0. \tag{8}$$

For the temperature gradient $(\partial_i T)$, the finite difference of the temperature (as predicted by the NN) of the voxel under consideration and its adjacent voxel (in the direction of the normal n_i) is inserted. If the two voxels have differing k values a piece-wise linear interpolation of the temperature was used. The bottom of the PCB is kept at a fixed temperature. On all other outer surfaces the flux boundary condition is included by evaluating the first term with:

$$-k\frac{\partial T}{\partial n} = \alpha(T - T_{\text{ext}}) \,. \tag{9}$$

This heat equation error should measure the local violation of the steady-state heat equation. If the net heat flux of a voxel is non-zero this implies that the temperature obtained from the NN does not represent a steady state.

The local L_1 loss values are compared with the local heat equation error for a particular sample (Figure 6) for different sections of the system in the z-direction. The proposed estimator appears to qualitatively identify many of the regions where the error of the NN estimation of the temperature map is high compared to the FEM solution. It is also apparent, however, that not all voxels with high error values are identified.

Moreover, the estimator does not provide a clear quantitative measure of the error. The absolute values of the heat equation error are much larger in regions with high k than in regions with lower k. This could indicate that the piece-wise linear interpolation used for the approximation of the temperature is not a good estimation of the actual temperature distribution. However, at the current state, it is not possible to use higher order approximations since they would require more voxels per uniform-k region, which in turn conflicts with the limited RAM available on the GPU.

More research in the direction of finding a confidence estimator is clearly needed.

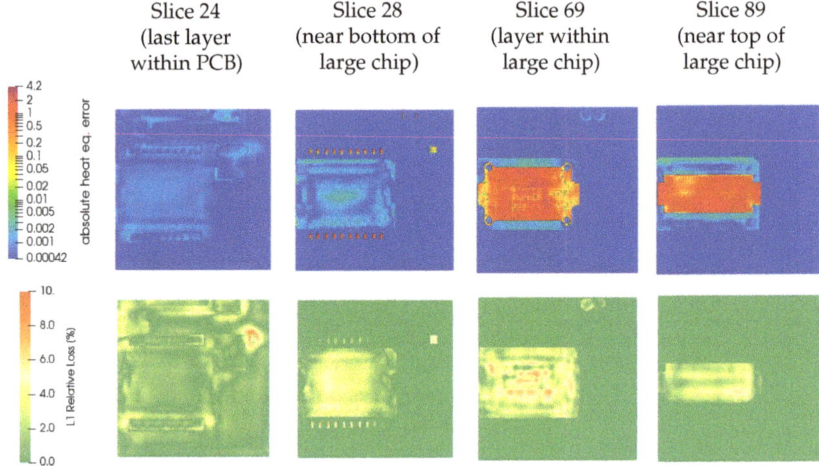

Figure 6. Comparison of the heat equation error (**top row**), which can be used as error predictor if no FE solution is available, and the L_1 error (**bottom row**) on selected slices from bottom to top (**left to right**). The heat equation error is able to indicate most of the regions with high error. It illustrates the checkerboard pattern expected from purely convolutional networks. Since the heat equation error is defined via the local imbalance of heat fluxes and sources, the detected errors can be slightly more localized compared to the actual error (e.g., orange points in slice 69, top compared to bottom).

4. Conclusions

An approach to quickly approximate the full-field temperature distribution of 3D electronic system using neural networks was proposed. The range of validity of the developed network spans a realistically wide range of material parameters and heat sources. Additionally, the network was trained to work with systems built with any number of the basic components at any position in the system.

The network proposed is able to provide a simulation result in an evaluation time of the order of milliseconds. The relative error achieved, when compared to a standard FEM simulation, is around 2% averaged over the whole system, with larger errors localized mostly on the surfaces of the components.

Finally we motivated a possible criterion for estimating the confidence of a prediction based on an integral version of the heat equation. Even though the criterion proposed does not allow for a quantitative estimate of the error of a given prediction, it may be used to identify regions with large estimation errors.

Author Contributions: Both authors contributed substantially to the conceptualization, methodology, and writing—original draft preparation, review and editing. All authors have read and agreed to the published version of the manuscript.

Funding: This work has been supported by Silicon Austria Labs GmbH (SAL), owned by the Republic of Austria, the Styrian Business Promotion Agency (SFG), the federal state of Carinthia, the Upper Austrian Research (UAR), and the Austrian Association for the Electric and Electronics Industry (FEEI).

Acknowledgments: We thank L. Ratschbacher for a critical reading of an early version of the manuscript and C. Mentin for support.

Conflicts of Interest: The authors declare no conflict of interest. The funders had no role in the design of the study; in the collection, analyses, or interpretation of data; in the writing of the manuscript, or in the decision to publish the results.

Appendix A. System Generation Details

Appendix A.1. System Generation

The main challenge we faced in generating systems was to obtain sufficiently diverse 3D geometries, and achieve enough variation of material properties and heat sources while ensuring that the resulting temperature fields stay within a physically plausible range. To simplify this task, we built the systems based on six different components. The components were designed manually (once) using `FreeCAD` (one large and one small IC, one large and one small capacitor, and copper layers of different shapes and sizes, see Figure 1a).

To generate a system, the components were randomly placed on a fixed-size PCB of dimensions 25 mm × 25 mm × 2 mm (Figure 1b). Such a random placing procedure does not produce functional electronic circuits. However, as long as the training dataset captures the impact that different component placements have on the heat transfer, the NN is expected to generalise also to (unseen) functional electronic circuits.

Most components (e.g., the ICs) have an internal structure (e.g., legs, die). Each of these sub-parts was assigned a typical material (e.g., copper for the legs). To account for variations in the material properties, a physically plausible range was specified for each material type as the average value given in Table A1 ±25%. For each system, the material properties of each part were chosen from a uniform distribution within those ranges.

Table A1. Average material properties: the actual values are chosen randomly from a range of [0.75 Avg, 1.25 Avg].

Property Unit	Avg. k (W/(m K))	Avg. c_P (J/(kg K))	Avg. ρ (kg/m^3)
Silicon	148	705	2330
Copper	384	385	8930
Epoxy	0.881	952	1682
FR_4	0.25	1200	1900
Al_2O_3	35	880	3890
Aluminum	148	128	1930

Appendix A.2. Finite Element Simulations

Once the systems were generated, the correct temperature distributions were obtained from FE simulations. Note, however, that instead of FE simulations any adequate simulation methods could have been used. In this proof-of-concept study we focused on the steady-state equation. Assuming that the material properties in the system do not change with temperature, the thermal behavior of the system is determined by the heat equation

$$- \vec{\nabla} \cdot \left(k(x) \vec{\nabla} T(x) \right) = \rho(x) h(x) \, , \tag{A1}$$

where ρ is the density, k the heat conductivity and h a heat source. ρ, k and h are assumed to be constant over time, i.e., independent of the current temperature. Three types of boundary conditions (BCs) were imposed on the systems for the Finite Element (FE) simulations: (1) a Dirichlet-type BC at the bottom of the PCB, which was set to a constant external temperature T_{ext}. The value of T_{ext} was assigned randomly to each of the systems we generated, from the range 300 K $\pm$ 25%. (2) a Neumann-type BC represented by a heat sink on the top of the large IC, modelled via a heat transfer coefficient α_{sink} (see Equation (9)). For each system, α_{sink} was randomly chosen from the range 2538.72 $W/(Km^2) \pm 25\%$. (3) on all other outer exposed surfaces a heat transfer coefficient α was prescribed, taken randomly from the range 14 $W/(Km^2) \pm 25\%$. Electronic losses were modelled as constant heat sources. The magnitudes of the volumetric heat sources, located at the silicon die of the ICs and at the core of the capacitors, were chosen randomly from the ranges specified in Table A2.

The material properties ρ and k depend on the position x, although in the generated systems they are constant over the different sub-parts of each component. ρ and h do not appear separately in Equation (A1), but rather as the product ρh. This implies that any solution of the equation will depend on this product only and, thus, it is sensible to use the product as input to the NN as explained below.

Next in the automatic workflow, a conformal mesh was generated using `gmsh` [41], which consisted on average of 3 million elements. Steady-state solutions were obtained with the open-source FEM solver `ElmerSolver` [42,43]. The main advantage of `ElmerSolver` is that a scriptable interface between `FreeCAD` (for automatic geometry generation), to `ElmerGrid/gmsh` (for automatic tetrahedral mesh generation) and `ElmerSolver` exists which enables an automatized workflow for the system generation.

Table A2. Range of applied heat sources for the different components of the systems (in W).

Component	Min	Max
Center of large capacitor	0.1	0.3
Silicon die of large chip	10	19
Silicon die of small chip	0.1	0.5

Appendix A.3. Voxelization

For the NN, the 3D-systems were represented as a stack of 3D-images. This enabled us to adapt NN methods previously developed for vision tasks to our problem (Figure 1d). During postprocessing, each system was converted to four 3D-images. One 3D-image for each of the physical properties (k, T_{ext}, ρh, α).

To create these 3D-images, the geometries of the basic components were voxelized once using a custom `FreeCAD`-python script. The voxel size was $0.19 \times 0.19 \times 0.05$ mm^3 so that a batch of ten images fits in the GPU memory for training, while the voxel size still resolves sufficient structural detail. The smaller resolution in z-direction was chosen in order to resolve the thin copper layers. During the system generation workflow, each component in the original system was then simply replaced by the previously voxelized representation. To create the four different images, the voxelized geometry was replaced part-by-part by the corresponding material parameter used in the FEM simulation. For our systems, this procedure led to images with $128 \times 128 \times 128$ voxels. The four 3D-images were then stacked to create a four channel input for the network.

As labels for the supervized learning procedure the temperature solution obtained by the FEM solver was used. The 3D temperature field was directly voxelized during post-processing by `Elmer`. The FEM solution is only available and of interest within the geometry (where a mesh is available). The outer image regions (we call them "air" regions) were excluded in the loss definition (see Section 2.2.6).

Appendix B. Introduction to ANNs

In this section, we give an overview of neural networks, highlighting the aspects that are most relevant to understand the rest of the paper; see, e.g., [44] for a detailed treatment. In this work, NNs are used as highly versatile and non-linear function approximators.

Generally speaking, a neural network can be visualized as a stack of layers, each of them computing a set of linear operations on their input data, subsequently applying a fixed non-linear operation and passing the result to the next layer. The input to the network in this work is a voxelized representation of a 3D electronic system (i.e., a 3D image), where the different properties of the system are passed as different channels (in analogy to the RGB channels of a standard 2D image). Convolutional neural networks (CNNs), consisting of a special type of layer called convolutional layer, are used for processing images.

The most important feature in a convolutional layer of a CNN is that not every pixel in the input image of a layer is connected to every pixel in the output, but rather each output pixel is determined by a small set of input pixels (the receptive field), the number of them is given by the size of the so-called convolutional kernel (or filter). To be specific, the equation defining the action of a 3D convolutional layer is:

$$z_{ijk;C} = b_C + \sum_{\alpha=0}^{k_h} \sum_{\beta=0}^{k_w} \sum_{\gamma=0}^{k_d} \sum_{F=0}^{n_{c'}} x_{i'j'k'F} \times w_{\alpha\beta\gamma;CF} , \tag{A2}$$

with

$$\begin{aligned} i' &= i \times s_h + d_h \times \alpha \\ j' &= j \times s_w + d_w \times \beta \\ k' &= k \times s_d + d_d \times \gamma \end{aligned} . \tag{A3}$$

where $z_{ijk;C}$ represents the value of the voxel at position i, j, k of the output 3D image, C labels the channel (or feature map) of the output image, s_h, s_w and s_d are natural numbers called strides and d_h, d_w and d_h are naturals called dilations (their meaning is explained in Section 2.2), k_h, k_w and k_d are the height, width and depth of the 3D convolutional kernels applied to the input image x, with $n_{c'}$ channels, b_C is a bias parameter for each feature map C and w are the weight parameters.

The next ideas to be presented are downsampling and upsampling. It is not necessarily the case that the size of the output and input images of a convolutional layer coincide. It can be inferred from Equation (A3), the output image size can be reduced by tuning the values of the stride parameters. For example, choosing $s_h = s_w = s_d = 2$, the size of the output image will be halved. One refers to this process as downsampling. In image processing, the purpose of downsampling is to *force* the CNN to extract relevant features from an image (e.g., edges)but, in this work, extracting features is not so relevant since they are, to some extent, manually implemented. Nevertheless, some degree of downsampling is necessary to reduce the memory requirements of the network. The inverse action to downsampling is upsampling. One can think of it as an interpolation procedure, where from one input pixel a higher number of pixels are generated as output. In practice, this is achieved by an operation called transposed convolution, which is mathematically very similar to the standard convolution in Equation (A2), with different weights and biases to be learned. Upsampling is necessary if downsampling is used in some of the layers of the network but the output of a CNN must be another image of the same size of the input image, as is our case. CNNs of this type are called fully-convolutional neural networks (FCNs).

Finally there is a different type of layer called pooling layer. They are parameter-free layers (non-learnable) that apply a fixed operation on their receptive field. Typical pooling layers include max-pooling and average-pooling layers, which give as output the maximum or the average value, respectively, of the pixel (resp. voxel) values in their receptive field.

The training of a CNN consists of finding the optimal values of the parameters w and b of each of the layers. This is achieved by minimising a certain objective function (loss) on a labelled dataset. For this work, as discuss in the main text, the dataset consists of a collection of simplified 3D circuits, each *labelled* by the result of a standard thermal simulation. The training is performed on a subset of the whole dataset (training set), whilst the rest are used to test the network (test set).

References

1. Langbauer, T.; Mentin, C.; Rindler, M.; Vollmaier, F.; Connaughton, A.; Krischan, K. Closing the Loop between Circuit and Thermal Simulation: A System Level Co-Simulation for Loss Related Electro-Thermal Interactions. In Proceedings of the 2019 25th International Workshop on Thermal Investigations of ICs and Systems (THERMINIC), Lecco, Italy, 25–27 September 2019; IEEE: Piscataway, NJ, USA, 2019; pp. 1–6. [CrossRef]
2. Zhao, S.; Blaabjerg, F.; Wang, H. An Overview of Artificial Intelligence Applications for Power Electronics. *Trans. Power Electron.* **2015**, *30*, 6791–6803. [CrossRef]
3. Wu, T.; Wang, Z.; Ozpineci, B.; Chinthavali, M.; Campbell, S. Automated heatsink optimization for air-cooled power semiconductor modules. *IEEE Trans. Power Electron.* **2018**, *34*, 5027–5031. [CrossRef]
4. Zhang, Y.; Wang, Z.; Wang, H.; Blaabjerg, F. Artificial Intelligence-Aided Thermal Model Considering Cross-Coupling Effects. *IEEE Trans. Power Electron.* **2020**, *35*, 9998–10002. [CrossRef]
5. Delaram, H.; Dastfan, A.; Norouzi, M. Optimal Thermal Placement and Loss Estimation for Power Electronic Modules. *IEEE Trans. Componen. Packag. Manuf. Technol.* **2018**, *8*, 236–243. [CrossRef]
6. Guillod, T.; Papamanolis, P.; Kolar, J.W. Artificial Neural Network (ANN) Based Fast and Accurate Inductor Modeling and Design. *IEEE Open J. Power Electron.* **2020**, *1*, 284–299. [CrossRef]
7. Chiozzi, D.; Bernardoni, M.; Delmonte, N.; Cova, P. A Neural Network Based Approach to Simulate Electrothermal Device Interaction in SPICE Environment. *IEEE Trans. Power Electron.* **2019**, *34*, 4703–4710. [CrossRef]
8. Xu, Z.; Gao, Y.; Wang, X.; Tao, X.; Xu, Q. Surrogate Thermal Model for Power Electronic Modules using Artificial Neural Network. In Proceedings of the 45th Annual Conference of the IEEE Industrial Electronics Society (IECON 2019), Lisbon, Portugal, 14–17 October 2019; IEEE: Piscataway, NJ, USA, 2019; pp. 3160–3165. [CrossRef]
9. Marie-Francoise, J.N.; Gualous, H.; Berthon, A. Supercapacitor thermal- and electrical-behaviour modelling using ANN. *IEEE Proc. Electr. Power Appl.* **2006**, *153*, 255. [CrossRef]
10. Kim, J.; Lee, C. Prediction of turbulent heat transfer using convolutional neural networks. *J. Fluid Mech.* **2020**, *882*, A18. [CrossRef]
11. Swischuk, R.; Mainini, L.; Peherstorfer, B.; Willcox, K. Projection-based model reduction: Formulations for physics-based machine learning. *Comput. Fluids* **2019**, *179*, 704–717. [CrossRef]
12. Gao, H.; Sun, L.; Wang, J.X. PhyGeoNet: Physics-Informed Geometry-Adaptive Convolutional Neural Networks for Solving Parametric PDEs on Irregular Domain. *arXiv* **2020**, arXiv:2004.13145.
13. Cai, S.; Wang, Z.; Lu, L.; Zaki, T.A.; Karniadakis, G.E. DeepM&Mnet: Inferring the electroconvection multiphysics fields based on operator approximation by neural networks. *arXiv* **2020**, arXiv:2009.12935.
14. He, Q.Z.; Barajas-Solano, D.; Tartakovsky, G.; Tartakovsky, A.M. Physics-informed neural networks for multiphysics data assimilation with application to subsurface transport. *Adv. Water Resour.* **2020**, *141*, 103610, [CrossRef]
15. Kadeethum, T.; Jørgensen, T.M.; Nick, H.M. Physics-informed neural networks for solving nonlinear diffusivity and Biot's equations. *PLoS ONE* **2020**, *15*, e0232683, [CrossRef]
16. Khan, A.; Ghorbanian, V.; Lowther, D. Deep learning for magnetic field estimation. *IEEE Trans. Magn.* **2019**, *55*, 7202304. [CrossRef]
17. Breen, P.G.; Foley, C.N.; Boekholt, T.; Zwart, S.P. Newton vs. the machine: Solving the chaotic three-body problem using deep neural networks. *Mon. Not. R. Astron. Soc.* **2019**, *494*, 2465–2470. [CrossRef]
18. Sanchez-Gonzalez, A.; Godwin, J.; Pfaff, T.; Ying, R.; Leskovec, J.; Battaglia, P.W. Learning to Simulate Complex Physics with Graph Networks. In Proceedings of the 37th International Conference on Machine Learning, ICML, Virtual Event, 13–18 July 2020; pp. 8459–8468.
19. Karniadakis, G.E.; Kevrekidis, I.G.; Lu, L.; Perdikaris, P.; Wang, S.; Yang, L. Physics-informed machine learning. *Nat. Rev. Phys.* **2021**, *3*, 422–440. [CrossRef]
20. Raissi, M.; Perdikaris, P.; Karniadakis, G.E. Physics-informed neural networks: A deep learning framework for solving forward and inverse problems involving nonlinear partial differential equations. *J. Comput. Phys.* **2019**, *378*, 686–707. [CrossRef]
21. Rasht-Behesht, M.; Huber, C.; Shukla, K.; Karniadakis, G.E. Physics-informed Neural Networks (PINNs) for Wave Propagation and Full Waveform Inversions. *arXiv* **2021**, arXiv:2108.12035.
22. Goswami, S.; Yin, M.; Yu, Y.; Karniadakis, G. A physics-informed variational DeepONet for predicting the crack path in brittle materials. *arXiv* **2021**, arXiv:2108.06905.
23. Lin, C.; Maxey, M.; Li, Z.; Karniadakis, G.E. A seamless multiscale operator neural network for inferring bubble dynamics. *J. Fluid Mech.* **2021**, *929*, A18. [CrossRef]
24. Kovacs, A.; Exl, L.; Kornell, A.; Fischbacher, J.; Hovorka, M.; Gusenbauer, M.; Breth, L.; Oezelt, H.; Praetorius, D.; Suess, D.; et al. Magnetostatics and micromagnetics with physics informed neural networks. *J. Magn. Magn. Mater.* **2022**, *548*, 168951, [CrossRef]
25. Jiang, J.; Zhao, J.; Pang, S.; Meraghni, F.; Siadat, A.; Chen, Q. Physics-informed deep neural network enabled discovery of size-dependent deformation mechanisms in nanostructures. *Int. J. Solids Struct.* **2022**, *236–237*, 111320. [CrossRef]
26. Di Leoni, P.C.; Lu, L.; Meneveau, C.; Karniadakis, G.; Zaki, T.A. DeepONet prediction of linear instability waves in high-speed boundary layers. *arXiv* **2021**, arXiv:2105.08697.
27. Cai, S.; Wang, Z.; Wang, S.; Perdikaris, P.; Karniadakis, G.E. Physics-informed neural networks for heat transfer problems. *J. Heat Transf.* **2021**, *143*, 060801. [CrossRef]

28. Battaglia, P.W.; Hamrick, J.B.; Bapst, V.; Sanchez-Gonzalez, A.; Zambaldi, V.; Malinowski, M.; Tacchetti, A.; Raposo, D.; Santoro, A.; Faulkner, R.; et al. Relational inductive biases, deep learning, and graph networks. *arXiv* **2018**, arXiv:1806.01261.
29. Blumberg, S.B.; Tanno, R.; Kokkinos, I.; Alexander, D.C. Deeper image quality transfer: Training low-memory neural networks for 3D images. In Proceedings of the International Conference on Medical Image Computing and Computer-Assisted Intervention, Granada, Spain, 16–20 September 2018; Springer: Berlin/Heidelberg, Germany, 2018; Volume 11070 LNCS, pp. 118–125. [CrossRef]
30. Ahmed, E.; Saint, A.; Shabayek, A.E.R.; Cherenkova, K.; Das, R.; Gusev, G.; Aouada, D.; Ottersten, B. A survey on Deep Learning Advances on Different 3D Data Representations. *arXiv* **2018**, arXiv:1808.01462.
31. Liu, W.; Sun, J.; Li, W.; Hu, T.; Wang, P. Deep Learning on Point Clouds and Its Application: A Survey. *Sensors* **2019**, *19*, 4188. [CrossRef]
32. Bello, S.A.; Yu, S.; Wang, C. Review: Deep learning on 3D point clouds. *Remote Sensing* **2020**, *12*, 1729. [CrossRef]
33. Rios, T.; Wollstadt, P.; Stein, B.V.; Back, T.; Xu, Z.; Sendhoff, B.; Menzel, S. Scalability of Learning Tasks on 3D CAE Models Using Point Cloud Autoencoders. In Proceedings of the 2019 IEEE Symposium Series on Computational Intelligence (SSCI), Xiamen, China, 6–9 December 2019; IEEE: Piscataway, NJ, USA, 2019; pp. 1367–1374. [CrossRef]
34. Riegler, G.; Ulusoy, A.O.; Geiger, A. OctNet: Learning Deep 3D Representations at High Resolutions. In Proceedings of the 30th IEEE Conference on Computer Vision and Pattern Recognition (CVPR 2017), Honolulu, HI, USA, 21–26 July 2017; pp. 6620–6629.
35. He, H.; Pathak, J. An unsupervised learning approach to solving heat equations on chip based on auto encoder and image gradient. *arXiv* **2020**, arXiv:2007.09684.
36. Köpüklü, O.; Kose, N.; Gunduz, A.; Rigoll, G. Resource efficient 3d convolutional neural networks. In Proceedings of the 2019 IEEE/CVF International Conference on Computer Vision Workshop (ICCVW), Seoul, Korea, 27–28 October 2019; IEEE: Piscataway, NJ, USA, 2019; pp. 1910–1919.
37. Klambauer, G.; Unterthiner, T.; Mayr, A.; Hochreiter, S. Self-normalizing neural networks. In *Advances in Neural Information Processing Systems*; MIT Press: Cambridge, MA, USA, 2017, pp. 971–980.
38. Maas, A.L.; Hannun, A.Y.; Ng, A.Y. *Rectifier Nonlinearities Improve Neural Network Acoustic Models*; Citeseer: Princeton, NJ, USA, 2013.
39. Iandola, F.N.; Han, S.; Moskewicz, M.W.; Ashraf, K.; Dally, W.J.; Keutzer, K. SqueezeNet: AlexNet-level accuracy with 50x fewer parameters and <0.5 MB model size. *arXiv* **2016**, arXiv:1602.07360.
40. Hinton, G.; Tieleman, T. Neural Networks for Machine Learning. Available online: http://www.cs.toronto.edu/~tijmen/csc321/slides/lecture_slides_lec6.pdf (accessed on 21 October 2021).
41. Geuzaine, C.; Remacle, J.F. Gmsh: A three-dimensional finite element mesh generator with built-in pre-and post-processing facilities. *Int. J. Numer. Methods Eng.* **2009**, *79*, 1309–1331. [CrossRef]
42. Malinen, M.; Råback, P. Elmer finite element solver for multiphysics and multiscale problems. *Multiscale Model. Methods Appl. Mater. Sci.* **2013**, *19*, 101–113.
43. Råback, P.; Malinen, M.; Ruokolainen, J.; Pursula, A.; Zwinger, T. *Elmer Models Manual*; Technical Report; CSC—IT Center for Science: Espoo, Finland, 2020.
44. Goodfellow, I.; Bengio, Y.; Courville, A.; Bengio, Y. *Deep Learning*; MIT Press: Cambridge, MA, USA, 2016; Volume 1.

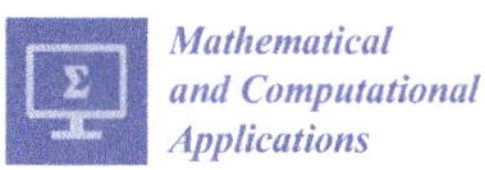
Mathematical and Computational Applications

MDPI

Article

Arbitrarily Accurate Analytical Approximations for the Error Function

Roy M. Howard

School of Electrical Engineering, Computing and Mathematical Sciences, Curtin University, GPO Box U1987, Perth 6845, Australia; r.howard@curtin.edu.au

Abstract: AbstractA spline-based integral approximation is utilized to define a sequence of approximations to the error function that converge at a significantly faster manner than the default Taylor series. The real case is considered and the approximations can be improved by utilizing the approximation $\mathrm{erf}(x) \approx 1$ for $|x| > x_o$ and with x_o optimally chosen. Two generalizations are possible; the first is based on demarcating the integration interval into m equally spaced subintervals. The second, is based on utilizing a larger fixed subinterval, with a known integral, and a smaller subinterval whose integral is to be approximated. Both generalizations lead to significantly improved accuracy. Furthermore, the initial approximations, and those arising from the first generalization, can be utilized as inputs to a custom dynamic system to establish approximations with better convergence properties. Indicative results include those of a fourth-order approximation, based on four subintervals, which leads to a relative error bound of 1.43×10^{-7} over the interval $[0, \infty]$. The corresponding sixteenth-order approximation achieves a relative error bound of 2.01×10^{-19}. Various approximations that achieve the set relative error bounds of 10^{-4}, 10^{-6}, 10^{-10}, and 10^{-16}, over $[0, \infty]$, are specified. Applications include, first, the definition of functions that are upper and lower bounds, of arbitrary accuracy, for the error function. Second, new series for the error function. Third, new sequences of approximations for $\exp(-x^2)$ that have significantly higher convergence properties than a Taylor series approximation. Fourth, the definition of a complementary demarcation function $e_C(x)$ that satisfies the constraint $e_C^2(x) + \mathrm{erf}^2(x) = 1$. Fifth, arbitrarily accurate approximations for the power and harmonic distortion for a sinusoidal signal subject to an error function nonlinearity. Sixth, approximate expressions for the linear filtering of a step signal that is modeled by the error function.

Keywords: error function; function approximation; spline approximation; Gaussian function

MSC: 33B20; 41A10; 41A15; 41A58

Citation: Howard, R.M. Arbitrarily Accurate Analytical Approximations for the Error Function. *Math. Comput. Appl.* **2022**, 27, 14. https://doi.org/10.3390/mca27010014

Received: 26 November 2021
Accepted: 27 January 2022
Published: 9 February 2022

Publisher's Note: MDPI stays neutral with regard to jurisdictional claims in published maps and institutional affiliations.

1. Introduction

The error function arises in many areas of mathematics, science, and scientific applications including diffusion associated with Brownian motion (Fick's second law), the heat kernel for the heat equation, e.g., [1], the modeling of magnetization, e.g., [2], the modelling of transitions between two levels, for example, with the modeling of smooth or soft limiters, e.g., [3] and the psychometric function, e.g., [4,5], the modeling of amplifier non-linearities, e.g., [6,7], and the modeling of rubber-like materials and soft tissue, e.g., [8,9]. It is widely used in the modeling of random phenomena as the error function defines the cumulative distribution of the Gaussian probability density function and examples include, the probability of error in signal detection, option pricing via the Black–Scholes formula, etc. Many other applications exist. In general, the error function is associated with a macro description of physical phenomena and the de Moivre–Laplace theorem is illustrative of the link between fundamental outcomes and a higher-level model.

The error function is defined on the complex plane according to

$$\mathrm{erf}(z) = \frac{2}{\sqrt{\pi}} \int_{\gamma} e^{-\lambda^2} \mathrm{d}\lambda, \, z \in \boldsymbol{C}, \tag{1}$$

where the path γ is between the points zero and z and is arbitrary. Associated functions are the complementary error function, the Faddeyeva function, the Voigt function, and Dawson's integral, e.g., [10]. The Faddeyeva function and the Voigt function, for example, have applications in spectroscopy, e.g., [11]. The error function can also be defined in terms of the spherical Bessel functions, e.g., Equation (7.6.8) of [10], and the incomplete Gamma function, e.g., Equation (7.11.1) of [10]. Marsaglia [12] provides a brief insight into the history of the error function.

For the real case, which is the case considered in this paper, the error function is defined by the integral

$$\mathrm{erf}(x) = \frac{2}{\sqrt{\pi}} \int_{0}^{x} e^{-\lambda^2} \mathrm{d}\lambda, \; x \in \boldsymbol{R}. \tag{2}$$

The widely used, and associated, cumulative distribution function for the standard normal distribution is defined according to

$$\Phi(x) = \frac{1}{\sqrt{2\pi}} \int_{-\infty}^{x} \exp\left[\frac{-\lambda^2}{2}\right] \mathrm{d}\lambda = 0.5 + 0.5\mathrm{erf}\left[\frac{x}{\sqrt{2}}\right]. \tag{3}$$

Being defined by an integral, which does not have an explicit analytical form, there is interest in approximations of the error function and, in recent decades, many approximations have been developed. Table 1 details indicative approximations for the real case and their relative errors are shown in Figure 1. Most of the approximations detailed in this table are custom and have a limited relative error bound with bounds in the range of 3.05×10^{-5} [13] to 7.07×10^{-3} [14]. It is preferable to have an approximation form that can be generalized to create approximations that converge to the error function. Examples include the standard Taylor series, the Bürmann series, and the approximation by Abrarov, which are defined in Table 1.

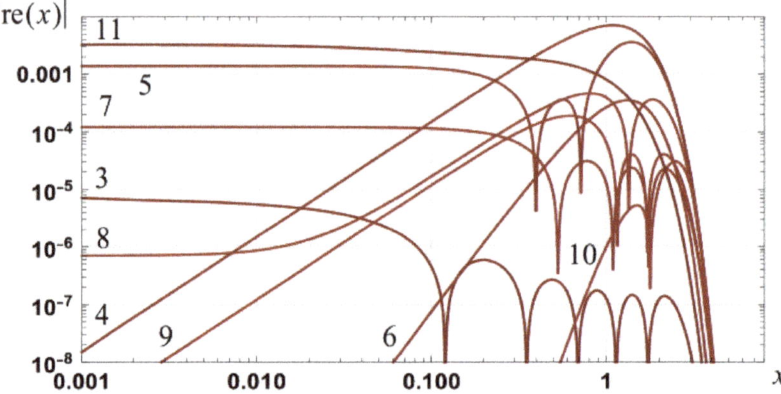

Figure 1. Graph of the magnitude of the relative error in the approximations, detailed in Table 1, for erf(x).

Many of the approximations detailed in Table 1 can be improved upon by approximating the associated residual function, denoted g, via a Padé approximant or a basis set decomposition. Examples of possible approximation forms, and the resulting residual functions, are given in Table 2. One example is that of a 4/2 Padé approximant for the function g_3, which leads to the approximation:

$$\mathrm{erf}(x) \approx \sqrt{1 - \exp\left[-x^2 \cdot \frac{4}{\pi}\left[1 + \frac{n_1 x_1 + n_2 x_1^2 + n_3 x_1^3 + n_4 x_1^4}{1 + d_1 x_1 + d_2 x_1^2}\right]\right]}, \; x_1 = \frac{x}{x+1}, x \geq 0, \quad (4)$$

$$\begin{aligned} n_1 &= \tfrac{279}{10{,}000{,}000}, n_2 = \tfrac{-303{,}923}{10{,}000{,}000}, n_3 = \tfrac{34{,}783}{5{,}000{,}000}, n_4 = \tfrac{40{,}793}{10{,}000{,}000} \\ d_1 &= \tfrac{-21{,}941{,}279}{10{,}000{,}000}, d_2 = \tfrac{3{,}329{,}407}{2{,}500{,}000}. \end{aligned} \quad (5)$$

The relative error bound in this approximation is 4.02×10^{-7}. Higher-order Padé approximants can be used to generate approximations with a lower relative error bound. Matic [15] provides a similar approximation with an absolute error of 5.79×10^{-6}.

An approximation of the error function can also be obtained by combining separate approximations, which are accurate, respectively, for $|x| \ll 1$ and $|x| \gg 1$, via a demarcation function d

$$\mathrm{erf}(x) = \frac{2x}{\sqrt{\pi}} d(x) + \left[1 - \frac{e^{-x^2}}{\sqrt{\pi}x}\right] \cdot [1 - d(x)], \quad x \geq 0, \quad (6)$$

where

$$d(x) = \frac{\sqrt{\pi}x\mathrm{erf}(x) - \sqrt{\pi}x + e^{-x^2}}{2x^2 - \sqrt{\pi}x + e^{-x^2}}. \quad (7)$$

Naturally, an approximation of d requires an approximation of the error function. Unsurprisingly, the relative error in the approximation for the error function equals the relative error in the approximation utilized to approximate the error function in d.

Finally, efficient numerical implementation of the error function is of interest and Chevillard [16] and De Schrijver [17] provide results and an overview. Highly accurate piecewise approximations have long since been defined, e.g., [18].

The two-point spline-based approximations for functions and integrals, detailed in [19], have recently been applied to find arbitrarily accurate approximations for the hyperbolic tangent function [20]. In this paper, the general two-point spline approximation form is applied to define a sequence of convergent approximations for the error function. The basic form of the approximation of order n, denoted f_n, is

$$\mathrm{erf}(x) \approx f_n(x) = p_{n,0}(x) + p_{n,1}(x)e^{-x^2}, \quad (8)$$

where $p_{n,1}$ is a polynomial of order n and $p_{n,0}$ is a polynomial of order less than n. Convergence of the sequence of approximations $f_1, f_2, \ldots$ to $\mathrm{erf}(x)$ is shown and the convergence is significantly better than the default Taylor series. For example, the second-order approximation

$$f_2(x) = \frac{x}{\sqrt{\pi}} \cdot \left[1 - \frac{x^2}{30}\right] + \frac{x}{\sqrt{\pi}} \cdot \left[1 + \frac{11x^2}{30} + \frac{x^4}{15}\right] e^{-x^2} \quad (9)$$

yields a relative error bound of 0.056 over the interval [0, 2] which is better than a fifteenth-order Taylor series approximation. The approximations can be improved by utilizing the approximation $\mathrm{erf}(x) \approx 1$ for $|x| \gg 1$ and thereby establishing approximations with a set relative error bound over the interval $[0, \infty)$.

Table 1. Examples of published approximations for $\mathrm{erf}(x)$, $0 < x < \infty$. For the third and second last approximations, the coefficient definitions are detailed in the associated reference. The stated relative error bounds arise from sampling the interval [0, 5] with 10,000 uniformly spaced points.

#	Reference	Approximation	Relative Error Bound
1	Taylor series	$T_n(x) = \frac{2}{\sqrt{\pi}} \cdot \left[x - \frac{x^3}{3 \cdot 1} + \frac{x^5}{5 \cdot 2!} - \ldots + \frac{(-1)^{(n-1)/2} x^n}{n \cdot [(n-1)/2]!}\right]$ $n \in \{1, 3, 5, \ldots\}$	
2	Abramowitz [21], p. 297, Equation (7.1.6)	$\frac{2}{\sqrt{\pi}}\left[x + \frac{2x^3}{1 \cdot 3} + \frac{2^2 x^5}{3 \cdot 5} + \frac{2^3 x^7}{3 \cdot 5 \cdot 7} + \ldots + \frac{2^n x^{2n+1}}{1 \cdot 3 \cdot 5 \cdot \ldots \cdot (2n+1)}\right] e^{-x^2}$	
3	Abramowitz [21], p. 299, Equation (7.1.26)	$1 - \left[\frac{a_1}{1+px} + \frac{a_2}{(1+px)^2} + \frac{a_3}{(1+px)^3} + \frac{a_4}{(1+px)^4} + \frac{a_5}{(1+px)^5}\right] e^{-x^2}$	8.09×10^{-6}
4	Menzel [14] and Nadagopal [22]	$\sqrt{1 - \exp\left[\frac{-4x^2}{\pi}\right]}$	7.07×10^{-3}
5	Bürmann series [23], Equation (33)	$\frac{2}{\sqrt{\pi}} \cdot \sqrt{1 - \exp(-x^2)} \cdot \left[\frac{\sqrt{\pi}}{2} + \frac{31}{200} e^{-x^2} - \frac{341}{8000} e^{-2x^2}\right]$	3.61×10^{-3}
6	Winitzki [24], Equation (3)	$\sqrt{1 - \exp\left[-x^2 \cdot \frac{4/\pi + ax^2}{1 + ax^2}\right]}, \quad a = \frac{8(\pi-3)}{3\pi(4-\pi)}$	3.50×10^{-4}
7	Soranzo [25], Equation (1)	$\sqrt{1 - \exp\left[-x^2 \cdot \frac{a_1 + a_2 x^2}{1 + b_2 x^2 + b_3 x^4}\right]}, \quad \begin{cases} a_1 = 1.2735457 \\ a_2 = 0.1487936 \end{cases}$ $b_2 = 0.1480931 \quad b_3 = 5.160 \times 10^{-4}$	1.20×10^{-4}
8	Vedder [26], Equation (5)	$\tanh\left[\frac{167x}{148} + \frac{11x^3}{109}\right]$	4.65×10^{-3}
9	Vazquez-Leal [27], Equation (3.1)	$\tanh\left[\frac{39x}{2\sqrt{\pi}} - \frac{111}{2} \cdot \tan^{-1}\left[\frac{35x}{111\sqrt{\pi}}\right]\right]$	1.88×10^{-4}
10	Sandoval-Hernandez [13], Equation (23)	$\frac{2}{1 + \exp[\alpha_1 x + \alpha_3 x^3 + \alpha_5 x^5 + \alpha_7 x^7 + \alpha_9 x^9]} - 1$	3.05×10^{-5}
11	Abrarov [28], Equation (16)	$1 - e^{-x^2}\left[\frac{1 - e^{-\tau_m x}}{\tau_m x} + \frac{\tau_m^2 x}{\sqrt{\pi}} \cdot \sum_{n=1}^{N} \frac{a_n[1 - (-1)^n e^{-\tau_m x}]}{n^2\pi^2 + \tau_m^2 x^2}\right]$ $a_n = \frac{2\sqrt{\pi}}{\tau_m} \cdot \exp[-n^2\pi^2/\tau_m^2], \quad \tau_m = 12$	3.27×10^{-3} ($N = 6$)

Table 2. Residual functions associated with approximations for $\mathrm{erf}(x)$, $0 < x < \infty$.

#	Error Function	Residual Function
1	$\tanh\left[\frac{2x}{\sqrt{\pi}}\right] + g_1(x)$	$g_1(x) = \mathrm{erf}(x) - \tanh\left[\frac{2x}{\sqrt{\pi}}\right]$
2	$\tanh\left[\frac{2x}{\sqrt{\pi}}[1 + g_2(x)]\right]$	$g_2(x) = \frac{\sqrt{\pi}}{2x} \cdot \mathrm{atanh}[\mathrm{erf}(x)] - 1$
3	$\sqrt{1 - \exp\left[-x^2 \cdot \frac{4}{\pi}[1 + g_3(x)]\right]}$	$g_3(x) = \frac{-\pi}{4x^2} \cdot \ln\left[1 - \mathrm{erf}(x)^2\right] - 1$

Two generalizations are detailed. The first is of the form

$$\mathrm{erf}(x) \approx p_0(x) + p_1(x) e^{-k_1 x^2} + \ldots + p_m(x) e^{-k_m x^2} \tag{10}$$

and is based on utilizing approximations associated with m equally spaced subintervals of the interval $[0, x]$. The second is based on utilizing a fixed subinterval within $[0, x]$ and then approximating the error function over the remainder of the interval. Both generalizations lead to significantly improved accuracy. For example, a fourth-order approximation based on four variable subintervals, when used with the approximation $\mathrm{erf}(x) \approx 1$ for $x \gg 1$, has a relative error bound of 1.43×10^{-7} over the interval $[0, \infty]$. The corresponding sixteenth-order approximation has a relative error bound of 2.01×10^{-19}. Finally, by utilizing the solutions of a custom dynamical system, approximations with better convergence properties can be established.

Applications of the proposed approximations for the error function include, first, the definition of functions that are upper and lower bounds, of arbitrary accuracy, for the error function. Second, new series for the error function. Third, new sequences of approximations for $\exp(-x^2)$ that have significantly higher convergence properties than a Taylor series approximation. Fourth, the definition of a complementary demarcation function $e_C(x)$ that satisfies the constraint $e_C^2(x) + \mathrm{erf}^2(x) = 1$. Fifth, arbitrarily accurate approximations for the power and harmonic distortion for a sinusoidal signal subject to an error function nonlinearity. Sixth, approximate expressions for the linear filtering of a step signal modeled by the error function.

Section 2 details the spline-based approximation for the error function and its convergence. Improved approximations, obtained by utilizing the nature of the error function for large arguments, are detailed in Section 3. Two generalizations, with the potential for lower relative error bounds, are detailed in Sections 4 and 5. Section 6 details how the initial approximations, and the approximations arising from the first generalization, can be utilized as inputs to a custom dynamical system to establish approximations with better convergence properties. Applications are specified in Section 7. Conclusions are stated in Section 8.

1.1. Notes and Notation

As $\mathrm{erf}(-x) = -\mathrm{erf}(x)$, it is sufficient to consider approximations for the interval $[0, \infty)$.

For a function f defined over the interval $[\alpha, \beta]$, an approximating function f_A has a relative error, at a point x_1, defined according to $\mathrm{re}(x_1) = 1 - f_A(x_1)/f(x_1)$. The relative error bound for the approximating function over the interval $[\alpha, \beta]$ is defined according to

$$\mathrm{re_B} = \max\{|re(x_1)|: x_1 \in [\alpha, \beta]\}. \tag{11}$$

The notation $f^{(k)}(x) = \frac{d^k}{dx^k} f(x)$ is used. The symbol u denotes the unit step function. Mathematica has been used to facilitate the analysis and to obtain numerical results.

1.2. Background Results

The following result underpins the bounds proposed for the error function:

Lemma 1. *Upper and Lower Functional Bounds. A positive approximation f_A to a positive function f over the interval $[\alpha, \beta]$, with a relative error bound*

$$-\varepsilon_B < 1 - \frac{f_A(x)}{f(x)} < \varepsilon_B, \quad x \in [\alpha, \beta], \varepsilon_B > 0, \tag{12}$$

leads to the following upper and lower bounded functions:

$$\frac{f_A(x)}{1+\varepsilon_B} < f(x) < \frac{f_A(x)}{1-\varepsilon_B}, \quad x \in [\alpha, \beta]. \tag{13}$$

The relative error bounds, over the interval $[\alpha, \beta]$, for the upper and lower bounded functions, respectively, are:

$$\frac{2\varepsilon_B}{1+\varepsilon_B}, \frac{2\varepsilon_B}{1-\varepsilon_B}. \tag{14}$$

Proof. The definition of the relative error bound, as specified by Equation (12), leads to

$$1 - \varepsilon_B < \frac{f_A(x)}{f(x)} < 1 + \varepsilon_B, \tag{15}$$

which implies

$$\frac{1-\varepsilon_B}{1+\varepsilon_B} < \frac{f_A(x)/(1+\varepsilon_B)}{f(x)} < 1, 1 < \frac{f_A(x)/(1-\varepsilon_B)}{f(x)} < \frac{1+\varepsilon_B}{1-\varepsilon_B} \tag{16}$$

and the relative error bounds:

$$\begin{aligned} &0 < 1 - \frac{f_A(x)/(1+\varepsilon_B)}{f(x)} < 1 - \frac{1-\varepsilon_B}{1+\varepsilon_B} = \frac{2\varepsilon_B}{1+\varepsilon_B} \\ &1 - \frac{1+\varepsilon_B}{1-\varepsilon_B} = \frac{-2\varepsilon_B}{1-\varepsilon_B} < 1 - \frac{f_A(x)/(1-\varepsilon_B)}{f(x)} < 0 \end{aligned} \tag{17}$$

□

Convergent Integral Approximations

One application of the proposed approximations for the error function requires knowledge of when function convergence implies convergence of associated integrals.

Lemma 2. *Convergent Integral Approximation. If a sequence of functions $f_1, f_2, \ldots$ converges, over the interval $[0, x]$, to a bounded and integrable function f, then point-wise convergence is sufficient for the associated integrals to be convergent, i.e., for*

$$\lim_{n\to\infty} \int_0^x f_n(\lambda)d\lambda = \int_0^x f(\lambda)d\lambda. \tag{18}$$

Proof. The required result follows if it is possible to interchange the order of limit and integration, i.e.,

$$\lim_{n\to\infty} \int_0^x f_n(\lambda)d\lambda = \int_0^x \lim_{n\to\infty} f_n(\lambda)d\lambda = \int_0^x f(\lambda)d\lambda. \tag{19}$$

Standard conditions for when the interchange is valid are specified by the monotone and dominated convergence theorems, e.g., [29], (p. 26). Sufficient conditions for a valid interchange include point-wise function convergence, and for f to be integrable and bounded. □

2. Spline-Based Approximations for Error Function

2.1. Spline Approximation for Error Function

The following nth-order, two-point spline-based approximation for an integral has been detailed in Equation (48) of [19], for a function f that is at least (n + 1)th-order differentiable:

$$\begin{aligned} \int_\alpha^x f(\lambda)d\lambda = \sum_{k=0}^{n} c_{n,k}(x-\alpha)^{k+1}\left[f^{(k)}(\alpha) + (-1)^k f^{(k)}(x)\right] + R_n(\alpha, x) \\ n \in \{0, 1, 2, \ldots\} \end{aligned} \tag{20}$$

where

$$c_{n,k} = \frac{n!}{(n-k)!(k+1)!} \cdot \frac{(2n+1-k)!}{2(2n+1)!}, \quad k \in \{0, 1, \ldots, n\}, \tag{21}$$

$$\begin{aligned} R_n^{(1)}(\alpha, x) = -\sum_{k=0}^{n} c_{n,k}(k+1)(x-\alpha)^k\left[f^{(k)}(\alpha) + (-1)^{k+1} \cdot \frac{n+1}{n-k+1} \cdot f^{(k)}(x)\right] + \\ c_{n,n}(-1)^{n+1}(x-\alpha)^{n+1} f^{(n+1)}(x). \end{aligned} \tag{22}$$

Direct application of this result to the integral defining the error function leads to the results stated in Theorem 1:

Theorem 1. *Spline-Based Integral Approximation for Error Function. The error function can be defined according to*

$$\operatorname{erf}(x) = f_n(x) + \varepsilon_n(x), \tag{23}$$

where f_n is the nth-order spline-based integral approximation defined according to

$$f_n(x) = \frac{2}{\sqrt{\pi}} \cdot \sum_{k=0}^{n} c_{n,k} x^{k+1} \left[p(k,0) + (-1)^k p(k,x) e^{-x^2} \right] \tag{24}$$

and $\varepsilon_n(x)$ is the associated residual function whose derivative is

$$\begin{aligned} \varepsilon_n^{(1)}(x) = & \frac{2e^{-x^2}}{\sqrt{\pi}} - \frac{2}{\sqrt{\pi}} \cdot \sum_{k=0}^{n} c_{n,k}(k+1)x^k p(k,0) - \\ & \frac{2e^{-x^2}}{\sqrt{\pi}} \cdot \sum_{k=0}^{n} c_{n,k} x^k (-1)^k \left[(k+1-2x^2) p(k,x) + x p^{(1)}(k,x) \right] \end{aligned} \tag{25}$$

In these equations,

$$p(k,x) = p^{(1)}(k-1,x) - 2xp(k-1,x), \ \ p(0,x) = 1. \tag{26}$$

A more general approximation is

$$\begin{aligned} \mathrm{erf}(x) - \mathrm{erf}(\alpha) & = \frac{2}{\sqrt{\pi}} \cdot \sum_{k=0}^{n} c_{n,k}(x-\alpha)^{k+1} \left[p(k,\alpha) e^{-\alpha^2} + (-1)^k p(k,x) e^{-x^2} \right] + \varepsilon_n(\alpha, x) \\ & = f_n(\alpha, x) + \varepsilon_n(\alpha, x). \end{aligned} \tag{27}$$

Proof. The proof is detailed in Appendix A. □

2.1.1. Note

The polynomial function $p(k,x)$ is equivalently defined by the kth-order Hermite polynomial function ([21], p. 775, Equation (23.3.10)) and an explicit form is

$$p(k,x) = \sum_{i=0}^{\lfloor k/2 \rfloor} \frac{(-1)^{i+k} k!}{i!(k-2i)!} \cdot 2^{k-2i} x^{k-2i}. \tag{28}$$

A change of variable $r = k - 2i$, and noting that $i \in \{0, 1, \ldots, \lfloor k/2 \rfloor\}$ implies $r \in \{k, k-2, \ldots, k - 2\lfloor k/2 \rfloor\}$, leads to the alternative form

$$p(k,x) = \sum_{r=k-2\lfloor k/2 \rfloor}^{k} d_{k,r} x^r, \quad d_{k,r} = \frac{(-1)^{(3k-r)/2} \left[\frac{1+(-1)^{r-\lfloor k-2\lfloor k/2 \rfloor \rfloor}}{2} \right] k! 2^r}{[(k-r)/2]! r!}. \tag{29}$$

Substitution of this form into Equation (24) leads to the direct polynomial form for the nth-order approximation to the error function:

$$f_n(x) = \frac{2}{\sqrt{\pi}} \cdot \sum_{k=0}^{n} c_{n,k} p(k,0) x^{k+1} + \frac{2e^{-x^2}}{\sqrt{\pi}} \cdot \sum_{k=0}^{n} \left[(-1)^k c_{n,k} \sum_{r=k-2\lfloor k-2 \rfloor}^{k} d_{k,r} x^{r+k+1} \right]. \tag{30}$$

2.1.2. Approximations

The polynomial functions p, as defined by Equations (26), (28), and (29), have the explicit forms:

$$\begin{aligned} & p(0,x) = 1, \quad p(1,x) = -2x, \quad p(2,x) = -2[1-2x^2], \\ & p(3,x) = 12x[1-2x^2/3], \quad p(4,x) = 12[1-4x^2+4x^4/3], \quad \ldots \end{aligned} \tag{31}$$

Approximations for the error function, as defined by Equation (24) and for orders zero to five, are:

$$f_0(x) = \frac{x}{\sqrt{\pi}} + \frac{x}{\sqrt{\pi}} \cdot e^{-x^2} \tag{32}$$

$$f_1(x) = \frac{x}{\sqrt{\pi}} + \frac{x}{\sqrt{\pi}}\left[1 + \frac{x^2}{3}\right]e^{-x^2} \tag{33}$$

$$f_2(x) = \frac{x}{\sqrt{\pi}}\left[1 - \frac{x^2}{30}\right] + \frac{x}{\sqrt{\pi}}\left[1 + \frac{11x^2}{30} + \frac{x^4}{15}\right]e^{-x^2} \tag{34}$$

$$f_3(x) = \frac{x}{\sqrt{\pi}}\left[1 - \frac{x^2}{21}\right] + \frac{x}{\sqrt{\pi}}\left[1 + \frac{8x^2}{21} + \frac{17x^4}{210} + \frac{x^6}{105}\right]e^{-x^2} \tag{35}$$

$$f_4(x) = \frac{x}{\sqrt{\pi}}\left[1 - \frac{x^2}{18} + \frac{x^4}{1260}\right] + \frac{x}{\sqrt{\pi}}\left[1 + \frac{7x^2}{18} + \frac{37x^4}{420} + \frac{4x^6}{315} + \frac{x^8}{945}\right]e^{-x^2} \tag{36}$$

$$f_5(x) = \frac{x}{\sqrt{\pi}}\left[1 - \frac{2x^2}{33} + \frac{x^4}{660}\right] + \frac{x}{\sqrt{\pi}}\left[1 + \frac{13x^2}{33} + \frac{61x^4}{660} + \frac{67x^6}{4620} + \frac{16x^8}{10,395} + \frac{x^{10}}{10,395}\right]e^{-x^2} \tag{37}$$

2.2. Results

The relative error in the zero- to tenth-order spline-based series approximations, along with the relative error in Taylor series approximations of orders 1–15, are detailed in Figure 2. The clear superiority, in terms of convergence, of the spline-based series relative to the Taylor series is evident. The relative error in the spline approximations, of orders 16, 20, 24, 28, and 32, are shown in Figure 3.

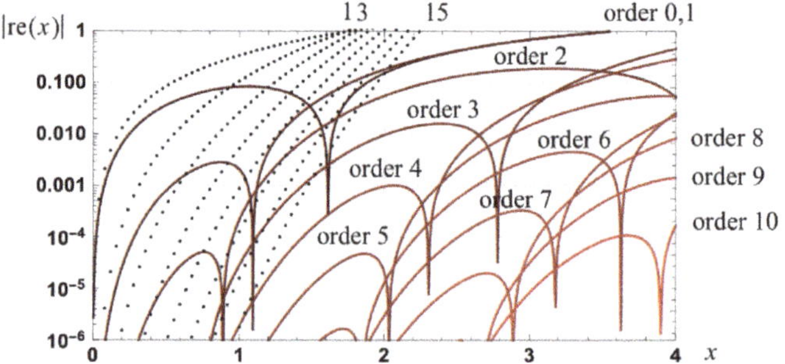

Figure 2. Graph of the magnitude of the relative errors in approximations to erf(x): zero to tenth order integral spline based series and first, third, ..., fifteenth order Taylor series (dotted).

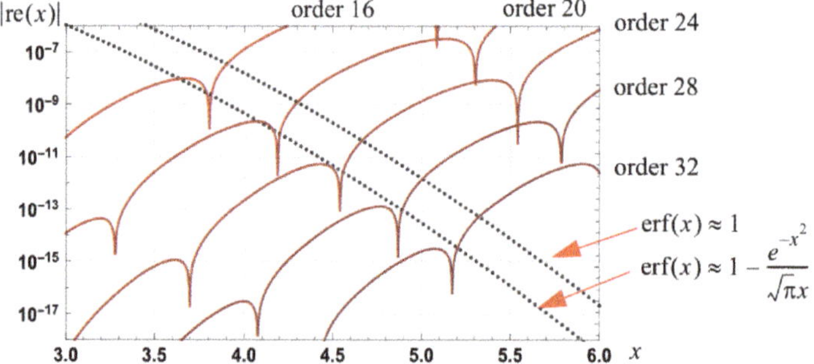

Figure 3. Graph of the magnitude of the relative errors associated with the approximation $\text{erf}(x) \approx 1$ and $\text{erf}(x) \approx 1 - \exp(-x^2)/\sqrt{\pi}x$ along with the relative error in spline approximations of orders 16, 20, 24, 28 and 32.

The Mathematica code underpinning the results shown in Figure 2, is detailed in Supplementary Material. Such code is indicative of the code underpinning the results detailed in the paper.

2.3. Approximation for Large Arguments

Zero- and first-order approximations for the error function, and for the case of $x \gg 1$, are

$$\operatorname{erf}(x) \approx 1, \quad \operatorname{erf}(x) \approx 1 - \frac{e^{-x^2}}{\sqrt{\pi}x}. \tag{38}$$

The relative errors in such approximations, respectively, are

$$\operatorname{re}(x) = 1 - \frac{1}{\operatorname{erf}(x)}, \quad \operatorname{re}(x) \approx 1 - \frac{1 - e^{-x^2}/\sqrt{\pi}x}{\operatorname{erf}(x)}, \tag{39}$$

and their graphs are shown in Figure 3.

2.4. Convergence

To prove the convergence of the sequence of functions $f_0, f_1, f_2, \ldots$, defined by Theorem 1, to the error function, it is sufficient to prove that the corresponding sequence of residual functions $\varepsilon_0, \varepsilon_1, \varepsilon_2, \ldots$ converge to zero. This can be shown by considering the derivatives of the residual functions defined by Equation (25). The derivatives of the residual functions of orders zero, one, and two, respectively, are:

$$\varepsilon_0^{(1)}(x) = \frac{1}{\sqrt{\pi}}\left[1 + 2x^2\right]e^{-x^2} - \frac{1}{\sqrt{\pi}} \tag{40}$$

$$\varepsilon_1^{(1)}(x) = \frac{1}{\sqrt{\pi}}\left[1 + x^2 + \frac{2x^4}{3}\right]e^{-x^2} - \frac{1}{\sqrt{\pi}} \tag{41}$$

$$\varepsilon_2^{(1)}(x) = \frac{1}{\sqrt{\pi}}\left[1 + \frac{9x^2}{10} + \frac{2x^4}{5} + \frac{2x^6}{15}\right]e^{-x^2} - \frac{1}{\sqrt{\pi}}\left[1 - \frac{x^2}{10}\right]. \tag{42}$$

Theorem 2. *Convergence of Spline-Based Approximations. For all fixed values of x, the derivatives of the residual functions converge to zero as the order increases, i.e., for all fixed values of x it is the case that*

$$\lim_{n \to \infty} \varepsilon_n^{(1)}(x) = 0, \quad x > 0. \tag{43}$$

This is sufficient for the convergence of the residual functions, i.e., $\lim_{n \to \infty} \varepsilon_n(x) = 0$, $x > 0$, *and, hence, for x fixed:*

$$\lim_{n \to \infty} f_n(x) = \operatorname{erf}(x), \quad x > 0. \tag{44}$$

The convergence is nonuniform.

Proof. The proof is detailed in Appendix B. □

2.5. Improved Approximation Based on Iteration

Consider the general result

$$\int_0^x \operatorname{erf}(\lambda)d\lambda = \frac{1}{\sqrt{\pi}x}\left[1 - e^{-x^2}\right] + x\operatorname{erf}(x) \tag{45}$$

By using the approximations, f_n, as defined in Theorem 1, in the integral, improved approximations for the error function can be defined.

Theorem 3. *Improved Approximation via Iteration. An improved approximation, of order n, for the error function is*

$$\begin{aligned} F_n(x) = & \frac{1}{\sqrt{\pi}x}\left[1 - e^{-x^2}\right] + \frac{2}{\sqrt{\pi}} \cdot \sum_{k=0}^{n} \frac{c_{n,k}p(k,0)}{k+2} \cdot x^{k+1} + \\ & \frac{2}{x\sqrt{\pi}} \cdot \sum_{k=0}^{n} \left[(-1)^k c_{n,k} \sum_{r=k-2\lfloor k/2 \rfloor}^{k} \frac{d_{k,r}}{2}\left[\frac{r+k}{2}\right]! \left[1 - e^{-x^2} \sum_{i=0}^{\frac{r+k}{2}} \frac{x^{2i}}{i!}\right]\right] \end{aligned} \tag{46}$$

where $c_{n,k}$ and $d_{k,r}$ are defined, respectively, by Equations (21) and (29).

Proof. From Equation (45), it follows that

$$\mathrm{erf}(x) \approx \frac{1}{\sqrt{\pi}x}\left[1 - e^{-x^2}\right] + \frac{1}{x} \cdot \int_0^x f_n(\lambda) d\lambda. \tag{47}$$

As

$$\int_0^x x^u e^{-\lambda^2} d\lambda = \frac{1}{2}\left[\frac{u-1}{2}\right]! \left[1 - e^{-x^2} \sum_{i=0}^{(u-1)/2} \frac{x^{2i}}{i!}\right], \quad u \in \{1, 3, 5, \ldots\}, \tag{48}$$

it follows, from the form for f_n detailed in Equation (30), that

$$\begin{aligned} \mathrm{erf}(x) \approx & \frac{1}{\sqrt{\pi}x}\left[1 - e^{-x^2}\right] + \frac{2}{\sqrt{\pi}} \cdot \sum_{k=0}^{n} \frac{c_{n,k}p(k,0)}{k+2} \cdot x^{k+1} + \\ & \frac{2}{x\sqrt{\pi}} \cdot \sum_{k=0}^{n} \left[(-1)^k c_{n,k} \sum_{r=k-2\lfloor k-2 \rfloor}^{k} \frac{d_{k,r}}{2}\left[\frac{r+k}{2}\right]! \left[1 - e^{-x^2} \sum_{i=0}^{\frac{r+k}{2}} \frac{x^{2i}}{i!}\right]\right] \end{aligned} \tag{49}$$

which is the required result. □

2.5.1. Explicit Approximations

Approximations to the error function, of orders zero to five, are:

$$f_0(x) = \frac{3}{2\sqrt{\pi}x}\left[1 + \frac{x^2}{3}\right] - \frac{3e^{-x^2}}{2\sqrt{\pi}x} \tag{50}$$

$$f_1(x) = \frac{5}{3\sqrt{\pi}x}\left[1 + \frac{3x^2}{10}\right] - \frac{5}{3\sqrt{\pi}x}\left[1 + \frac{x^2}{10}\right]e^{-x^2} \tag{51}$$

$$f_2(x) = \frac{7}{4\sqrt{\pi}x}\left[1 + \frac{2x^2}{7} - \frac{x^4}{210}\right] - \frac{7}{4\sqrt{\pi}x}\left[1 + \frac{x^2}{7} + \frac{2x^4}{105}\right]e^{-x^2} \tag{52}$$

$$f_3(x) = \frac{9}{5\sqrt{\pi}x}\left[1 + \frac{5x^2}{18} - \frac{5x^4}{756}\right] - \frac{9}{5\sqrt{\pi}x}\left[1 + \frac{x^2}{6} + \frac{23x^4}{756} + \frac{x^6}{378}\right]e^{-x^2} \tag{53}$$

$$\begin{aligned} f_4(x) = & \frac{11}{6\sqrt{\pi}x}\left[1 + \frac{3x^2}{11} - \frac{x^4}{132} + \frac{x^6}{13{,}860}\right] - \\ & \frac{11}{6\sqrt{\pi}x}\left[1 + \frac{2x^2}{11} + \frac{5x^4}{132} + \frac{16x^6}{3465} + \frac{x^8}{3465}\right]e^{-x^2} \end{aligned} \tag{54}$$

$$\begin{aligned} f_5(x) = & \frac{13}{7\sqrt{\pi}x}\left[1 + \frac{7x^2}{26} - \frac{7x^4}{858} + \frac{7x^6}{51{,}480}\right] - \\ & \frac{13}{7\sqrt{\pi}x}\left[1 + \frac{5x^2}{26} + \frac{37x^4}{858} + \frac{313x^6}{51{,}480} + \frac{7x^8}{12{,}850} + \frac{x^{10}}{38{,}610}\right]e^{-x^2} \end{aligned} \tag{55}$$

Note that integration of these expressions leads to functions defined, in part, by the Gamma function that is an integral. This makes further iteration impractical.

2.5.2. Results

The relative errors in even order approximations, of orders 0–10, are shown in Figure 4. A comparison of the results detailed in Figures 2 and 4 show the clear improvement in the approximations specified by Equation (46).

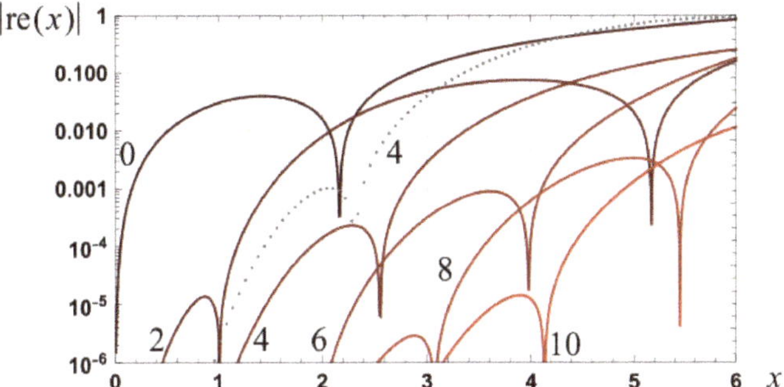

Figure 4. Graph of the magnitude of the relative errors in the approximations to erf(x), of even orders, as specified by Equation (46). The dotted results are for the fourth order approximation specified by Theorem 1 (Equation (36)).

3. Improved Approximations

3.1. Improved Approximation for Error Function

An improved approximation for the error function can be achieved by noting, as illustrated in Figure 3, that the approximation $\mathrm{erf}(x) \approx 1$ is increasingly accurate for the case of $x \gg 1$ and for x increasing. By switching at a suitable point x_o, as illustrated in Figure 5, from a spline-based approximation to the approximation $\mathrm{erf}(x) \approx 1$, an improved approximation is achieved. Naturally, it is possible to switch to the approximation $\mathrm{erf}(x) \approx 1 - e^{-x^2}/\sqrt{\pi}x$, or higher-order approximations, in a similar manner.

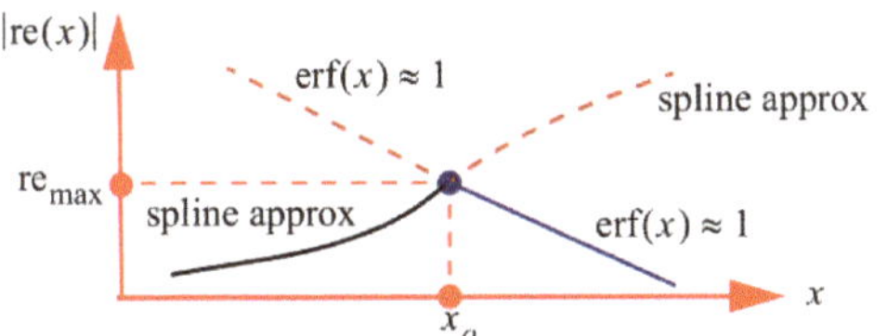

Figure 5. Illustration of the crossover point where the magnitude of the relative error in the approximation $\mathrm{erf}(x) \approx 1$ equals the magnitude of the relative error in a set order spline approximation.

Theorem 4. *Improved Approximation for Error Function. Improved approximations for the error function, based on the nth-order approximations detailed in Theorems 1 and 3, and consistent with the illustration shown in* Figure 5, *are*

$$\begin{aligned} \mathrm{erf}(x) &\approx f_n(x)u[x_o(n) - x] + [1 - u[x_o(n) - x]] \\ \mathrm{erf}(x) &\approx F_n(x)u[x_o(n) - x] + [1 - u[x_o(n) - x]] \end{aligned} \quad (56)$$

where the transition points, respectively, are defined according to

$$\begin{aligned} x_o(n) &= x : \left|1 - \tfrac{1}{\mathrm{erf}(x)}\right| = \left|1 - \tfrac{f_n(x)}{\mathrm{erf}(x)}\right| \\ x_o(n) &= x : \left|1 - \tfrac{1}{\mathrm{erf}(x)}\right| = \left|1 - \tfrac{F_n(x)}{\mathrm{erf}(x)}\right| \end{aligned} \quad (57)$$

Proof. The improved approximation results follow from optimally switching, as illustrated in Figure 5 and at the point specified by Equation (57), to the approximation $\text{erf}(x) \approx 1$, which has a lower relative error magnitude. □

Transition Points and Relative Error Bounds

The transition points, for various orders of spline approximation, are specified in Table 3. The relationship between the transition point and order is shown in Figure 6 for the case of the approximations f_n. This relationship can be approximated, with a second-order polynomial, according to

$$x_o(n) = 1.3607 + 0.20511n - 0.002932n^2, \quad 0 \leq n \leq 24. \tag{58}$$

Table 3. The transition points, x_o, and the resulting relative error bounds for the spline-based approximations specified by Equation (56). The transition points are based on sampling the interval [0, 5] with 10,000 points.

Approx. Order: n	Transition Point for f_n	Relative Error Bound for f_n	Transition Point for F_n	Relative Error Bound for F_n
0	1.3085	0.0851	1.465	0.0400
1	1.492	0.0362	1.769	0.0126
2	1.658	1.95×10^{-2}	1.929	6.42×10^{-3}
3	1.8975	7.36×10^{-3}	2.1725	2.13×10^{-3}
4	2.3715	1.03×10^{-3}	2.6305	2.28×10^{-4}
6	2.4715	4.75×10^{-4}	2.73	1.13×10^{-4}
8	2.963	2.79×10^{-5}	3.1855	6.69×10^{-6}
10	3.0785	1.35×10^{-5}	3.324	2.59×10^{-6}
12	3.4625	9.78×10^{-7}	3.67	2.12×10^{-7}
14	3.5845	4.00×10^{-7}	3.8205	6.57×10^{-8}
16	3.9025	3.44×10^{-8}	4.101	6.66×10^{-9}
18	4.0285	1.22×10^{-8}	4.257	1.75×10^{-9}
20	4.300	1.20×10^{-9}	4.493	2.11×10^{-10}
22	4.429	3.76×10^{-10}	4.652	4.75×10^{-11}
24	4.6655	4.18×10^{-11}	4.854	6.70×10^{-12}

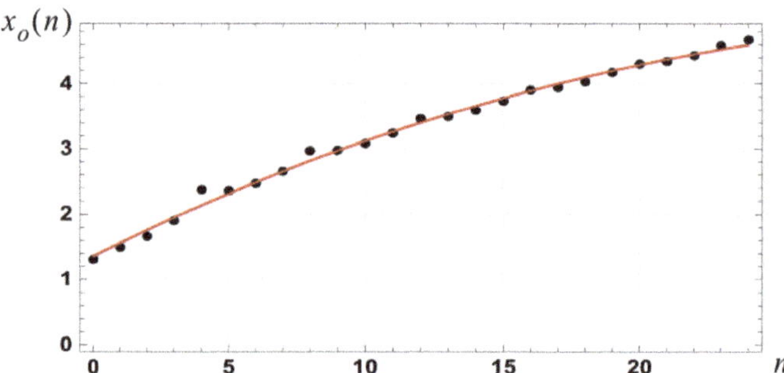

Figure 6. Graph of the relationship between the optimum transition point $x_o(n)$, as defined by Equation (57) for the case of f_n and the order of the spline approximation.

However, as small variations in $x_o(n)$ can lead to significant changes in the maximum relative error in the approximation for the error function, precise values for $x_o(n)$ are preferable.

The graphs of the relative errors in the approximations f_n to erf(x), as specified by Equation (56), are shown in Figure 7 for orders 2, 4, 6, ..., 20. The relative error bounds that can be achieved over the interval $[0, \infty]$, using the optimally chosen transition points, are detailed in Table 3.

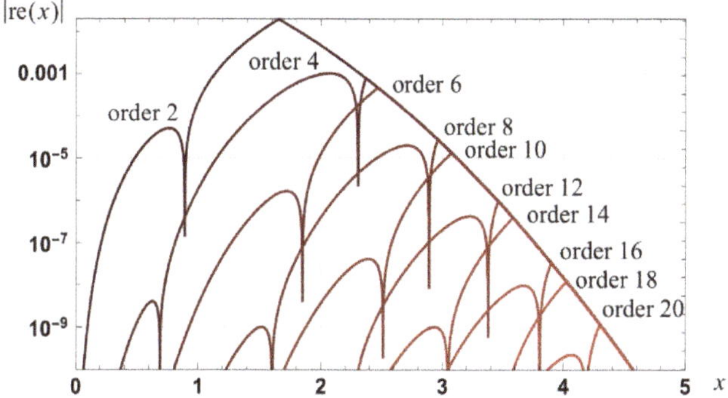

Figure 7. Graph of the relative errors in the approximations, f_n, to erf(x), of orders 2, 4, 6, ..., 20, based on utilizing the approximation $\mathrm{erf}(x) \approx 1$ in an optimum manner.

3.2. Improved Approximation for Taylor Series

The approximation of $\mathrm{erf}(x) \approx 1$ for $x \gg 1$ can be utilized to improve the relative error bound for a Taylor series approximation according to

$$\mathrm{erf}(x) \approx T_n(x)u[x_o(n) - x] + [1 - u[x_o(n) - x]], \tag{59}$$

where T_n is a nth-order Taylor series, with n odd, as specified in Table 1. The optimum transition points and relative error bounds, for selected orders, are detailed in Table 4. The variation of the relative errors, with order, are shown in Figure 8. The change in the optimum transition point can be approximated according to

$$x_o(n) \approx 0.932 + 0.0560n - 0.0003503n^2, \ 3 \le n \le 61, \tag{60}$$

but, again, as small variations in $x_o(n)$ can lead to significant changes in the maximum relative error in the approximation for the error function, precise values for $x_o(n)$ are preferable. The clear superiority in the convergence of the spline-based series is evident from a visual comparison of the relative errors shown in Figures 7 and 8.

Table 4. The transition points, and resulting relative error bounds, for Taylor series approximations specified by Equation (59). The transition points are based on sampling the interval [0, 4] with 10,000 points.

Order: n	Transition Point : $x_o(n)$	Relative Error Bound in T_n
1	0.8864	0.266
3	1.078	0.146
5	1.222	0.0917
7	1.344	0.0609
9	1.4532	0.0416
13	1.6452	0.0204
17	1.8144	0.0105
21	1.9672	5.44×10^{-3}
25	2.1084	2.89×10^{-3}
29	2.24	1.55×10^{-3}
37	2.4812	4.53×10^{-4}
45	2.70	1.35×10^{-4}
53	2.902	4.09×10^{-5}
61	3.09	1.24×10^{-5}

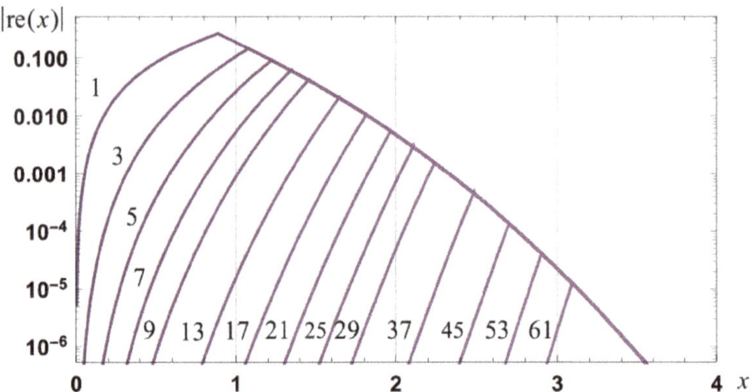

Figure 8. Graph of the magnitude of the relative error in Taylor series approximations to $\text{erf}(x)$ that utilize an optimized change to the approximation $\text{erf}(x) \approx 1$.

4. Variable Subinterval Approximations for Error Function

An improved analytic approximation for the error function can be achieved by demarcating the interval $[0, x]$ into variable subintervals, e.g., the subintervals $[0, x/4]$, $[x/4, x/2]$, $[x/2, 3x/4]$, and $[3x/4, x]$ for the four-subintervals case, and by utilizing spline-based integral approximations for each subinterval. Chiani [30] utilized subintervals to enhance approximations for the complementary error function.

Theorem 5. *Variable Subinterval Approximations for Error Function. The nth-order spline-based approximation to the error function, based on m equal-width subintervals, is*

$$f_{n,m}(x) = \frac{2}{\sqrt{\pi}} \sum_{i=0}^{m-1} \left[\sum_{k=0}^{n} c_{n,k} \left[\frac{x}{m}\right]^{k+1} \left[\begin{array}{c} p\left[k, \frac{ix}{m}\right] \exp\left[-\frac{i^2x^2}{m^2}\right] + \\ (-1)^k p\left[k, \frac{(i+1)x}{m}\right] \exp\left[-\frac{(i+1)^2x^2}{m^2}\right] \end{array} \right] \right] \tag{61}$$

where

$$p(k,x) = p^{(1)}(k-1,x) - 2xp(k-1,x), \;\; p(0,x) = 1. \tag{62}$$

An alternative form is

$$f_{n,m}(x) = \frac{2}{\sqrt{\pi}} \left[p_{n,0}(x) + \sum_{i=1}^{m-1} p_{n,i}(x) \exp\left[-\frac{i^2x^2}{m^2}\right] + p_{n,m}(x)\exp(-x^2) \right] \tag{63}$$

where

$$\begin{aligned} p_{n,0}(x) &= \sum_{k=0}^{n} \tfrac{c_{n,k}}{m^{k+1}} \cdot p(k,0)x^{k+1} \\ p_{n,i}(x) &= \sum_{k=0}^{n} \tfrac{c_{n,k}}{m^{k+1}} \cdot \left[1+(-1)^k\right] p(k, \tfrac{ix}{m})x^{k+1} \\ p_{n,m}(x) &= \sum_{k=0}^{n} \tfrac{c_{n,k}}{m^{k+1}} \cdot (-1)^k p(k,x)x^{k+1}. \end{aligned} \tag{64}$$

Proof. The first result follows by applying Equation (27) in Theorem 1 to the subintervals $[0, x/m]$, $[x/m, 2x/m]$, ..., $\left[x - \frac{x}{m}, x\right]$. The alternative form arises by expanding the outer summation in Equation (61) and collecting terms of similar form. □

4.1. Explicit Expressions

A first-order approximation, based on m subintervals, is

$$f_{1,m}(x) = \frac{x}{m\sqrt{\pi}}\left[1 + 2\sum_{i=1}^{m-1}\exp\left[\frac{-i^2x^2}{m^2}\right] + \exp(-x^2)\right] + \frac{x^3}{3m^2\sqrt{\pi}}\cdot\exp(-x^2). \quad (65)$$

For the four-subintervals case, explicit expressions are

$$f_{0,4}(x) = \frac{x}{4\sqrt{\pi}}\left[1 + 2\exp\left[\frac{-x^2}{16}\right] + 2\exp\left[\frac{-x^2}{4}\right] + 2\exp\left[\frac{-9x^2}{16}\right] + \exp(-x^2)\right] \quad (66)$$

$$f_{1,4}(x) = \frac{x}{4\sqrt{\pi}}\left[1 + 2\exp\left[\frac{-x^2}{16}\right] + 2\exp\left[\frac{-x^2}{4}\right] + 2\exp\left[\frac{-9x^2}{16}\right] + \exp(-x^2)\right] + \frac{x^3}{48\sqrt{\pi}}\cdot\exp(-x^2) \quad (67)$$

Using the alternative form, a fourth-order expression is

$$\begin{aligned} f_{4,4}(x) = \; & \tfrac{x}{4\sqrt{\pi}}\left[1 - \tfrac{x^2}{288} + \tfrac{x^4}{322{,}560}\right] + \\ & \tfrac{x}{2\sqrt{\pi}}\cdot\exp\left[\tfrac{-x^2}{16}\right]\left[1 - \tfrac{x^2}{288} + \tfrac{47x^4}{107{,}520} - \tfrac{x^6}{1{,}290{,}040} + \tfrac{x^8}{61{,}931{,}520}\right] + \\ & \tfrac{x}{2\sqrt{\pi}}\cdot\exp\left[\tfrac{-x^2}{4}\right]\left[1 - \tfrac{x^2}{288} + \tfrac{187x^4}{107{,}520} - \tfrac{x^6}{322{,}560} + \tfrac{x^8}{3{,}870{,}720}\right] + \\ & \tfrac{x}{2\sqrt{\pi}}\cdot\exp\left[\tfrac{-9x^2}{16}\right]\left[1 - \tfrac{x^2}{288} + \tfrac{1261x^4}{322{,}560} - \tfrac{x^6}{143{,}360} + \tfrac{3x^8}{2{,}293{,}760}\right] + \\ & \tfrac{x}{4\sqrt{\pi}}\cdot\exp(-x^2)\left[1 + \tfrac{31x^2}{288} + \tfrac{101x^4}{15{,}360} - \tfrac{19x^6}{80{,}640} + \tfrac{x^8}{241{,}920}\right] + \end{aligned} \quad (68)$$

A fourth-order spline approximation, which utilizes 16 subintervals, is detailed in Appendix C. This expression, when utilized with the transition point $x_o = 7.1544$, yields an approximation with a relative error bound of 4.82×10^{-16}.

Results

The relative errors in the spline approximations of orders one to six, and for the case of four equal subintervals $[0, x/4]$, $[x/4, x/2]$, $[x/2, 3x/4]$, and $[3x/4, x]$, are shown in Figure 9.

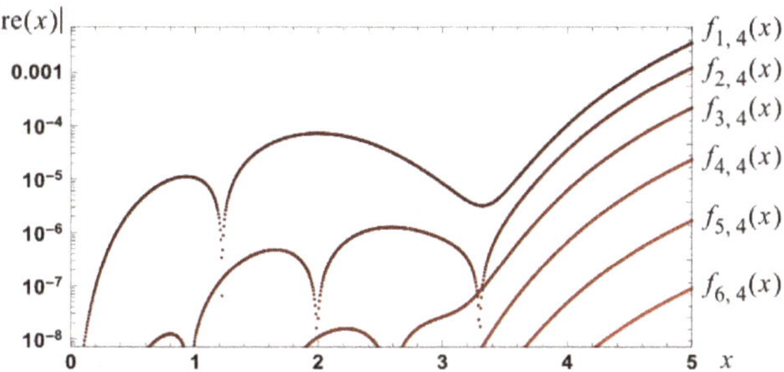

Figure 9. Graph of the relative errors in spline approximations to $\text{erf}(x)$, of orders one to six and based on four variable sub-intervals of equal width.

4.2. Improved Approximation

The spline approximations utilizing variable subintervals can be improved by using the transition to the approximation $\text{erf}(x) \approx 1$ at a suitable point, as specified by Equation (56). The relative error in the spline approximations of orders one to seven, and for the case of four equal subintervals $[0, x/4]$, $[x/4, x/2]$, $[x/2, 3x/4]$, and $[3x/4, x]$, are updated in Figure 10 to show the improvement associated with utilizing the optimum transition point to the approximation $\text{erf}(x) \approx 1$. The relative error bounds, and transition points, are detailed in Tables 5 and 6 for the cases of four and 16 subintervals.

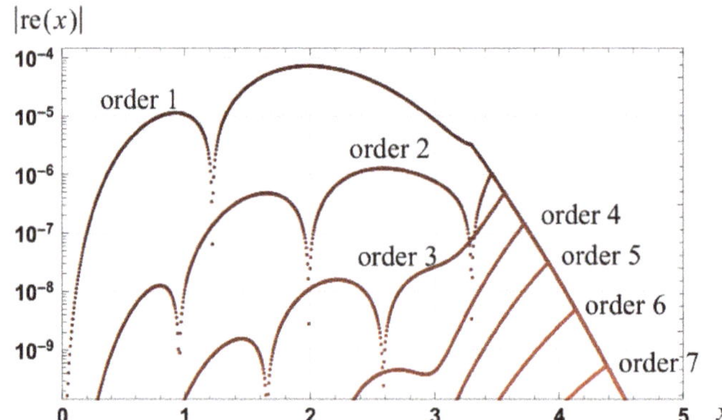

Figure 10. Graph of the relative errors in approximations to $\text{erf}(x)$: first to seventh order spline based series based on four sub-intervals of equal width and with utilization of the approximation $\text{erf}(x) \approx 1$ at the optimum transition point.

Table 5. Transition point and relative error bound for the four equal subintervals case. The transition points are based on sampling the interval [0, 8] with 10,000 points.

Spline Order	Transition Point	Relative Error Bound
0	2.7016	5.32×10^{-3}
1	3.292	7.21×10^{-5}
2	3.4544	1.27×10^{-6}
4	3.7208	1.43×10^{-7}
8	4.6616	4.34×10^{-11}
12	5.6784	9.75×10^{-16}
16	6.3736	2.01×10^{-19}
20	7.1544	4.62×10^{-24}
24	7.7136	1.06×10^{-27}

Table 6. Transition point and relative error bound for the 16 equal subintervals case. The transition points are based on sampling the interval [0, 12] with 10,000 points.

Spline Order	Transition Point	Relative Error Bound
0	5.5008	3.32×10^{-4}
1	6.8796	2.82×10^{-7}
2	7.0224	3.14×10^{-10}
4	7.1544	4.82×10^{-16}
8	7.5996	6.22×10^{-27}
12	8.2032	4.16×10^{-31}
16	8.9244	1.66×10^{-36}
20	9.7284	4.68×10^{-43}
24	10.584	1.21×10^{-50}

Examples

A first-order approximation, based on m subintervals, as specified by Equation (65), yields the relative error bound of 7.21×10^{-5} for four subintervals and the transition point $x_o = 3.2928$; 4.51×10^{-6} for eight subintervals with a transition point $x_o = 4.784$; 2.82×10^{-7} for 16 subintervals and a transition point of $x_o = 6.88$; and 1.10×10^{-9} for 64 subintervals and a transition point $x_o = 15.7888$.

The fourth-order approximation, based on four equal subintervals, as specified by Equation (68), leads to the relative error bound of 1.43×10^{-7} when used with the transition point $x_o = 3.7208$. A sixteenth-order approximation, based on four equal subintervals, leads to an error bound of 2.01×10^{-19} when used with the optimum transition point of $x_o = 6.3736$.

5. Dynamic Constant plus Spline Approximation

Consider the demarcation of the areas, as illustrated in Figure 11 and based on a resolution Δ, that define the error function. It follows that

$$\text{erf}(x) = \sum_{k=0}^{\lfloor x/\Delta \rfloor} c_k + \frac{2}{\sqrt{\pi}} \int_{\Delta\lfloor x/\Delta \rfloor}^{x} e^{-\lambda^2} d\lambda \quad \begin{cases} c_0 = 0 \\ c_k = \text{erf}(k\Delta) - \text{erf}[(k-1)\Delta], k \geq 1. \end{cases} \tag{69}$$

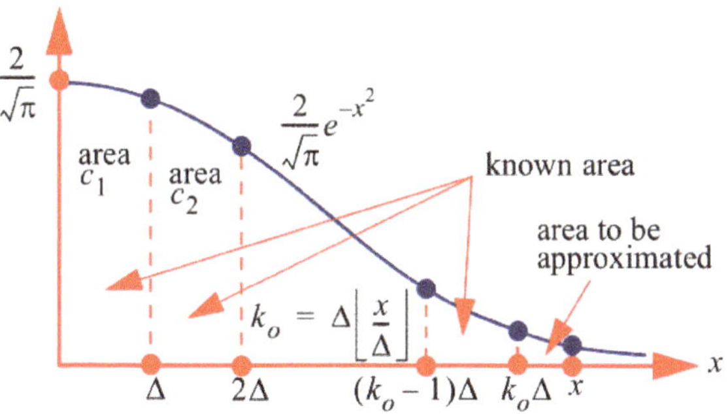

Figure 11. Illustration of areas comprising erf(x).

For the general case of nonuniformly spaced intervals, as defined by the set of monotonically increasing points $\{x_0, x_1, x_2, \ldots, x_m\}$, and where it is not necessarily the case that $x > x_m$, the error function is defined according to

$$\text{erf}(x) = \sum_{k=1}^{m} c_k u(x - x_k) + \frac{2}{\sqrt{\pi}} \int_{x_S}^{x} e^{-\lambda^2} d\lambda, \tag{70}$$

where $c_0 = 0$, $x_0 = 0$ and

$$c_k = \text{erf}(x_k) - \text{erf}[x_{k-1}], \quad x_S = \sum_{k=1}^{m} [x_k - x_{k-1}] u(x - x_k). \tag{71}$$

A spline-based approximation, as defined by Equation (27), can be utilized for the unknown integrals in Equations (69) and (70). This leads to the results stated in Theorem 6:

Theorem 6. *Error Function Approximation: Dynamic Constant Plus a Spline Approximation. The error function, as defined by Equations (69) and (70), can be approximated, respectively, by*

$$f_{n,\Delta}(x) = \sum_{k=0}^{\lfloor x/\Delta \rfloor} c_k + \frac{2}{\sqrt{\pi}} \left[\sum_{k=0}^{n} c_{n,k} \left[x - \Delta \left\lfloor \frac{x}{\Delta} \right\rfloor \right]^{k+1} \left[\begin{array}{c} p\left[k, \Delta \left\lfloor \frac{x}{\Delta} \right\rfloor\right] \exp\left[-\Delta^2 \left\lfloor \frac{x}{\Delta} \right\rfloor^2 \right] + \\ (-1)^k p(k,x) exp(-x^2) \end{array} \right] \right] \tag{72}$$

$$\text{erf}(x) \approx \sum_{k=1}^{m} c_k u(x - x_k) + \frac{2}{\sqrt{\pi}} \left[\sum_{k=0}^{n} c_{n,k} (x - x_S)^{k+1} \left[\begin{array}{c} p(k, x_S) exp(-x_S^2) + \\ (-1)^k p(k,x) exp(-x^2) \end{array} \right] \right] \tag{73}$$

Proof. These results arise from spline approximation of order n, as defined by Equation (27), for the integrals, respectively, over the intervals $[\Delta\lfloor x/\Delta \rfloor, x]$ and $[x_S, x]$. □

5.1. Approximations of Orders Zeros to Four

Approximations of orders zero to four arising from Theorem 6 are:

$$f_{0,\Delta}(x) = \sum_{k=0}^{\lfloor x/\Delta \rfloor} c_k + \frac{x - \Delta\lfloor x/\Delta \rfloor}{\sqrt{\pi}} \cdot \left[e^{-\Delta^2 \lfloor x/\Delta \rfloor^2} + e^{-x^2}\right] \quad (74)$$

$$\begin{aligned} f_{1,\Delta}(x) = & \sum_{k=0}^{\lfloor x/\Delta \rfloor} c_k + \frac{x-\Delta\lfloor x/\Delta \rfloor}{\sqrt{\pi}} \cdot \left[e^{-\Delta^2 \lfloor x/\Delta \rfloor^2} + e^{-x^2}\right] - \\ & \frac{(x-\Delta\lfloor x/\Delta \rfloor)^2}{3\sqrt{\pi}} \cdot \left[\Delta\lfloor x/\Delta \rfloor e^{-\Delta^2 \lfloor x/\Delta \rfloor^2} - xe^{-x^2}\right] \end{aligned} \quad (75)$$

$$\begin{aligned} f_{2,\Delta}(x) = & \sum_{k=0}^{\lfloor x/\Delta \rfloor} c_k + \frac{x-\Delta\lfloor x/\Delta \rfloor}{\sqrt{\pi}} \cdot \left[e^{-\Delta^2 \lfloor x/\Delta \rfloor^2} + e^{-x^2}\right] - \\ & \frac{2(x-\Delta\lfloor x/\Delta \rfloor)^2}{5\sqrt{\pi}} \cdot \left[\Delta\lfloor x/\Delta \rfloor e^{-\Delta^2 \lfloor x/\Delta \rfloor^2} - xe^{-x^2}\right] - \\ & \frac{(x-\Delta\lfloor x/\Delta \rfloor)^3}{30\sqrt{\pi}} \cdot \left[\left[1 - 2\Delta^2\lfloor x/\Delta \rfloor^2\right] e^{-\Delta^2 \lfloor x/\Delta \rfloor^2} + \left[1 - 2x^2\right] e^{-x^2}\right] \end{aligned} \quad (76)$$

$$\begin{aligned} f_{3,\Delta}(x) = & \sum_{k=0}^{\lfloor x/\Delta \rfloor} c_k + \frac{x-\Delta\lfloor x/\Delta \rfloor}{\sqrt{\pi}} \cdot \left[e^{-\Delta^2 \lfloor x/\Delta \rfloor^2} + e^{-x^2}\right] - \\ & \frac{3(x-\Delta\lfloor x/\Delta \rfloor)^2}{7\sqrt{\pi}} \cdot \left[\Delta\lfloor x/\Delta \rfloor e^{-\Delta^2 \lfloor x/\Delta \rfloor^2} - xe^{-x^2}\right] - \\ & \frac{(x-\Delta\lfloor x/\Delta \rfloor)^3}{21\sqrt{\pi}} \cdot \left[\left[1 - 2\Delta^2\lfloor x/\Delta \rfloor^2\right] e^{-\Delta^2 \lfloor x/\Delta \rfloor^2} + \left[1 - 2x^2\right] e^{-x^2}\right] + \\ & \frac{(x-\Delta\lfloor x/\Delta \rfloor)^4}{70\sqrt{\pi}} \cdot \left[\Delta\lfloor x/\Delta \rfloor \left[1 - \frac{2\Delta^2\lfloor x/\Delta \rfloor^2}{3}\right] e^{-\Delta^2 \lfloor x/\Delta \rfloor^2} - x\left[1 - \frac{2x^2}{3}\right] e^{-x^2}\right] \end{aligned} . \quad (77)$$

$$\begin{aligned} f_{4,\Delta}(x) = & \sum_{k=0}^{\lfloor x/\Delta \rfloor} c_k + \frac{x-\Delta\lfloor x/\Delta \rfloor}{\sqrt{\pi}} \cdot \left[e^{-\Delta^2 \lfloor x/\Delta \rfloor^2} + e^{-x^2}\right] - \\ & \frac{4(x-\Delta\lfloor x/\Delta \rfloor)^2}{9\sqrt{\pi}} \cdot \left[\Delta\lfloor x/\Delta \rfloor e^{-\Delta^2 \lfloor x/\Delta \rfloor^2} - xe^{-x^2}\right] - \\ & \frac{(x-\Delta\lfloor x/\Delta \rfloor)^3}{18\sqrt{\pi}} \cdot \left[\left[1 - 2\Delta^2\lfloor x/\Delta \rfloor^2\right] e^{-\Delta^2 \lfloor x/\Delta \rfloor^2} + \left[1 - 2x^2\right] e^{-x^2}\right] + \\ & \frac{(x-\Delta\lfloor x/\Delta \rfloor)^4}{42\sqrt{\pi}} \cdot \left[\Delta\lfloor x/\Delta \rfloor \left[1 - \frac{2\Delta^2\lfloor x/\Delta \rfloor^2}{3}\right] e^{-\Delta^2 \lfloor x/\Delta \rfloor^2} - x\left[1 - \frac{2x^2}{3}\right] e^{-x^2}\right] + \\ & \frac{(x-\Delta\lfloor x/\Delta \rfloor)^5}{1260\sqrt{\pi}} \cdot \left[\left[1 - 4\Delta^2\lfloor x/\Delta \rfloor^2 + \frac{4\Delta^4\lfloor x/\Delta \rfloor^4}{3}\right] e^{-\Delta^2 \lfloor x/\Delta \rfloor^2} + \left[1 - 4x^2 + \frac{4x^4}{3}\right] e^{-x^2}\right] \end{aligned} \quad (78)$$

5.2. Results

For a resolution of $\Delta = 1/2$, the coefficients are tabulated in Table 7.

A resolution of $\Delta = 1/2$ yields a relative error bound of 1.16×10^{-5} for a second-order approximation, a relative error bound of 1.35×10^{-9} for a fourth-order approximation, a relative error bound of 7.15×10^{-14} for a sixth-order approximation and a relative error bound of 9.03×10^{-37} for a sixteenth-order approximation. These bounds are based on 10,000 equally spaced samples in the interval [0, 8].

Table 7. Coefficient values for the case of $\Delta = 1/2$.

k	Definition for c_k	c_k
1	erf(1/2)	$5.204998778 \times 10^{-1}$
2	erf(1) − erf(1/2)	$3.222009151 \times 10^{-1}$
3	erf(3/2) − erf(1)	$1.234043535 \times 10^{-1}$
4	erf(2) − erf(3/2)	$2.921711854 \times 10^{-2}$
5	erf(5/2) − erf(2)	$4.270782964 \times 10^{-3}$
6	erf(3) − erf(5/2)	$3.848615204 \times 10^{-4}$
7	erf(7/2) − erf(3)	$2.134739863 \times 10^{-5}$
8	erf(4) − erf(7/2)	$7.276811144 \times 10^{-7}$
9	erf(9/2) − erf(4)	$1.522064186 \times 10^{-8}$
10	erf(5) − erf(9/2)	$1.950785844 \times 10^{-10}$
11	erf(11/2) − erf(5)	$1.530101947 \times 10^{-12}$
12	erf(6) − erf(11/2)	$7.336328181 \times 10^{-15}$

The variation of the relative error bound with resolution, and order, is detailed in Figure 12. The nature of the variation of the relative error, for orders two, three, and four, is shown in Figure 13 for a resolution of 0.5. It is possible to obtain better results by using nonuniformly spaced intervals but the improvement, in general, does not warrant the increase in complexity.

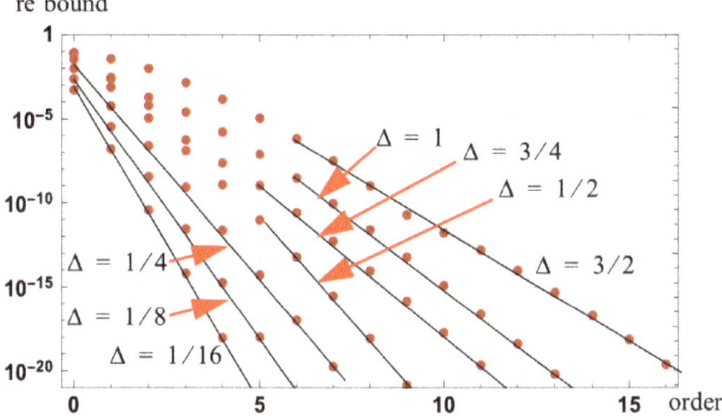

Figure 12. Graph of the relative error bound, versus the order of approximation, for various set resolutions.

Figure 13. Graph of the relative errors, based on a resolution of $\Delta = 0.5$, in second to fourth order approximations to erf(x).

6. A Dynamic System to Yield Improved Approximations

It is possible to utilize the approximations detailed in Theorems 1 and 5 as the basis for determining new approximations with a lower relative error. The approach is indirect and based on considering the feedback system illustrated in Figure 14, which has varying feedback. The differential equation characterizing the system is

$$y\prime(t) + f_M(t)y(t) = x(t). \tag{79}$$

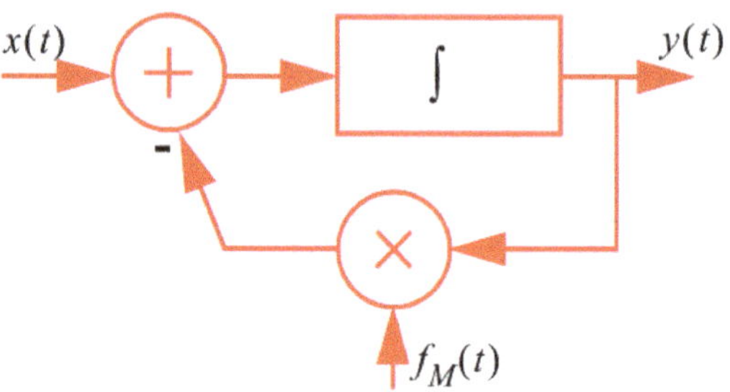

Figure 14. Feedback system with dynamically varying (modulated) feedback.

For specific input, x, and modulated feedback, f_M, signals the output has a known form. For example, for the case of $x(t) = f_M(t) = \mathrm{erf}(t)u(t)$, the output signal, assuming zero initial conditions, is

$$y(t) = 1 - exp\left[\frac{1}{\sqrt{\pi}} \cdot \left[1 - e^{-t^2}\right] - t\mathrm{erf}(t)\right], \quad t \geq 0. \tag{80}$$

For the case of

$$x(t) = f_M(t) = \frac{4}{\sqrt{\pi}} e^{-t^2} \mathrm{erf}(t)u(t) \tag{81}$$

the output signal, assuming zero initial conditions, is

$$y(t) = 1 - exp\left[-\mathrm{erf}^2(t)\right], \quad t \geq 0. \tag{82}$$

This case facilitates approximations for the error function, which can be made arbitrarily accurate and which are valid for the positive real line.

Theorem 7. *Dynamical System Approximations for Error Function. Based on the differential equation specified by Equation (79), a nth-order approximation to* $\mathrm{erf}(x)$*, for the case of* $x \geq 0$*, can be defined according to*

$$f_n(x) = \sqrt{p_{n,0} + p_{n,1}(x)e^{-x^2} + p_{n,2}(x)e^{-2x^2}}, \tag{83}$$

where, for the case of $n \geq 2$*:*

$$\begin{aligned} p_{n,1}(x) &= \alpha_0 + \alpha_2 x^2 + \ldots + \alpha_m x^m, \quad m = \begin{cases} n-1 & m \text{ odd} \\ n & m \text{ even} \end{cases} \\ p_{n,2}(x) &= \beta_0 + \beta_2 x^2 + \ldots + \beta_{2n} x^{2n} \\ p_{n,0}(x) &= -(\alpha_0 + \beta_0) \end{aligned} \tag{84}$$

and with

$$\begin{aligned}
\alpha_m &= \tfrac{-4}{\pi} \cdot c_{n,m} a_{m,0} \\
\alpha_{m-2i} &= \tfrac{(m-2i+2)\alpha_{m-2i+2}}{2} - \tfrac{4}{\pi} \cdot c_{n,m-2i} a_{m-2i,0}, \ i \in \left\{1, \ldots, \tfrac{m}{2} - 1\right\} \\
\alpha_0 &= \alpha_2 - \tfrac{4}{\pi} \cdot c_{n,0} a_{0,0}
\end{aligned} \tag{85}$$

$$\begin{aligned}
\beta_{2n} &= \tfrac{-2}{\pi} \cdot (-1)^n c_{n,n} a_{n,n} \\
\beta_{2n-2i} &= \tfrac{(n-i+1)\beta_{2n-2i+2}}{2} - \tfrac{2}{\pi} \sum_{k=n-i}^{\min\{2n-2i,n\}} (-1)^k c_{n,k} a_{k,2(n-i)-k}, \ i \in \{1, \ldots, n-1\} \\
\beta_0 &= \tfrac{\beta_2}{2} - \tfrac{2}{\pi} \cdot c_{n,0} a_{0,0}
\end{aligned} \tag{86}$$

Here, the coefficients $a_{i,j}$, $i, j \in \{0, 1, \ldots, n\}$ are defined by the expansion

$$p(k, x) = a_{k,0} + a_{k,1}x + a_{k,2}x^2 + \ldots + a_{k,k}x^k, \quad k \in \{0, 1, \ldots, n\}, \tag{87}$$

arising from the polynomials (Equation (26))

$$p(k, x) = p^{(1)}(k-1, x) - 2xp[k-1, x], \quad p(0, x) = 1 \tag{88}$$

Finally, it is the case that

$$\lim_{n \to \infty} f_n(x) = \operatorname{erf}(x), \quad x \geq 0, \tag{89}$$

with the convergence being uniform.

Proof. The proof is detailed in Appendix D. □

6.1. Explicit Approximations

Explicit approximations for orders zero to four for $\operatorname{erf}(x)$, $x \geq 0$ are:

$$f_0(x) = \frac{1}{\sqrt{\pi}} \sqrt{3 - 2e^{-x^2} - e^{-2x^2}} \tag{90}$$

$$f_1(x) = \frac{1}{\sqrt{\pi}} \sqrt{\frac{19}{6} - 2e^{-x^2} - \frac{7e^{-2x^2}}{6}\left[1 + \frac{2x^2}{7}\right]} \tag{91}$$

$$f_2(x) = \frac{1}{\sqrt{\pi}} \sqrt{\frac{63}{20} - \frac{29e^{-x^2}}{15}\left[1 - \frac{x^2}{29}\right] - \frac{73e^{-2x^2}}{60}\left[1 + \frac{26x^2}{73} + \frac{4x^4}{73}\right]} \tag{92}$$

$$f_3(x) = \frac{1}{\sqrt{\pi}} \sqrt{\frac{22}{7} - \frac{40e^{-x^2}}{21}\left[1 - \frac{x^2}{20}\right] - \frac{26e^{-2x^2}}{21}\left[1 + \frac{10x^2}{26} + \frac{x^4}{13} + \frac{x^6}{130}\right]} \tag{93}$$

$$f_4(x) = \frac{1}{\sqrt{\pi}} \sqrt{\begin{array}{l} \frac{377}{120} - \frac{596e^{-x^2}}{315}\left[1 - \frac{17x^2}{298} + \frac{x^4}{1192}\right] - \\ \frac{3149e^{-2x^2}}{2520}\left[1 + \frac{1258x^2}{3149} + \frac{278x^4}{3149} + \frac{112x^6}{9447} + \frac{8x^8}{9447}\right] \end{array}} \tag{94}$$

6.2. Results

The relative error bounds associated with the approximations to $\operatorname{erf}(x)$ are detailed in Table 8. The graphs of the relative errors in the approximations are shown in Figure 15. The clear advantage of the proposed approximations is evident, with the improvement increasing with the order of the initial approximation (i.e., a function with an initial lower relative error bound leads to an increasingly lower relative error bound). The other clear advantage of the approximations, as is evident in Figure 15, is that the relative error is bounded as $x \to \infty$.

Table 8. Relative error bounds, over the interval $(0,\infty)$, for approximations to $\mathrm{erf}(x)$ as defined in Theorem 7.

Order of Approx.	Relative Error Bound: Original Series—Optimum Transition Point (Table 3)	Relative Error Bound: Approx. Defined by Equation (83)
0	0.0851	2.68×10^{-2}
1	0.0362	3.98×10^{-3}
2	1.95×10^{-2}	1.34×10^{-3}
3	7.36×10^{-3}	2.03×10^{-4}
4	1.03×10^{-3}	1.82×10^{-5}
6	4.75×10^{-4}	9.20×10^{-7}
8	2.79×10^{-5}	1.69×10^{-8}
10	1.35×10^{-5}	7.43×10^{-10}
12	9.78×10^{-7}	1.67×10^{-11}
14	4.00×10^{-7}	6.47×10^{-13}
16	3.44×10^{-8}	1.68×10^{-14}
18	1.22×10^{-8}	5.90×10^{-16}
20	1.20×10^{-9}	1.73×10^{-17}
22	3.76×10^{-10}	5.56×10^{-19}
24	4.18×10^{-11}	1.79×10^{-20}

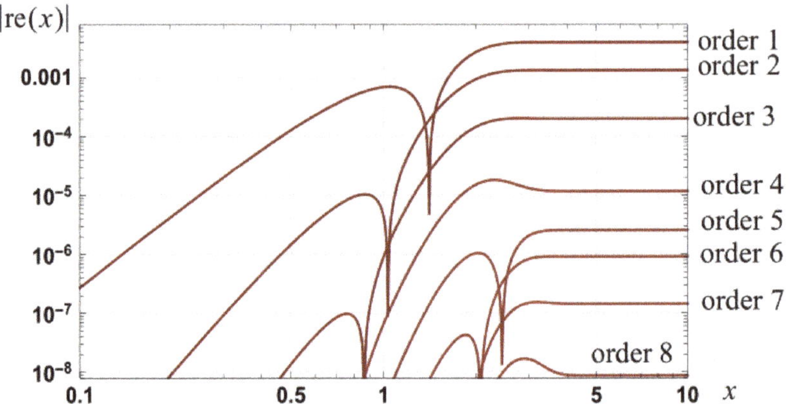

Figure 15. Graph of the relative errors in approximations, of orders one to eight, to erf(x) as defined in Theorem 7.

6.3. Extension

By utilizing the approximations detailed in Theorem 5, similar approximations can be detailed, with lower relative error. For example, the first-order approximation, $f_{1,4}$, which is based on four equal subintervals and is defined by Equation (67), yields the approximation

$$f_{1,4}(x) = \frac{1}{\sqrt{\pi}}\sqrt{\frac{128{,}177}{40{,}800} - \frac{e^{-x^2}}{2} - \frac{16e^{-17x^2/16}}{17} - \frac{4e^{-5x^2/4}}{5} - \frac{16e^{-25x^2/16}}{25} - \frac{25e^{-2x^2}}{96}\left[1 + \frac{2x^2}{25}\right]} \quad (95)$$

which has a relative error bound of 2.83×10^{-6}. With an optimum transition point of 3.292, the original approximation has a relative error bound of 7.21×10^{-5} (see Table 5).

6.4. Notes

First, the constants $p_{n,0}$, $n \in \{0, 1, \ldots\}$, as defined in Equation (84), form a series that in the limit converges to 1. It then follows that the corresponding series converges to π:

$$3, \frac{19}{6}, \frac{63}{20}, \frac{22}{7}, \frac{377}{120}, \frac{174,169}{55,440}, \frac{4,528,409}{1,441,440}, \ldots. \tag{96}$$

Second, the square root functional structure has been utilized for approximations to the error function, as is evident from the approximations detailed in Table 1. It is easy to conclude that the form

$$f_n(x) = \sqrt{p_{n,0} - p_{n,1}(x)e^{-k_1 x^2} - p_{n,2}(x)e^{-k_2 x^2} - \ldots} \tag{97}$$

is well suited for approximating the error function.

7. Applications

This section details indicative applications for the approximations to the error function that have been detailed above.

The distinct analytical forms specified in Theorems 1, 3, 5 and 7, for approximations to the error function, in general, facilitate analysis for different applications. For example, the form detailed in Theorem 1 underpins approximations to $\exp(-x^2)$, as detailed in Section 7.2. The form detailed in Theorem 7 underpins analytical approximations for the power associated with the output of a nonlinearity modeled by the error function when subject to a sinusoidal signal. The approximations are detailed in Section 7.6.

For applications, where a set relative error bound over a set interval is required, the suitable approximation will depend, in part, on the domain over which an approximation is required as well as the level of the relative error bound that is acceptable. For example, the approximations detailed in Theorems 1 and 3 lead to simple analytical forms and modest relative error bounds over $[0, \infty)$ when used with an appropriate transition point for the approximation of $\text{erf}(x) \approx 1$. Without the use of a transition point, such approximations are likely to be best suited for a restricted domain—for example, the domain $\left[0, \frac{3}{\sqrt{2}}\right]$, which is consistent with the three sigma case arising from a Gaussian distribution. The fourth-order approximations, as specified by Equations (36) and (54), have relative error bounds, respectively, of 1.02×10^{-3} and 1.90×10^{-4} over $\left[0, \frac{3}{\sqrt{2}}\right]$. Section 7.1 provides examples of approximations that are consistent with set relative error bounds, over the interval $[0, \infty)$, of 10^{-4}, 10^{-6}, 10^{-10}, and 10^{-16}.

7.1. Error Function Approximations: Set Relative Error Bounds

Consider the case where an approximation for the error function, with a relative error bound over the positive real line of 10^{-4}, is required. A 47th-order Taylor series, with a transition point of $x_o = 2.752$, yields a relative bound of 1.00×10^4.

An eighth-order spline approximation, with a transition point of $x_o = 2.963$, yields a relative error bound of 2.79×10^{-5}. The approximation, according to Equation (56), is

$$f_8(x) = \frac{x}{\sqrt{\pi}} \cdot u[x_o - x] \cdot \left[\begin{array}{l} 1 - \frac{7x^2}{102} + \frac{x^4}{340} - \frac{x^6}{18,564} + \frac{x^8}{5,250,960} + \\ \left[\begin{array}{l} 1 + \frac{41x^2}{102} + \frac{101x^4}{1020} + \frac{1591x^6}{92,820} + \frac{4793x^8}{2,162,160} + \\ \frac{2017x^{10}}{9,189,180} + \frac{38x^{12}}{2,297,295} + \frac{31x^{14}}{34,459,425} + \\ \frac{x^{16}}{34,459,425} \end{array} \right] e^{-x^2} \end{array} \right] + \\ [1 - u[x_o - u]] \tag{98}$$

A seventh-order approximation, with a transition point of $x_o = 2.65$, yields a relative error bound of 1.79×10^{-4}.

A first-order spline approximation, based on four equal subintervals $[0, x/4]$, $[x/4, x/2]$, $[x/2, 3x/4]$ and $[3x/4, x]$, is defined according to

$$f_{1,4}(x) = \left[\begin{array}{r} \frac{x}{4\sqrt{\pi}} \cdot \left[1 + 2\exp\left[\frac{-x^2}{16}\right] + 2\exp\left[\frac{-x^2}{4}\right] + 2\exp\left[\frac{-9x^2}{16}\right] + \exp\left[-x^2\right]\right] + \\ \frac{x^3}{48\sqrt{\pi}} \cdot \exp\left[-x^2\right] \end{array} \right] u[x_o - x] + \tag{99}$$
$$[1 - u[x_o - u]]$$

and yields a relative error bound of 7.21×10^{-5} with the transition point $x_o = 3.292$.

A dynamic constant plus a spline approximation of order 2, and based on a resolution of $\Delta = 19/20$, achieves a relative error bound of 8.33×10^{-5} (10,000 points in the interval $[0, 5]$). The approximation is

$$\begin{aligned} f_{2,\Delta}(x) = {} & \sum_{k=0}^{\lfloor 20x/19 \rfloor} c_k + \frac{1}{\sqrt{\pi}}\left[x - \frac{19}{20} \cdot \left\lfloor \frac{20x}{19} \right\rfloor\right]\left[\exp\left[\frac{-361}{400}\left\lfloor \frac{20x}{19} \right\rfloor^2\right] + e^{-x^2}\right] - \\ & \frac{2}{5\sqrt{\pi}}\left[x - \frac{19}{20} \cdot \left\lfloor \frac{20x}{19} \right\rfloor\right]^2 \left[\frac{19}{20} \cdot \left\lfloor \frac{20x}{19} \right\rfloor \exp\left[\frac{-361}{400}\left\lfloor \frac{20x}{19} \right\rfloor^2\right] - xe^{-x^2}\right] - \\ & \frac{1}{30\sqrt{\pi}}\left[x - \frac{19}{20} \cdot \left\lfloor \frac{20x}{19} \right\rfloor\right]^3 \cdot \left[\left[1 - \frac{361}{200} \cdot \left\lfloor \frac{20x}{19} \right\rfloor^2\right] \exp\left[\frac{-361}{400}\left\lfloor \frac{20x}{19} \right\rfloor^2\right] + (1 - 2x^2)e^{-x^2}\right] \end{aligned} \tag{100}$$

where

$$\begin{aligned} & c_0 = 0, \quad c_k = \operatorname{erf}\left[\frac{19k}{20}\right] - \operatorname{erf}\left[\frac{19(k-1)}{20}\right], \quad k \in \{1, 2, \ldots\}, \\ & c_1 = 0.82089081, \quad c_2 = 0.17189962, \\ & c_3 = 7.1539145 \times 10^{-3}, \quad c_4 = 5.5579 \times 10^{-5}, \quad c_5 = c_6 = \ldots = 0. \end{aligned} \tag{101}$$

Here, the approximation of $\operatorname{erf}(x) \approx 1$, for $x \geq 57/20$ (after three intervals) can be utilized without impacting the relative error bound.

Utilizing a fourth-order spline approximation and iteration consistent with Theorem 7, the approximation

$$f_4(x) = \frac{1}{\sqrt{\pi}} \sqrt{\begin{array}{l} \frac{377}{120} - \frac{596e^{-x^2}}{315}\left[1 - \frac{17x^2}{298} + \frac{x^4}{1192}\right] - \\ \frac{3149e^{-2x^2}}{2520}\left[1 + \frac{1258x^2}{3149} + \frac{278x^4}{3149} + \frac{112x^6}{9447} + \frac{8x^8}{9447}\right] \end{array}} \tag{102}$$

yields a relative error bound of 1.82×10^{-5}.

Details of approximations that are consistent with higher-order relative error bounds are detailed in Table 9.

Table 9. Approximations that are consistent with a set relative error bound. The actual relative error bound is specified by re_B.

Relative Error Bound	Spline Approx: Theorem 4	Variable Interval Approx: Theorem 5	Dynamic Constant Plus Spline Approx: Theorem 6	Iterative Approx: Theorem 7
10^{-6}	order = 12 $x_o = 3.4625$ $re_B = 9.78 \times 10^{-7}$	order = 5 3 subintervals $x_o = 3.51$ $re_B = 6.96 \times 10^{-7}$	order = 3 resolution = 3/4 $re_B = 5.53 \times 10^{-7}$	order = 6 $re_B = 9.20 \times 10^{-7}$
10^{-10}	order = 23 $x_o = 4.581$ $re_B = 9.31 \times 10^{-11}$	order = 8 4 subintervals $x_o = 4.6616$ $re_B = 4.34 \times 10^{-11}$	order = 4 resolution = 3/8 $re_B = 9.12 \times 10^{-11}$	order = 11 $re_B = 1.34 \times 10^{-10}$ order = 12 $re_B = 1.67 \times 10^{-11}$
10^{-16}	order = 39 $x_o = 5.9017$ $re_B = 7.21 \times 10^{-17}$	order = 11 6 subintervals $x_o = 5.98$ $re_B = 2.75 \times 10^{-17}$	order = 6 resolution = 1/4 $re_B = 1.01 \times 10^{-17}$	order = 19 $re_B = 1.18 \times 10^{-16}$ order = 20 $re_B = 1.73 \times 10^{-17}$

7.2. Approximation for Exp($-x^2$)

A nth-order approximation to the Gaussian function $\exp(-x^2)$ is detailed in the following theorem:

Theorem 8. *Approximation for Gaussian Function. A nth-order approximation, g_n, to the Gaussian function $\exp(-x^2)$ is*

$$g_n(x) = \frac{\sum_{k=0}^{n} c_{n,k}(k+1)x^k p(k,0)}{1 + \sum_{k=0}^{n} c_{n,k}(-1)^{k+1} x^k \left[p(k,x)[k+1-2x^2] + xp^{(1)}(k,x)\right]} \tag{103}$$

where $c_{n,k}$ is defined by Equation (21) and $p(k,x)$ is defined by Equation (26).

Proof. The proof is detailed in Appendix E. □

7.2.1. Approximations

Approximations to $\exp(-x^2)$, of orders zero to five, are:

$$g_0(x) = \frac{1}{1+2x^2} \quad g_1(x) = \frac{1}{1+x^2+\frac{2x^4}{3}} \quad g_2(x) = \frac{1-x^2/10}{1+\frac{9x^2}{10}+\frac{2x^4}{5}+\frac{2x^6}{15}} \tag{104}$$

$$\begin{aligned} g_3(x) &= \frac{1-x^2/7}{1+\frac{6x^2}{7}+\frac{5x^4}{14}+\frac{2x^6}{21}+\frac{2x^8}{105}} \\ g_4(x) &= \frac{1-x^2/6+x^4/252}{1+\frac{5x^2}{6}+\frac{85x^4}{252}+\frac{11x^6}{126}+\frac{x^8}{63}+\frac{2x^{10}}{945}} \end{aligned} \tag{105}$$

$$g_5(x) = \frac{1-2x^2/11+x^4/132}{1+\frac{9x^2}{11}+\frac{43x^4}{132}+\frac{x^6}{12}+\frac{x^8}{66}+\frac{x^{10}}{495}+\frac{2x^{12}}{10{,}395}} \tag{106}$$

7.2.2. Results

The relative errors in the above defined approximations to $\exp(-x^2)$ are detailed in Figure 16 for approximations of order 0, 2, 4, 6, 8, 10, and 12, along with the relative error in Taylor series for orders $1, 3, 5, \ldots, 15$. The clear superiority of the defined approximations is evident.

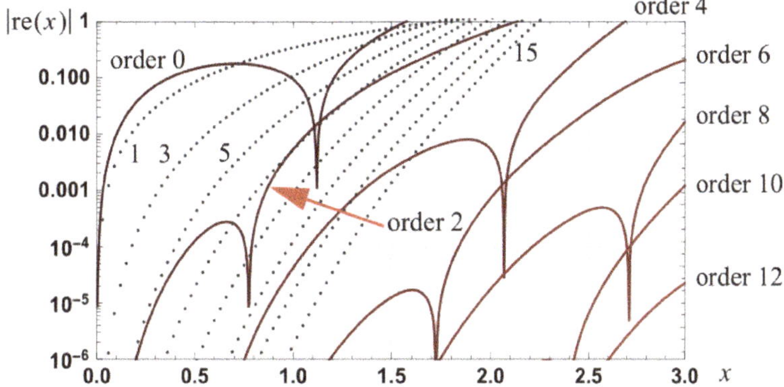

Figure 16. Graph of the magnitude of the relative errors in approximations to $\exp(-x^2)$, as defined by Equation (103), of orders 0, 2, 4, 6, 8, 10 and 12. The dotted curves are the relative errors associated with Taylor series of orders 1, 3, 5, 7, 9, 11, 13 and 15.

7.2.3. Comparison

The following nth-order approximation for $\exp(-x^2)$ has been proposed in Equation (77) of [19]:

$$h_n(x) = \frac{1 - c_{n,0}x^2 + c_{n,1}x^4 - c_{n,2}x^6 + \ldots + c_{n,n}(-1)^{n+1}x^{2n+2}}{1 + c_{n,0}x^2 + c_{n,1}x^4 + c_{n,2}x^6 + \ldots + c_{n,n}x^{2n+2}} \quad (107)$$

where $c_{n,k}$ is defined by Equation (21). The relative error bounds over the interval $\left[0,\ 3/\sqrt{2}\right]$ (the three sigma bound case for Gaussian probability distributions) for this approximation, and the approximation defined by Equation (103), are detailed in Table 10. The tabulated results clearly show that this approximation is more accurate than the approximation detailed in Equation (103). The improvement is consistent with the higher-order Padé approximant being used.

Table 10. Relative error bounds for approximations over the interval $\left[0,\ 3/\sqrt{2}\right]$.

Order of Approx.	Relative Error Bound: Equation (103)	Relative Error Bound: Equation (77) of [19]
0	8.00	35.6
1	3.74	6.98
2	0.957	0.767
3	1.25×10^{-1}	5.25×10^{-2}
4	8.04×10^{-3}	2.42×10^{-3}
5	7.71×10^{-3}	7.98×10^{-5}
6	2.09×10^{-3}	1.97×10^{-6}
7	3.72×10^{-4}	3.75×10^{-8}
8	5.02×10^{-5}	5.69×10^{-10}
10	4.54×10^{-7}	7.22×10^{-14}
12	1.09×10^{-9}	4.62×10^{-18}

The following approximations (seventh- and fifth-order) yield relative error bounds of less than 0.001 over the interval $\left[0,\ 3/\sqrt{2}\right]$:

$$g_7(x) = \frac{1 - \frac{x^2}{5} + \frac{x^4}{78} - \frac{x^6}{4290}}{1 + \frac{4x^2}{5} + \frac{61x^4}{195} + \frac{34x^6}{429} + \frac{83x^8}{5270} + \frac{x^{10}}{495} + \frac{7x^{12}}{32{,}175} + \frac{4x^{14}}{225{,}225} + \frac{2x^{16}}{2{,}027{,}025}} \quad (108)$$

$$h_5(x) = \frac{1 - \frac{x^2}{2} + \frac{5x^4}{44} - \frac{x^6}{66} + \frac{x^8}{792} - \frac{x^{10}}{15{,}840} + \frac{x^{12}}{665{,}280}}{1 + \frac{x^2}{2} + \frac{5x^4}{44} + \frac{x^6}{66} + \frac{x^8}{792} + \frac{x^{10}}{15{,}840} + \frac{x^{12}}{665{,}280}} \quad (109)$$

A twenty-seventh-order Taylor series approximation yields a relative bound of 1.03×10^{-3}.

7.3. Upper and Lower Bounded Approximations to Error Function

Establishing bounds for erf(x) has received modest research interest, e.g., [31], and published bounds for erf(x) for the case of $x > 0$ include Chu [32]:

$$\sqrt{1 - \exp[-px^2]} \leq \mathrm{erf}(x) \leq \sqrt{1 - \exp[-qx^2]}, \quad p \in (0,1], q \in [4/\pi, \infty]. \quad (110)$$

Corollary 4.2 of Neuman [33]:

$$\frac{2x}{\sqrt{\pi}} \exp\left[\frac{-x^2}{3}\right] \leq \mathrm{erf}(x) \leq \frac{4x}{3\sqrt{\pi}}\left[1 + \frac{\exp(-x^2)}{2}\right] \quad (111)$$

and refinements to the form proposed by Chu [32], e.g., Yang [34,35], Corollary 3.4 of [35]:

$$\sqrt{1 - \tfrac{20}{3\pi}\left[1 - \tfrac{\pi}{4}\right]\exp\left[\tfrac{-8x^2}{5}\right] - \tfrac{8}{3}\left[1 - \tfrac{5}{2\pi}\right]\exp(-x^2)} \leq \mathrm{erf}(x) \leq \sqrt{1 - \lambda(p_0)\exp(-p_0 x^2) - [1 - \lambda(p_0)]\exp[-\mu(p_0)x^2]} \tag{112}$$

where

$$p_0 = \tfrac{21\pi - 60 + \sqrt{3(147\pi^2 - 920\pi + 1440)}}{30(\pi - 3)}, \quad \lambda(p) = \tfrac{4[7\pi - 20 - 5(\pi - 3)p]}{\pi(15p^2 - 40p + 28)}, \quad \mu(p) = \tfrac{4(5p - 7)}{5(3p - 4)}. \tag{113}$$

The relative error in these bounds is detailed in Figure 17.

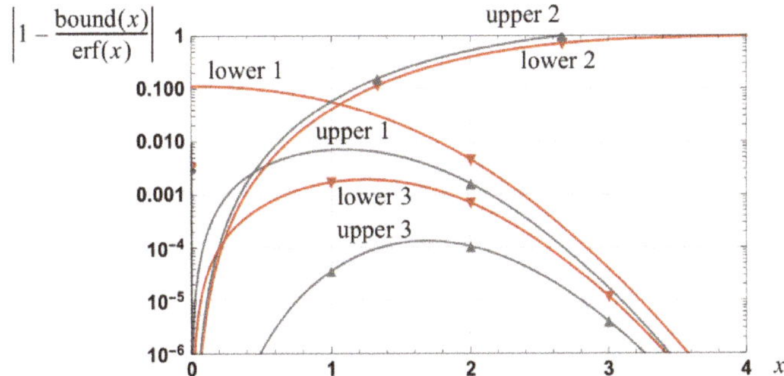

Figure 17. Relative error in upper and lower bounds to erf(x) as, respectively, defined by Equation (110)–(112). The parameters p = 1 and q = $\pi/4$ have been used for the bounds defined by Equation (110).

Utilizing the results of Lemma 1, it follows that any of the approximations detailed in Theorems 4, 5, 6, or 7 can be utilized to create upper and lower bounded functions for erf(x), $x > 0$, of arbitrary accuracy and with an arbitrary relative error bound. For example, the approximation $f_{1,4}$ specified by Equation (99) yields the functional bounds:

$$\frac{f_{1,4}(x)}{1+\varepsilon_B} \leq \mathrm{erf}(x) \leq \frac{f_{1,4}(x)}{1-\varepsilon_B}, \quad \varepsilon_B = 7.21 \times 10^{-5}, \quad x > 0, \tag{114}$$

with a relative error bound of 8.33×10^{-5} for the lower bounded function and 1.44×10^{-4} for the upper bounded function. Such accuracy is better than the bounds underpinning the results shown in Figure 17. The sixteenth-order approximation, $f_{4,16}$, based on four equal subintervals, specified in Appendix C and when used with a transition point $x_o = 7.1544$, leads to the functional bounds

$$\frac{f_{4,16}(x)}{1+\varepsilon_B} \leq \mathrm{erf}(x) \leq \frac{f_{4,16}(x)}{1-\varepsilon_B}, \quad \varepsilon_B = 4.82 \times 10^{-16}, \quad x > 0, \tag{115}$$

with a relative error bound of less than 9.64×10^{-16} for the lower bounded function and 9.32×10^{-16} for the upper bounded function.

7.4. New Series for Error Function

Consider the exact results

$$\mathrm{erf}(x) = f_n(x) + \varepsilon_n(x), \quad n \in \{0, 1, 2, \ldots\}, \tag{116}$$

where f_n is specified by Equation (24) and $\varepsilon_n^{(1)}(x)$ is specified by Equation (25). By utilizing a Taylor series approximation for $\exp(-x^2)$ in $\varepsilon_n^{(1)}(x)$ and then integrating, an approximation for $\varepsilon_n(x)$ can be established. This leads to new series for the error function.

Theorem 9. *New Series for Error Function. Based on zero-, first-, and second-order approximations, the following series for the error function are valid:*

$$\begin{aligned}\text{erf}(x) = &\frac{x}{\sqrt{\pi}} + \frac{x}{\sqrt{\pi}} \cdot e^{-x^2} + \\ &\frac{x^3}{\sqrt{\pi}}\left[\frac{1}{3} - \frac{3x^2}{2\cdot 5} + \frac{5x^4}{6\cdot 7} - \frac{7x^6}{24\cdot 9} + \frac{9x^8}{120\cdot 11} - \ldots + \frac{(-1)^k(2k+1)x^{2k}}{(2k+3)\cdot(k+1)!} + \ldots\right]\end{aligned} \tag{117}$$

$$\begin{aligned}\text{erf}(x) = &\frac{x}{\sqrt{\pi}} + \frac{x}{\sqrt{\pi}}\left[1 + \frac{x^2}{3}\right]e^{-x^2} + \\ &\frac{x^5}{6\sqrt{\pi}}\left[\frac{1}{5} - \frac{3x^2}{(3/2)\cdot 7} + \frac{5x^4}{4\cdot 9} - \frac{7x^6}{15\cdot 11} + \frac{9x^8}{72\cdot 13} - \ldots + \frac{(-1)^k(2k+1)(k+1)!x^{2k}}{(2k+5)\prod\limits_{r=1}^{k}\left[\sum\limits_{i=1}^{r} 2i+1\right]} + \ldots\right]\end{aligned} \tag{118}$$

$$\begin{aligned}\text{erf}(x) = &\frac{x}{\sqrt{\pi}}\left[1 - \frac{x^2}{30}\right] + \frac{x}{\sqrt{\pi}}\left[1 + \frac{11x^2}{30} + \frac{x^4}{15}\right]e^{-x^2} + \\ &\frac{1}{\sqrt{\pi}} \cdot \frac{x^7}{60}\left[\begin{array}{c}\frac{1\cdot 3}{1\cdot 3\cdot 7} - \frac{3\cdot 5x^2}{2\cdot 3\cdot 9} + \frac{5\cdot 7x^4}{4\cdot 5\cdot 11} - \frac{7\cdot 9x^6}{9\cdot 10\cdot 13} + \ldots + \\ \frac{(-1)^k(2k+1)(2k+3)(k+1)!x^{2k}}{3(2k+7)\prod\limits_{r=2}^{k+1}\left[2\sum\limits_{i=2}^{r} i\right]} + \ldots\end{array}\right]\end{aligned} \tag{119}$$

Further series, based on higher-order approximations, can also be established.

Proof. The proof is detailed in Appendix F. □

Results

The relative errors associated with the zero- and second-order series are shown in Figures 18 and 19. Clearly, the relative error improves as the number of terms used in the series expansion increases. The significant improvement in the relative error, for small values of x, is evident. A comparison with the relative errors associated with Taylor series approximations, as shown in Figure 2, shows the improved performance.

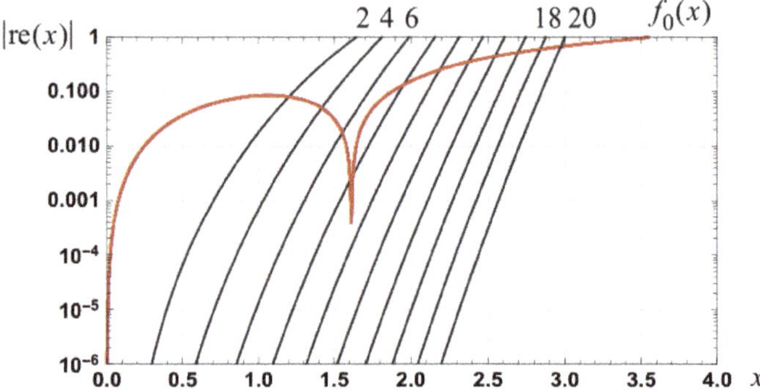

Figure 18. Relative error in the approximations $f_0(x)$ and $f_0(x) + \varepsilon_0(x)$ to erf(x) where the residual function $\varepsilon_0(x)$ is approximated by the stated order.

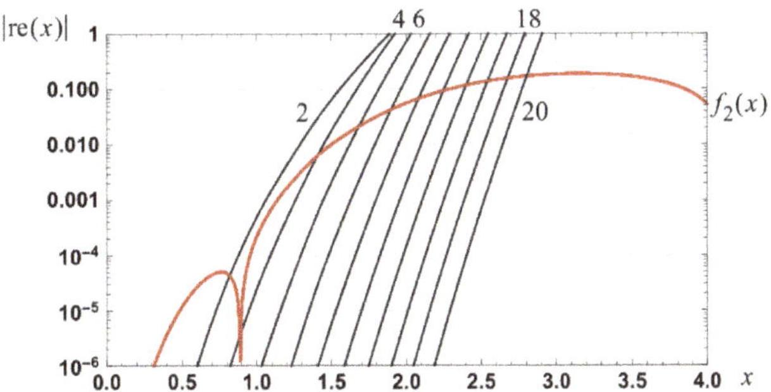

Figure 19. Relative error in the approximations $f_2(x)$ and $f_2(x) + \varepsilon_2(x)$ to erf(x) where the residual function $\varepsilon_2(x)$ is approximated by the stated order.

The second-order approximation arising from Equation (118), i.e.,

$$\text{erf}(x) = \frac{x}{\sqrt{\pi}} + \frac{x}{\sqrt{\pi}}\left[1 + \frac{x^2}{3}\right]e^{-x^2} + \frac{x^5}{6\sqrt{\pi}}\left[\frac{1}{5} - \frac{2x^2}{7}\right] \tag{120}$$

yields a relative error bound of less than 0.001 for the interval [0, 0.87] and less than 0.01 for the interval [0, 1.1].

7.5. Complementary Demarcation Functions

Consider a complementary function e_C which is such that

$$e_C^2(x) + \text{erf}^2(x) = 1,\ x \geq 0. \tag{121}$$

With the approximation detailed in Theorem 7 (and by noting that $\lim_{n\to\infty} p_{n,0} = 1$—see Equation (96)), it is the case that

$$e_C^2(x) = \lim_{n\to\infty}\left[p_{n,1}(x)e^{-x^2} + p_{n,2}(x)e^{-2x^2}\right] \tag{122}$$

and, thus, $e_C(x)$ can be defined independently of the error function. This function is shown in Figure 20 along with $\text{erf}(x)^2$. These two functions act as complementary demarcation functions for the interval $[0, \infty)$. The transition point is $x_o = 0.74373198514677$ as

$$\text{erf}(x)|_{x=0.74373198514677} = \frac{1}{\sqrt{2}}. \tag{123}$$

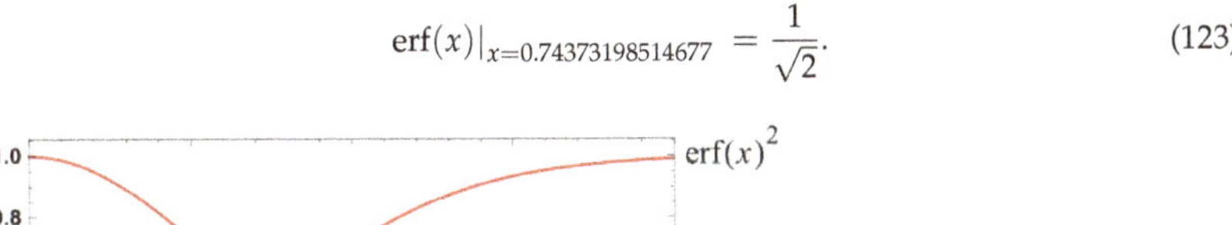

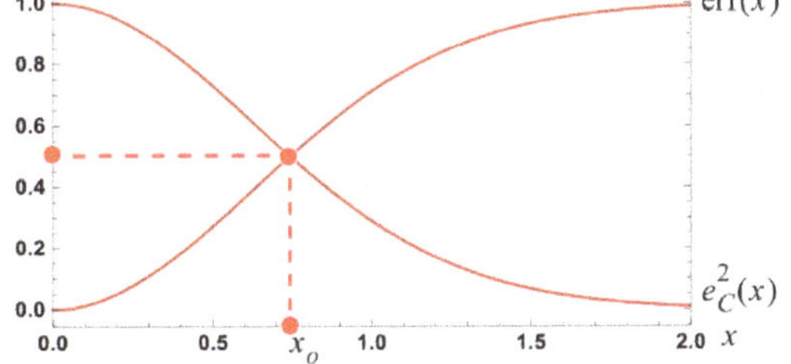

Figure 20. Graph of the signals $e_C^2(x)$ and $\text{erf}(x)^2$.

7.6. Power and Harmonic Distortion: Erf Modeled Nonlinearity

The error function is often used to model nonlinearities and the harmonic distortion created by such a nonlinearity is of interest. Examples include the harmonic distortion in magnetic recording, e.g., [2,36], and the harmonic distortion arising, in a communication context, by a power amplifier, e.g., [7]. For these cases, the interest was in obtaining, with a sinusoidal input signal defined by $a\sin(2\pi f_o t)$, the harmonic distortion created by an error function nonlinearity over the input amplitude range of $[-2, 2]$.

Consider the output signal of a nonlinearity modelled by the error function:

$$y(t) = \text{erf}[a\sin(2\pi f_o t)]. \tag{124}$$

For such a case, the output power is defined according to

$$P = \frac{1}{T}\int_0^T y^2(t)dt = \frac{1}{T}\int_0^T \text{erf}^2[a\sin(2\pi f_o t)]dt, \quad T = 1/f_o, \tag{125}$$

and the output amplitude associated with the kth harmonic is

$$\frac{\sqrt{2}}{\sqrt{T}}\int_0^T \text{erf}[a\sin(2\pi f_o t)]\sin(2\pi k f_o t)dt. \tag{126}$$

To determine an analytical approximation to the output power, the approximations stated in Theorem 7 lead to relatively simple expressions. Consider the third-order approximation, as specified by Equation (93), which has a relative error bound of 2.03×10^{-4} for the positive real line. For such a case, the output signal is approximated according to

$$y_3(t) = \frac{1}{\sqrt{\pi}}\left[\begin{array}{l} \frac{22}{7} - \frac{40e^{-[a\sin(2\pi f_o t)]^2}}{21}\left[1 - \frac{[a\sin(2\pi f_o t)]^2}{20}\right] - \\ \frac{26e^{-2[a\sin(2\pi f_o t)]^2}}{21}\left[\begin{array}{l} 1 + \frac{10[a\sin(2\pi f_o t)]^2}{26} + \frac{[a\sin(2\pi f_o t)]^4}{13} + \\ \frac{[a\sin(2\pi f_o t)]^6}{130} \end{array}\right] \end{array}\right]^{1/2} \tag{127}$$

and is shown in Figure 21.

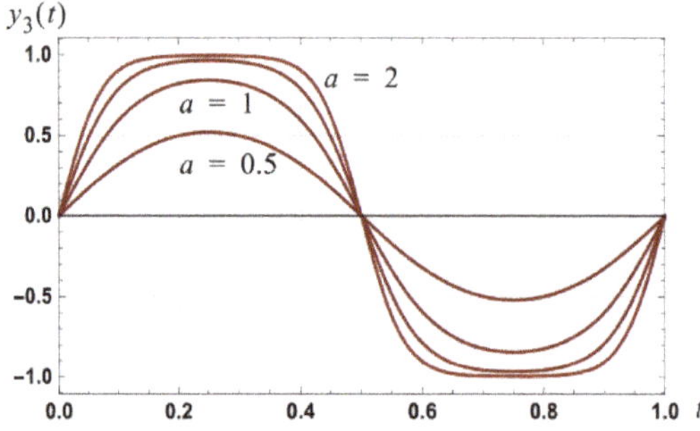

Figure 21. Graph of $y_3(t)$ for the case of $f_o = 1$ and for amplitudes of $a = 0.5$, $a = 1$, $a = 1.5$ and $a = 2$.

The power in y_3 can be readily determined (e.g., via Mathematica) and it then follows that an approximation to the true power is

$$\begin{array}{rl} P(a) \approx & \frac{22}{7\pi} - \frac{40}{21\pi}I_0\left[\frac{a^2}{2}\right]\left[1 - \frac{a^2}{40}\right]e^{-a^2/2} - \frac{26}{21\pi}I_0\left[a^2\right]\left[1 + \frac{5a^2}{26} + \frac{41a^4}{1040} + \frac{a^6}{260}\right]e^{-a^2} - \\ & \frac{a^2}{21\pi}I_1\left[\frac{a^2}{2}\right]e^{-a^2/2} + \frac{37a^2}{140\pi}I_1\left[a^2\right]\left[1 + \frac{43a^2}{222} + \frac{2a^4}{111}\right]e^{-a^2} \end{array} \tag{128}$$

where I_0 and I_1, respectively, are the zero- and first-order Bessel functions of the first kind. The variation in output power is shown in Figure 22.

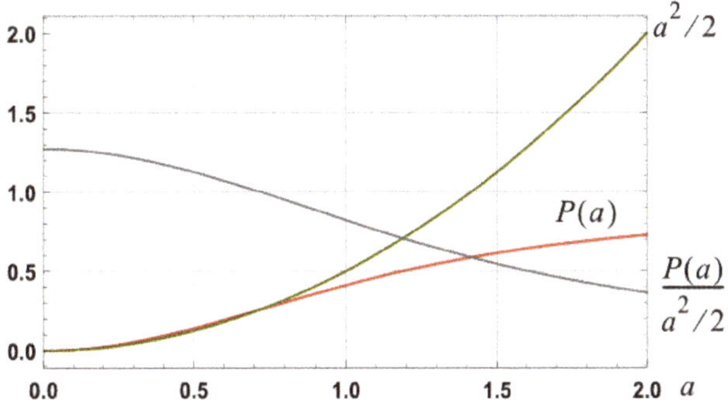

Figure 22. Graph of the input power, output power and ratio of output power to input power as the amplitude of the input signal varies.

Harmonic Distortion

To establish analytical approximations for the harmonic distortion, the functional forms detailed in Theorem 7 are not suitable. However, the functional forms detailed in Theorem 1 do lead to analytical approximations that are valid over a restricted domain. Consider a fourth-order spline approximation, as specified by Equation (36), which approximates the error function over the range $[-2, 2]$ with a relative error bound that is better than 0.001 and leads to the approximation

$$\begin{aligned} y_4(t) = & \frac{a\sin(2\pi f_o t)}{\sqrt{\pi}}\left[1 - \frac{a^2\sin(2\pi f_o t)^2}{18} + \frac{a^4\sin(2\pi f_o t)^4}{1260}\right] + \\ & \frac{a\sin(2\pi f_o t)}{\sqrt{\pi}}\left[\begin{array}{c} 1 + \frac{7a^2\sin(2\pi f_o t)^2}{18} + \frac{37a^4\sin(2\pi f_o t)^4}{420} + \\ \frac{4a^6\sin(2\pi f_o t)^6}{315} + \frac{a^8\sin(2\pi f_o t)^8}{945} \end{array}\right] e^{-a^2\sin(2\pi f_o t)^2} \end{aligned} \quad (129)$$

The amplitude of the kth harmonic in such a signal is given by

$$c_{4,k} = \frac{\sqrt{2}}{\sqrt{T}}\int_0^T y_4[a\sin(2\pi f_o t)]\sin(2\pi k f_o t)dt = \sqrt{2T}\int_0^1 y_4[a\sin(2\pi\lambda)]\sin(2\pi k\lambda)d\lambda \quad (130)$$

where the change of variable $\lambda = f_o t$ has been used. The first, third, fifth, and seventh harmonic levels are:

$$\begin{aligned} \frac{c_{4,1}}{\sqrt{T}} = & \frac{\sqrt{2}a}{2\sqrt{\pi}}\left[1 - \frac{a^2}{24} + \frac{a^4}{2016}\right] + \frac{\sqrt{2}a}{2\sqrt{\pi}} \cdot I_0\left[\frac{a^2}{2}\right]\left[1 + \frac{11a^2}{24} + \frac{11a^4}{105} + \frac{a^6}{70} + \frac{a^8}{945}\right]e^{-a^2/2} - \\ & \frac{5\sqrt{2}a}{6\sqrt{\pi}} \cdot I_1\left[\frac{a^2}{2}\right]\left[1 + \frac{1481a^2}{4200} + \frac{38a^4}{525} + \frac{29a^6}{3150} + \frac{a^8}{1575}\right]e^{-a^2/2} \end{aligned} \quad (131)$$

$$\begin{aligned} \frac{c_{4,3}}{\sqrt{T}} = & \frac{\sqrt{2}a^3}{144\sqrt{\pi}}\left[1 - \frac{a^2}{56}\right] - \frac{115\sqrt{2}a}{84\sqrt{\pi}} \cdot I_0\left[\frac{a^2}{2}\right]\left[1 + \frac{403a^2}{1380} + \frac{6a^4}{115} + \frac{31a^6}{5175} + \frac{2a^8}{5175}\right]e^{-a^2/2} + \\ & \frac{115\sqrt{2}}{21a\sqrt{\pi}} \cdot I_1\left[\frac{a^2}{2}\right]\left[1 + \frac{8a^2}{23} + \frac{163a^4}{1840} + \frac{76a^6}{5175} + \frac{11a^8}{6900} + \frac{a^{10}}{10{,}350}\right]e^{-a^2/2} \end{aligned} \quad (132)$$

$$\begin{aligned} \frac{c_{4,5}}{\sqrt{T}} = & \frac{\sqrt{2}a^5}{40{,}320\sqrt{\pi}} + \frac{262\sqrt{2}}{15a\sqrt{\pi}} \cdot I_0\left[\frac{a^2}{2}\right]\left[1 + \frac{1943a^2}{7336} + \frac{1485a^4}{29{,}344} + \frac{73a^6}{11{,}004} + \frac{13a^8}{22{,}008} + \frac{a^{10}}{33{,}012}\right]e^{-a^2/2} - \\ & \frac{1048\sqrt{2}}{15a^3\sqrt{\pi}} \cdot I_1\left[\frac{a^2}{2}\right]\left[1 + \frac{1943a^2}{7336} + \frac{1201a^4}{14{,}672} + \frac{5125a^6}{352{,}128} + \frac{5a^8}{2751} + \frac{41a^{10}}{264{,}096} + \frac{a^{12}}{132{,}048}\right]e^{-a^2/2} \end{aligned} \quad (133)$$

$$\begin{aligned}\frac{c_{4,7}}{\sqrt{T}} = &\frac{-6784\sqrt{2}}{21a^3\sqrt{\pi}} \cdot I_0\left[\frac{a^2}{2}\right]\left[1 + \frac{779a^2}{3392} + \frac{631a^4}{13{,}568} + \frac{2047a^6}{325{,}632} + \frac{25a^8}{40{,}704} + \frac{17a^{10}}{407{,}040} + \frac{a^{12}}{610{,}560}\right]e^{-a^2/2} - \\ &\frac{27{,}136\sqrt{2}}{21a^5\sqrt{\pi}} \cdot I_1\left[\frac{a^2}{2}\right]\left[\begin{array}{r} 1 + \frac{779a^2}{3392} + \frac{1055a^4}{13{,}568} + \frac{137a^6}{10{,}176} + \frac{2269a^8}{1{,}302{,}528} + \frac{67a^{10}}{407{,}040} + \\ \frac{a^{12}}{92{,}160} + \frac{a^{14}}{2{,}442{,}240} \end{array}\right]e^{-a^2/2}\end{aligned} \quad (134)$$

The variation, with the input signal amplitude, of the harmonic distortion, as defined by $c_{4,k}^2/c_{4,1}^2$, is shown in Figure 23.

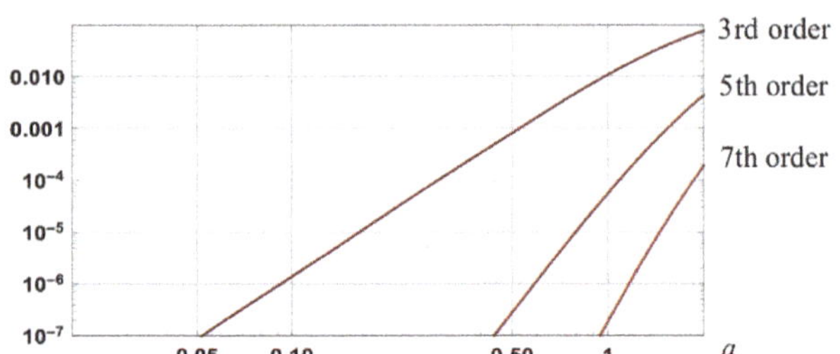

Figure 23. Graph of the variation of harmonic distortion with amplitude.

7.7. Linear Filtering of an Error Function Step Signal

Consider the case of a practical step input signal that is modeled by the error function, $\mathrm{erf}(t/\gamma)$ and the case where such a signal is input to a 2nd-order linear filter with a transfer function defined by

$$H(s) = \frac{1}{\left[1 + \frac{s}{2\pi f_p}\right]^2} \Leftrightarrow h(t) = \frac{te^{-t/\tau}}{\tau^2} \cdot u(t), \;\; \tau = \frac{1}{2\pi f_p}. \quad (135)$$

Theorem 10. *Linear Filtering of an Error Function Signal. The output of a second-order linear filter, defined by Equation (135), to an error function input signal, defined by* $\mathrm{erf}(t/\gamma)$*, is*

$$\begin{aligned} y(t) = &\mathrm{erf}\left[\frac{t}{\gamma}\right]u(t) + \pi\theta \\ &\frac{e^{-t/\tau}}{\tau} \cdot \left[\begin{array}{r} \left[\frac{\gamma^2}{2\tau} - (t+\tau)\right]\exp\left[\frac{\gamma^2}{4\tau^2}\right]\left[\mathrm{erf}\left[\frac{\gamma}{2\tau}\right] - \mathrm{erf}\left[\frac{\gamma}{2\tau} - \frac{t}{\gamma}\right]\right] - \\ \frac{\gamma}{\sqrt{\pi}}\exp\left[\frac{\gamma^2}{4\tau^2}\right]\exp\left[-\left[\frac{t}{\gamma} - \frac{\gamma}{2\tau}\right]^2\right] + \frac{\gamma}{\sqrt{\pi}} \end{array}\right]u(t) \end{aligned} \quad (136)$$

and can be approximated by the nth-order signal

$$\begin{aligned} y_n(t) = &f_n\left[\frac{t}{\gamma}\right]u(t) + \\ &\frac{e^{-t/\tau}}{\tau} \cdot \left[\begin{array}{r} \left[\frac{\gamma^2}{2\tau} - (t+\tau)\right]\exp\left[\frac{\gamma^2}{4\tau^2}\right]\left[f_n\left[\frac{\gamma}{2\tau}\right] - f_n\left[\frac{\gamma}{2\tau} - \frac{t}{\gamma}\right]\right] - \\ \frac{\gamma}{\sqrt{\pi}}\exp\left[\frac{\gamma^2}{4\tau^2}\right]\exp\left[-\left[\frac{t}{\gamma} - \frac{\gamma}{2\tau}\right]^2\right] + \frac{\gamma}{\sqrt{\pi}} \end{array}\right]u(t) \end{aligned} \quad (137)$$

where f_n *is defined by one of the approximations detailed in Theorems 4, 5, 6, or 7. It is the case that*

$$\lim_{n\to\infty} y_n(t) = y(t), \;\; t \in (0,\infty). \quad (138)$$

Proof. The proof is detailed in Appendix G. □

Results

For an input signal $\mathrm{erf}(t/\gamma)u(t)$, $\gamma = 1/2$, input into a second-order linear filter with $f_p = 1$, the output signal is shown in Figure 24. The relative errors in the approximations to the output signal are shown in Figure 25 for the case of approximations as specified by Equation (56) and with the use of optimum transition points.

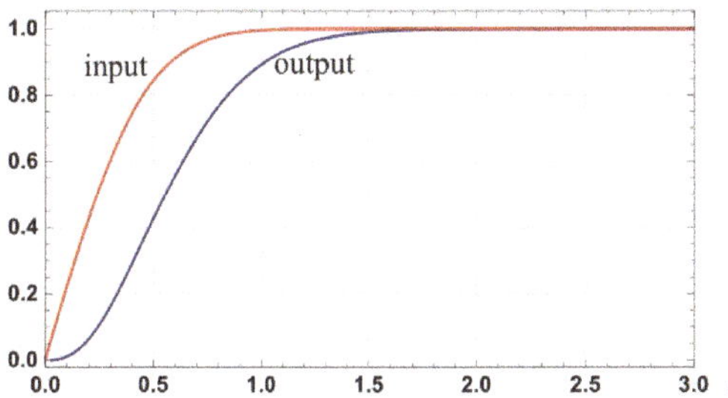

Figure 24. Graph of the input signal $\mathrm{erf}(t/\gamma)$, $\gamma = 1/2$ and the corresponding approximation to the output of a second order linear filter with $f_p = 1$, $\tau = 1/2\pi$.

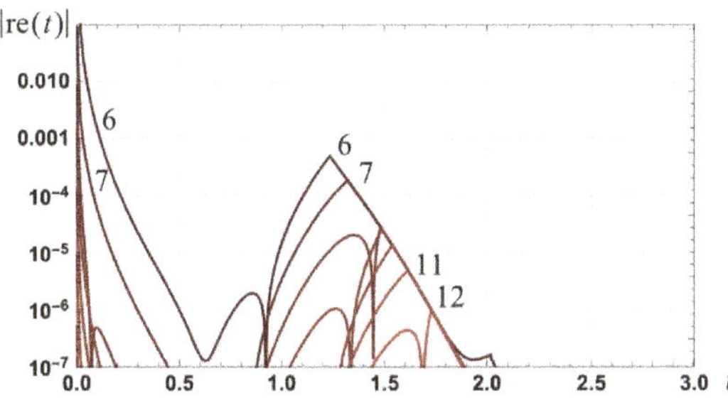

Figure 25. Graph of the relative errors associated with the output signal, shown in Figure 24, for approximations to the error function (Equation (56)) of orders six to twelve which utilize optimum transition points.

7.8. Extension to Complex Case

By definition, the error function, for the general complex case, is defined according to

$$\mathrm{erf}(z) = \frac{2}{\sqrt{\pi}} \int_{\gamma} e^{-\lambda^2} d\lambda, \ z \in C, \tag{139}$$

where the path γ is between the points zero and z and is arbitrary. For the case of $z = x + jy$, and a path along the x axis to the point $(x, 0)$ and then to the point z, the error function is defined according to Equation (5) of [37]:

$$\begin{aligned} \mathrm{erf}(x + jy) = & \ \tfrac{2}{\sqrt{\pi}} \int_0^x e^{-\lambda^2} d\lambda + \tfrac{2j}{\sqrt{\pi}} \int_0^y e^{-(x+j\lambda)^2} d\lambda \\ & = \mathrm{erf}(x) + \tfrac{2e^{-x^2}}{\sqrt{\pi}} \cdot \int_0^y e^{\lambda^2} \sin(2x\lambda) d\lambda + \tfrac{2je^{-x^2}}{\sqrt{\pi}} \cdot \int_0^y e^{\lambda^2} \cos(2x\lambda) d\lambda \end{aligned} \tag{140}$$

Explicit approximations for $\mathrm{erf}(x + jy)$ then arise when integrable approximations for the two-dimensional surfaces $\exp(y^2)\sin(2xy)$ and $\exp(y^2)\cos(2xy)$ over $[0, x] \times [0, y]$ are available. Naturally, significant existing research exists, e.g., [28,37].

7.9. Approximation for the Inverse Error Function

There are many applications where the inverse error function is required and accurate approximations for this function are of interest. From the research underpinning this paper, the author's view is that finding approximations to the inverse error function is best treated directly and as a separate problem, rather than approaching it via finding the inverse of an approximation to the error function.

8. Conclusions

This paper has detailed analytical approximations for the real case of the error function, underpinned by a spline-based integral approximation, which have significantly better convergence than the default Taylor series. The original approximations can be improved by utilizing the approximation $\text{erf}(x) \approx 1$ for $x > x_o$, with x_o being dependent on the order of approximation. The fourth-order approximations arising from Theorems 1 and 3, with respective transition points of $x_o = 2.3715$ and $x_o = 2.6305$, achieve relative error bounds over the interval $[0, \infty]$, respectively, of 1.03×10^{-3} and 2.28×10^{-4}. The respective sixteenth-order approximations, with $x_o = 3.9025$ and $x_o = 4.101$, have relative error bounds of 3.44×10^{-8} and 6.66×10^{-9}.

Further improvements were detailed via two generalizations. The first was based on utilizing integral approximations for each of the m equally spaced subintervals in the required interval of integration. The second was based on utilizing a fixed subinterval within the interval of integration with a known tabulated area, and then utilizing an integral approximation over the remainder of the interval. Both generalizations lead to significantly improved accuracy. For example, a fourth-order approximation based on four subintervals, with $x_o = 3.7208$, achieves a relative error bound of 1.43×10^{-7} over the interval $[0, \infty]$. A sixteenth-order approximation, with $x_o = 6.3726$, has a relative error bound of 2.01×10^{-19}.

Finally, it was shown that a custom feedback system, with inputs defined by either the original error function approximations or approximations based on the use of subintervals, leads to analytical approximations with improved accuracy that are valid over the positive real line without utilizing the approximation $\text{erf}(x) \approx 1$ for suitably large values of x. The original fourth-order error function approximation yields an approximation with a relative error bound of 1.82×10^{-5} over the interval $[0, \infty]$. The original sixteenth-order approximation yields an approximation with a relative error bound of 1.68×10^{-14}.

Applications of the approximations include, first, approximations to achieve the specified error bounds of 10^{-4}, 10^{-6}, 10^{-10}, and 10^{-16} over the positive real line. Second, the definitions of functions that are upper and lower bounds, of arbitrary accuracy, for the error function. Third, new series for the error function. Fourth, new sequences of approximations for $\exp(-x^2)$ that have significantly higher convergence properties than a Taylor series approximation. Fifth, a complementary demarcation function satisfying the constraint $e_C^2(x) + erf^2(x) = 1$ was defined. Sixth, arbitrarily accurate approximations for the power and harmonic distortion for a sinusoidal signal subject to an error function nonlinearity. Seventh, approximate expressions for the linear filtering of a step signal that is modeled by the error function.

Supplementary Materials: The following supporting information can be downloaded at: https://www.mdpi.com/article/10.3390/mca27010014/s1, File S1. Mathematica Code for Figure 2 Results.

Funding: This research received no external funding.

Acknowledgments: The support of Abdelhak M. Zoubir, SPG, Technische Universität Darmstadt, Darmstadt, Germany, who hosted a visit during which the research for, and writing of, this paper was completed, is gratefully acknowledged. An anonymous reviewer provided a comment, which, after consideration, led to the improved approximations detailed in Theorem 3.

Conflicts of Interest: The authors declare no conflict of interest.

Appendix A. Proof of Theorem 1

Consider $f(x) = \exp(-x^2)$. Successive differentiation of this function leads to the iterative formula

$$f^{(k)}(x) = p(k,x)\exp(-x^2), \tag{A1}$$

where

$$p(k,x) = p^{(1)}(k-1,x) - 2xp(k-1,x), \;\; p(0,x) = 1. \tag{A2}$$

It then follows from Equation (20) that

$$\begin{aligned} \frac{2}{\sqrt{\pi}} \cdot \int_{\alpha}^{x} \exp(-\lambda^2)d\lambda &\approx \frac{2}{\sqrt{\pi}} \cdot \sum_{k=0}^{n} c_{n,k}(x-\alpha)^{k+1}\left[f^{(k)}(\alpha) + (-1)^k f^{(k)}(x)\right] \\ &= \frac{2}{\sqrt{\pi}} \cdot \sum_{k=0}^{n} c_{n,k}(x-\alpha)^{k+1}\left[\begin{array}{c} p(k,\alpha)\exp(-\alpha^2)+ \\ (-1)^k p(k,x)\exp(-x^2) \end{array}\right] \end{aligned} \tag{A3}$$

The result for the case of $\alpha = 0$ then yields the nth-order approximation for the error function:

$$f_n(x) = \frac{2}{\sqrt{\pi}} \cdot \sum_{k=0}^{n} c_{n,k}x^{k+1}\left[p(k,0) + (-1)^k p(k,x)e^{-x^2}\right] \tag{A4}$$

To determine $\varepsilon_n^{(1)}(x)$ consider the equality $\mathrm{erf}(x) = f_n(x) + \varepsilon_n(x)$. Differentiation yields

$$\begin{aligned} \varepsilon_n^{(1)}(x) = \; & \frac{2e^{-x^2}}{\sqrt{\pi}} - \frac{2}{\sqrt{\pi}} \cdot \sum_{k=0}^{n} c_{n,k}(k+1)x^k\left[p(k,0) + (-1)^k p(k,x)e^{-x^2}\right] - \\ & \frac{2e^{-x^2}}{\sqrt{\pi}} \cdot \sum_{k=0}^{n} c_{n,k}x^{k+1}(-1)^k\left[p^{(1)}(k,x) - 2xp(k,x)\right] \end{aligned} \tag{A5}$$

and the required result:

$$\begin{aligned} \varepsilon_n^{(1)}(x) = \; & \frac{2e^{-x^2}}{\sqrt{\pi}} - \frac{2}{\sqrt{\pi}} \cdot \sum_{k=0}^{n} c_{n,k}(k+1)x^k p(k,0) - \\ & \frac{2e^{-x^2}}{\sqrt{\pi}} \cdot \sum_{k=0}^{n} c_{n,k}x^k(-1)^k\left[(k+1-2x^2)p(k,x) + xp^{(1)}(k,x)\right] \end{aligned} \tag{A6}$$

Appendix B. Proof of Theorem 2

The use of a Taylor series expansion for $\exp(-x^2)$ in the definitions of $\varepsilon_0^{(1)}(x)$ and $\varepsilon_1^{(1)}(x)$, as defined by Equations (40) and (41), yields:

$$\begin{aligned} \varepsilon_0^{(1)}(x) &= \frac{x^2}{\sqrt{\pi}} \cdot \left[1 - \frac{3x^2}{2} + \frac{5x^4}{6} - \frac{7x^6}{24} + \frac{3x^8}{40} - \frac{11x^{10}}{720} + \frac{13x^{12}}{5040} - \frac{x^{14}}{2688} + \ldots\right] \\ &= \frac{x^2}{\sqrt{\pi}} \cdot \left[c_{0,0} - c_{0,1}x^2 + \ldots + (-1)^k c_{0,k}x^{2k} + \ldots\right], \quad c_{0,k} = \frac{2}{k!} - \frac{1}{(k+1)!}, k \geq 0 \end{aligned} \tag{A7}$$

$$\varepsilon_1^{(1)}(x) = \frac{x^4}{6\sqrt{\pi}} \cdot \left[1 - 2x^2 + \frac{5x^4}{4} - \frac{7x^6}{16} + \frac{x^8}{8} - \frac{11x^{10}}{420} \cdots\right] \tag{A8}$$

From a consideration of $\varepsilon_0^{(1)}(x)$ and $\varepsilon_1^{(1)}(x)$, as well as higher-order residual functions, it can be readily seen that the polynomial terms of order zero to $2n + 1$ in $\varepsilon_1^{(1)}(x)$ have coefficients of zero. It then follows that $\varepsilon_n^{(1)}(x)$ can be written as

$$\varepsilon_n^{(1)}(x) = \frac{1}{\sqrt{\pi}} \cdot \frac{x^{2n+2}}{x_{n,0}} \cdot g_n(x), \quad n \in \{0, 1, 2, \ldots\}, \tag{A9}$$

where

$$g_n(x) = 1 - d_{n,1}x^2 + d_{n,2}x^4 - \ldots + (-1)^k d_{n,k}x^{2k} + \ldots \tag{A10}$$

and where it can readily be shown that

$$x_{n,0} = 2^n \prod_{i=0}^{n}(2i+1) \geq 2^n n!, \ \ n \in \{0,1,2,\ldots\}. \tag{A11}$$

Graphs of the magnitude of the residual functions, $\varepsilon_n^{(1)}(x)$, for orders zero, two, four, six, and eight, are shown in Figure A1. The magnitude of the functions defined by g_n, for the same orders, are shown in Figure A2 and it is evident that

$$|g_n(x)| \leq k_o, \ \ x \geq 0, n \in \{0,1,2,\ldots\}, \tag{A12}$$

for a fixed constant k_o, which is of the order of unity. Hence:

$$\left|\varepsilon_n^{(1)}(x)\right| \leq \frac{k_o}{\sqrt{\pi}} \cdot \frac{x^{2n+2}}{x_{n,0}}, \ \ x \geq 0, n \in \{0,1,2,\ldots\}. \tag{A13}$$

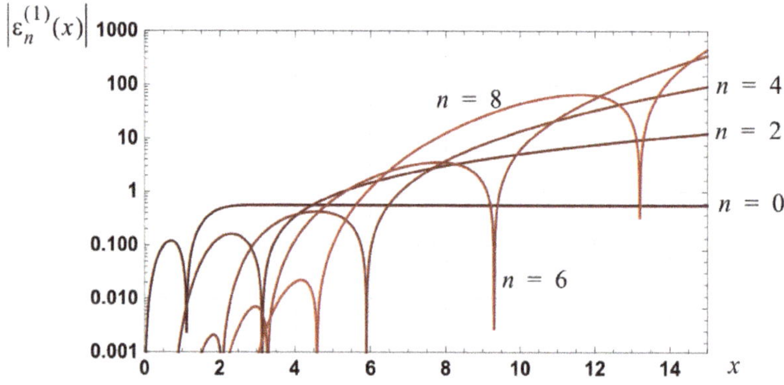

Figure A1. Graphs of $\left|\varepsilon_n^{(1)}(x)\right|$ for orders zero, two, four, six and eight.

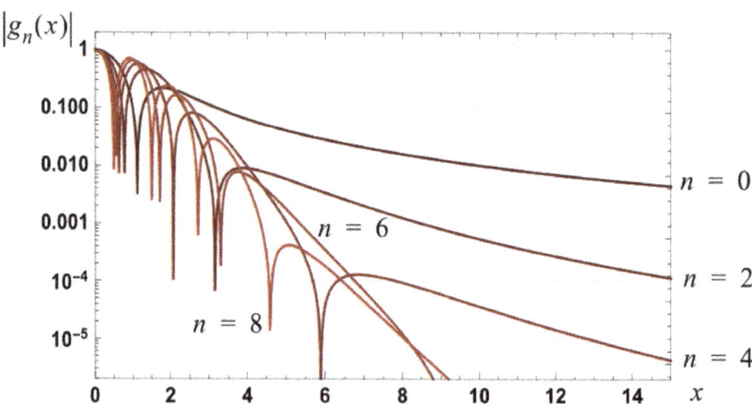

Figure A2. Graphs of $|g_n(x)|$ for orders zero, two, four, six and eight.

Further, as $x_{n,0} \geq 2^n n!$, it follows, for all fixed values of x, that

$$\lim_{n\to\infty} \varepsilon_n^{(1)}(x) = 0, \ \ x \geq 0. \tag{A14}$$

The convergence is not uniform.

It then follows, for all fixed values of x, that there exists an order of approximation, n, such that the error in the approximation $\varepsilon_n^{(1)}(x)$ can be made arbitrarily small, i.e., $\forall\varepsilon_o > 0$ there exists a number $N(x)$ such that

$$\left|\varepsilon_n^{(1)}(x)\right| < \varepsilon_o, \quad \forall n > N(x). \tag{A15}$$

In general, $N(x)$ increases with x. Thus, $\forall\varepsilon_o > 0$ there exists a number $N_{x_o}(x_o)$ such that

$$\left|\varepsilon_n^{(1)}(x)\right| < \varepsilon_o, \quad x \in [0, x_o], \forall n > N_{x_o}(x_o). \tag{A16}$$

Finally, as $\varepsilon_n^{(1)}(0) = 0$, for all n, it then follows, for x fixed, that

$$|\varepsilon_n(x)| = \left|\int_0^x \varepsilon_n^{(1)}(\lambda)d\lambda\right| < \varepsilon_o x, \quad \forall n > N_x(x), \tag{A17}$$

which proves convergence.

Appendix C. Fourth-Order Spline Approximation—The Sixteen-Subintervals Case

Consistent with Theorem 5, a fourth-order spline approximation, which utilizes 16 subintervals, is

$$\begin{aligned}
f_{4,16}(x) = {} & \frac{x}{16\sqrt{\pi}}\left[1 - \frac{16x^2}{73{,}728} + \frac{16x^4}{1{,}321{,}205{,}760}\right] + \\
& \frac{x}{8\sqrt{\pi}} \cdot \exp\left[\frac{-x^2}{256}\right]\left[1 - \frac{x^2}{4608} + \frac{47x^4}{27{,}525{,}120} - \frac{x^6}{5{,}284{,}823{,}040} + \frac{x^8}{4{,}058{,}744{,}094{,}720}\right] + \\
& \frac{x}{8\sqrt{\pi}} \cdot \exp\left[\frac{-x^2}{64}\right]\left[1 - \frac{x^2}{4608} + \frac{187x^4}{27{,}525{,}120} - \frac{x^6}{1{,}321{,}205{,}760} + \frac{x^8}{253{,}671{,}505{,}920}\right] + \\
& \frac{x}{8\sqrt{\pi}} \cdot \exp\left[\frac{-9x^2}{256}\right]\left[1 - \frac{x^2}{4608} + \frac{1261x^4}{82{,}575{,}360} - \frac{x^6}{587{,}202{,}560} + \frac{3x^8}{150{,}323{,}855{,}360}\right] + \\
& \frac{x}{8\sqrt{\pi}} \cdot \exp\left[\frac{-x^2}{16}\right]\left[1 - \frac{x^2}{4608} + \frac{249x^4}{9{,}175{,}040} - \frac{x^6}{330{,}301{,}440} + \frac{x^8}{15{,}854{,}469{,}120}\right] + \\
& \frac{x}{8\sqrt{\pi}} \cdot \exp\left[\frac{-25x^2}{256}\right]\left[1 - \frac{x^2}{4608} + \frac{389x^4}{9{,}175{,}040} - \frac{5x^6}{1{,}056{,}964{,}608} + \frac{125x^8}{811{,}748{,}818{,}944}\right] + \\
& \frac{x}{8\sqrt{\pi}} \cdot \exp\left[\frac{-9x^2}{64}\right]\left[1 - \frac{x^2}{4608} + \frac{5041x^4}{82{,}575{,}360} - \frac{x^6}{146{,}800{,}640} + \frac{3x^8}{9{,}395{,}240{,}960}\right] + \\
& \frac{x}{8\sqrt{\pi}} \cdot \exp\left[\frac{-49x^2}{256}\right]\left[1 - \frac{x^2}{4608} + \frac{2287x^4}{27{,}525{,}120} - \frac{7x^6}{754{,}974{,}720} + \frac{343x^8}{579{,}820{,}584{,}960}\right] + \\
& \frac{x}{8\sqrt{\pi}} \cdot \exp\left[\frac{-x^2}{4}\right]\left[1 - \frac{x^2}{4608} + \frac{2987x^4}{27{,}525{,}120} - \frac{x^6}{82{,}575{,}360} + \frac{x^8}{990{,}904{,}320}\right] + \\
& \frac{x}{8\sqrt{\pi}} \cdot \exp\left[\frac{-81x^2}{256}\right]\left[1 - \frac{x^2}{4608} + \frac{11341x^4}{82{,}575{,}360} - \frac{9x^6}{587{,}202{,}560} + \frac{243x^8}{150{,}323{,}855{,}360}\right] + \\
& \frac{x}{8\sqrt{\pi}} \cdot \exp\left[\frac{-25x^2}{64}\right]\left[1 - \frac{x^2}{4608} + \frac{4667x^4}{27{,}525{,}120} - \frac{5x^6}{264{,}241{,}152} + \frac{125x^8}{50{,}734{,}301{,}184}\right] + \\
& \frac{x}{8\sqrt{\pi}} \cdot \exp\left[\frac{-121x^2}{256}\right]\left[1 - \frac{x^2}{4608} + \frac{5647x^4}{27{,}525{,}120} - \frac{121x^6}{5{,}284{,}823{,}040} + \frac{14641x^8}{4{,}058{,}744{,}094{,}720}\right] + \\
& \frac{x}{8\sqrt{\pi}} \cdot \exp\left[\frac{-9x^2}{16}\right]\left[1 - \frac{x^2}{4608} + \frac{20161x^4}{82{,}575{,}360} - \frac{x^6}{36{,}700{,}160} + \frac{3x^8}{587{,}202{,}560}\right] + \\
& \frac{x}{8\sqrt{\pi}} \cdot \exp\left[\frac{-169x^2}{256}\right]\left[1 - \frac{x^2}{4608} + \frac{2629x^4}{9{,}175{,}040} - \frac{169x^6}{5{,}284{,}823{,}040} + \frac{28{,}561x^8}{4{,}058{,}744{,}094{,}720}\right] + \\
& \frac{x}{8\sqrt{\pi}} \cdot \exp\left[\frac{-49x^2}{64}\right]\left[1 - \frac{x^2}{4608} + \frac{3049x^4}{9{,}175{,}040} - \frac{7x^6}{188{,}743{,}680} + \frac{343x^8}{36{,}238{,}786{,}560}\right] + \\
& \frac{x}{8\sqrt{\pi}} \cdot \exp\left[\frac{-225x^2}{256}\right]\left[1 - \frac{x^2}{4608} + \frac{31501x^4}{82{,}575{,}360} - \frac{5x^6}{117{,}440{,}512} + \frac{375x^8}{30{,}064{,}771{,}072}\right] + \\
& \frac{x}{16\sqrt{\pi}} \cdot \exp(-x^2)\left[1 + \frac{127x^2}{4608} + \frac{3929x^4}{9{,}175{,}040} + \frac{79x^6}{20{,}643{,}840} + \frac{x^8}{61{,}931{,}520}\right]
\end{aligned} \tag{A18}$$

When this approximation is utilized with the transition point $x_o = 7.1544$, the relative error bound in the approximation to the error function, over the interval $(0, \infty)$, is 4.82×10^{-16}.

Appendix D. Proof of Theorem 7

Consider the differential Equation

$$y_n{\prime}(t) + g_n(t)y_n(t) = g_n(t), \; y_n(0) = 0, \tag{A19}$$

for the case where g_n is based on the nth-order approximation f_n to the error function, defined in Theorem 1, and is defined according to

$$g_n(t) = \frac{4}{\sqrt{\pi}}e^{-t^2}f_n(t) = \frac{8e^{-t^2}}{\pi} \cdot \sum_{k=0}^{n} c_{n,k}t^{k+1}\left[p(k,0) + (-1)^k p(k,t)e^{-t^2}\right], t \geq 0. \tag{A20}$$

To find a solution to the differential equation for such a driving signal, first note that the solution to the differential equation for the case of $g_n(t) = \frac{4}{\sqrt{\pi}}e^{-t^2}\mathrm{erf}(t)$ is

$$y_n(t) = 1 - \exp[-\mathrm{erf}^2(t)]. \tag{A21}$$

Second, with g_n defined by Equation (A20), the following signal form

$$y_n(t) = 1 - \exp\left[-\left[p_{n,0} + p_{n,1}(t)e^{-t^2} + p_{n,2}(t)e^{-2t^2}\right]\right], \; t \geq 0, \tag{A22}$$

has potential as a basis for finding solutions for the unknown polynomial functions $p_{n,1}$ and $p_{n,2}$ and the unknown constant $p_{n,0}$. With such a form, the initial condition of $y_n(0) = 0$ implies

$$p_{n,0} = -[p_{n,1}(0) + p_{n,2}(0)]. \tag{A23}$$

It is the case that

$$y_n{\prime}(t) = \left[p_{n,1}^{(1)}(t)e^{-t^2} - 2tp_{n,1}(t)e^{-t^2} + p_{n,2}^{(1)}(t)e^{-2t^2} - 4tp_{n,2}(t)e^{-2t^2}\right] \cdot [1 - y_n(t)]. \tag{A24}$$

Substitution of $y_n(t)$ and $y_n{\prime}(t)$ into the differential equation yields

$$\begin{gathered}\left[p_{n,1}^{(1)}(t)e^{-t^2} - 2tp_{n,1}(t)e^{-t^2} + p_{n,2}^{(1)}(t)e^{-2t^2} - 4tp_{n,2}(t)e^{-2t^2}\right] \cdot \exp\left[-\left[p_{n,0} + p_{n,1}(t)e^{-t^2} + p_{n,2}(t)e^{-2t^2}\right]\right] + \\ \frac{4}{\sqrt{\pi}}e^{-t^2}f_n(t) \cdot \left[1 - \exp\left[-\left[p_{n,0} + p_{n,1}(t)e^{-t^2} + p_{n,2}(t)e^{-2t^2}\right]\right]\right] = \frac{4}{\sqrt{\pi}}e^{-t^2}f_n(t)\end{gathered} \tag{A25}$$

which implies

$$p_{n,1}^{(1)}(t)e^{-t^2} - 2tp_{n,1}(t)e^{-t^2} + p_{n,2}^{(1)}(t)e^{-2t^2} - 4tp_{n,2}(t)e^{-2t^2} - \frac{4}{\sqrt{\pi}}e^{-t^2}f_n(t) = 0 \tag{A26}$$

and

$$\begin{gathered}p_{n,1}^{(1)}(t)e^{-t^2} - 2tp_{n,1}(t)e^{-t^2} + p_{n,2}^{(1)}(t)e^{-2t^2} - 4tp_{n,2}(t)e^{-2t^2} = \\ \frac{8}{\pi}e^{-t^2}\sum_{k=0}^{n} c_{n,k}t^{k+1}\left[p(k,0) + (-1)^k p(k,t)e^{-t^2}\right].\end{gathered} \tag{A27}$$

Thus:

$$\begin{aligned}p_{n,1}^{(1)}(t) - 2tp_{n,1}(t) &= \frac{8}{\pi}\sum_{k=0}^{n} c_{n,k}p(k,0)t^{k+1} \\ p_{n,2}^{(1)}(t) - 4tp_{n,2}(t) &= \frac{8}{\pi}\sum_{k=0}^{n} c_{n,k}(-1)^k p(k,t)t^{k+1}\end{aligned} \tag{A28}$$

To solve for the polynomials $p_{n,1}$ and $p_{n,2}$, first note (see Equation (26)) that

$$p(k,t) = a_{k,0} + a_{k,1}t + a_{k,2}t^2 + \ldots + a_{k,k}t^k, k \in \{0,1,\ldots,n\}, \tag{A29}$$

for appropriately defined coefficients $a_{k,j}$, $j \in \{0,1,\ldots,k\}$.

Appendix D.1. Solving for Coefficients of First Polynomial

Substitution of $p(k,0)$ from Equation (A29) into the differential equation defining $p_{n,1}$ yields

$$p_{n,1}^{(1)}(t) - 2tp_{n,1}(t) = \frac{8}{\pi}\sum_{k=0}^{n} c_{n,k}a_{k,0}t^{k+1}. \tag{A30}$$

With $a_{n,0} = 0$ for n odd, the maximum power for t on the right side of the differential equation is t^{n+1}, n even, and t^n for n odd. Thus, the form required for $p_{n,1}$ is

$$p_{n,1}(t) = \begin{cases} \alpha_0 + \alpha_1 t + \ldots + \alpha_{n-1}t^{n-1} & n \text{ odd} \\ \alpha_0 + \alpha_1 t + \ldots + \alpha_n t^n & n \text{ even} \end{cases} \tag{A31}$$

Substitution then yields

$$\begin{aligned} &[\alpha_1 + 2\alpha_2 t + \ldots + (n-1)\alpha_{n-1}t^{n-2}] - 2t[\alpha_0 + \alpha_1 t + \ldots + \alpha_{n-1}t^{n-1}] = \\ &\qquad \tfrac{8}{\pi}\sum_{k=0}^{n-1} c_{n,k}a_{k,0}t^{k+1} \quad n \text{ odd} \\ &[\alpha_1 + 2\alpha_2 t + \ldots + n\alpha_n t^{n-1}] - 2t[\alpha_0 + \alpha_1 t + \ldots + \alpha_n t^n] = \\ &\qquad \tfrac{8}{\pi}\sum_{k=0}^{n} c_{n,k}a_{k,0}t^{k+1} \quad n \text{ even} \end{aligned} \tag{A32}$$

For the case of n even, equating coefficients associated with set powers of t, yields:

$$\begin{aligned} &t^{n+1} : -2\alpha_n = \tfrac{8}{\pi}\cdot c_{n,n}a_{n,0} \Rightarrow \alpha_n = \tfrac{-4}{\pi}\cdot c_{n,n}a_{n,0} \\ &t^n : -2\alpha_{n-1} = \tfrac{8}{\pi}\cdot c_{n,n-1}a_{n-1,0} \Rightarrow \alpha_{n-1} = \tfrac{-4}{\pi}\cdot c_{n,n-1}a_{n-1,0} \\ &t^{n-1} : n\alpha_n - 2\alpha_{n-2} = \tfrac{8}{\pi}\cdot c_{n,n-2}a_{n-2,0} \Rightarrow \alpha_{n-2} = \tfrac{n}{2}\cdot\alpha_n - \tfrac{4}{\pi}\cdot c_{n,n-2}a_{n-2,0} \\ &\ldots \\ &t^2 : 3\alpha_3 - 2\alpha_1 = \tfrac{8}{\pi}\cdot c_{n,1}a_{1,0} \Rightarrow \alpha_1 = \tfrac{3\alpha_3}{2} - \tfrac{4}{\pi}\cdot c_{n,1}a_{1,0} \\ &t^1 : 2\alpha_2 - 2\alpha_0 = \tfrac{8}{\pi}\cdot c_{n,0}a_{0,0} \Rightarrow \alpha_0 = \alpha_2 - \tfrac{4}{\pi}\cdot c_{n,0}a_{0,0} \end{aligned} \tag{A33}$$

With the odd coefficients $a_{1,0}$, $a_{3,0}$, ..., $a_{n-1,0}$ being zero, it follows that the corresponding odd coefficients $\alpha_{n-1}, \alpha_{n-3}, \ldots, \alpha_1$ are also zero. For the even coefficients, the algorithm is:

$$\begin{aligned} \alpha_m &= \tfrac{-4}{\pi}\cdot c_{n,m}a_{m,0} \\ \alpha_{m-2i} &= \tfrac{(m-2i+2)\alpha_{m-2i+2}}{2} - \tfrac{4}{\pi}\cdot c_{n,m-2i}a_{m-2i,0}, \quad i \in \{1,\ldots,\tfrac{m}{2}-1\} \\ \alpha_0 &= \alpha_2 - \tfrac{4}{\pi}\cdot c_{n,0}a_{0,0} \end{aligned} \tag{A34}$$

where $m = n$. For the case of n being odd, the odd coefficients $\alpha_n, \alpha_{n-2}, \ldots, \alpha_1$ are again zero and the algorithm is the same as that specified in Equation (A34) with $m = n - 1$.

Appendix D.2. Solving for Coefficients of Second Polynomial

Substitution of $p(k,t)$ from Equation (A29) into the differential equation defining $p_{n,2}$ yields:

$$p_{n,2}^{(1)}(t) - 4tp_{n,2}(t) = \frac{8}{\pi}\cdot\sum_{k=0}^{n} c_{n,k}(-1)^k\left[\sum_{i=0}^{k} a_{k,i}t^{k+i+1}\right]. \tag{A35}$$

The coefficients $a_{k,i}$ that are associated with a given power of t are illustrated in Figure A3. It then follows, for a fixed power of t, say t^r, that the associated coefficients, $a_{k,i}$, are

$$\begin{aligned} &k \in \left\{\left\lfloor \tfrac{r}{2} \right\rfloor, \left\lfloor \tfrac{r}{2} \right\rfloor + 1, \ldots, \min\{r-1, n\}\right\} \\ &i = r - k - 1. \end{aligned} \tag{A36}$$

Thus:

$$p_{n,2}^{(1)}(t) - 4tp_{n,2}(t) = \frac{8}{\pi} \cdot \sum_{r=1}^{2n+1} \left[\sum_{k=\lfloor \frac{r}{2} \rfloor}^{\min\{r-1,n\}} c_{n,k}(-1)^k a_{k,r-k-1} \right] t^r. \tag{A37}$$

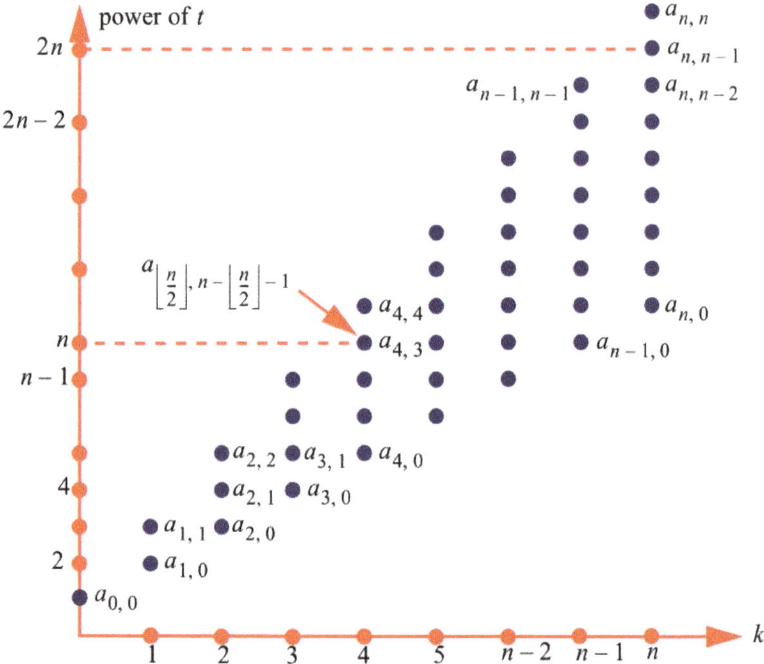

Figure A3. Illustration of the coefficients that potentially are non-zero for a set power of t. The illustration is for the case of $n = 8$.

With

$$p_{n,2}(t) = \beta_0 + \beta_1 t + \ldots + \beta_m t^m, \tag{A38}$$

the differential equation implies

$$\begin{aligned} &\left[\beta_1 + 2\beta_2 t + \ldots + m\beta_m t^{m-1}\right] - 4t\left[\beta_0 + \beta_1 t + \ldots + \beta_m t^m\right] \\ &\quad = \tfrac{8}{\pi} \cdot \sum_{r=1}^{2n+1} \left[\sum_{k=\lfloor \frac{r}{2} \rfloor}^{\min\{r-1,n\}} c_{n,k}(-1)^k a_{k,r-k-1} \right] t^r. \end{aligned} \tag{A39}$$

The requirement, thus, is for $m = 2n$. Equating coefficients (see Figure A3) yields:

$$
\begin{aligned}
&t^{2n+1}: \quad -4\beta_{2n} = \tfrac{8}{\pi}\cdot(-1)^n c_{n,n}a_{n,n} \Rightarrow \beta_{2n} = \tfrac{-2}{\pi}\cdot(-1)^n c_{n,n}a_{n,n} \\
&t^{2n}: \quad -4\beta_{2n-1} = \tfrac{8}{\pi}\left[(-1)^n c_{n,n}a_{n,n-1}\right] \Rightarrow \beta_{2n-1} = \tfrac{-2}{\pi}\cdot(-1)^n c_{n,n}a_{n,n-1} \\
&t^{2n-1}: \quad 2n\beta_{2n} - 4\beta_{2n-2} = \tfrac{8}{\pi}\left[(-1)^{n-1}c_{n,n-1}a_{n-1,n-1} + (-1)^n c_{n,n}a_{n,n-2}\right] \\
&\qquad \Rightarrow \beta_{2n-2} = \tfrac{2n\beta_{2n}}{4} - \tfrac{2}{\pi}\left[(-1)^{n-1}c_{n,n-1}a_{n-1,n-1} + (-1)^n c_{n,n}a_{n,n-2}\right] \\
&\qquad \ldots \\
&t^3: \quad 4\beta_4 - 4\beta_2 = \tfrac{8}{\pi}[-c_{n,1}a_{1,1} + c_{n,2}a_{2,0}] \Rightarrow \beta_2 = \beta_4 - \tfrac{2}{\pi}[-c_{n,1}a_{1,1} + c_{n,2}a_{2,0}] \\
&t^2: \quad 3\beta_3 - 4\beta_1 = \tfrac{-8}{\pi}\cdot c_{n,1}a_{1,0} \Rightarrow \beta_1 = \tfrac{3\beta_3}{4} + \tfrac{2}{\pi}\cdot c_{n,1}a_{1,0} \\
&t^1: \quad 2\beta_2 - 4\beta_0 = \tfrac{8}{\pi}\cdot c_{n,0}a_{0,0} \Rightarrow \beta_0 = \tfrac{2\beta_2}{4} - \tfrac{2}{\pi}\cdot c_{n,0}a_{0,0}
\end{aligned}
\tag{A40}
$$

As the coefficients $a_{n,n-1}, a_{n,n-3}, \ldots, a_{1,0}$ are zero, the algorithm is:

$$
\begin{aligned}
&\beta_{2n} = \tfrac{-2}{\pi}\cdot(-1)^n c_{n,n}a_{n,n} \\
&\beta_{2n-2i} = \tfrac{[2n-2i+2]\beta_{2n-2i+2}}{4} - \tfrac{2}{\pi}\sum_{k=n-i}^{\min\{2n-2i,n\}} (-1)^k c_{n,k}a_{k,2(n-i)-k},\ i \in \{1,\ldots,n-1\} \\
&\beta_0 = \tfrac{\beta_2}{2} - \tfrac{2}{\pi}\cdot c_{n,0}a_{0,0}
\end{aligned}
\tag{A41}
$$

Appendix E. Proof of Theorem 8

Consider the results stated in Theorem 1:

$$\text{erf}(x) = \frac{2}{\sqrt{\pi}}\cdot\int_0^x e^{-\lambda^2}\mathrm{d}\lambda \approx \frac{2}{\sqrt{\pi}}\cdot\sum_{k=0}^{n} c_{n,k}x^{k+1}\left[p(k,0) + (-1)^k p(k,x)e^{-x^2}\right]. \tag{A42}$$

Differentiation then yields

$$
\begin{aligned}
e^{-x^2} \approx\ &\sum_{k=0}^{n} c_{n,k}(k+1)x^k\left[p(k,0) + (-1)^k p(k,x)e^{-x^2}\right] + \\
&\sum_{k=0}^{n} c_{n,k}(-1)^k x^{k+1}\left[p^{(1)}(k,x)e^{-x^2} - 2xp(k,x)e^{-x^2}\right]
\end{aligned}
\tag{A43}
$$

which leads to the required result:

$$e^{-x^2} \approx \frac{\sum_{k=0}^{n} c_{n,k}(k+1)x^k p(k,0)}{1 + \sum_{k=0}^{n} c_{n,k}(-1)^{k+1}x^k\left[p(k,x)(k+1-2x^2) + xp^{(1)}(k,x)\right]} \tag{A44}$$

Appendix F. Proof of Theorem 9

Consider the exact result

$$\text{erf}(x) = f_0(x) + \varepsilon_0(x) \tag{A45}$$

where f_0 is specified by Equation (32) and the derivative of the error term, $\varepsilon_0^{(1)}(x)$, is specified by Equation (40). By utilizing a Taylor series approximation for $\exp(-x^2)$, $\varepsilon_0^{(1)}(x)$ can be written as

$$\varepsilon_0^{(1)}(x) = \frac{x^2}{\sqrt{\pi}}\cdot\left[1 - \frac{3x^2}{2} + \frac{5x^4}{6} - \frac{7x^6}{24} + \frac{9x^8}{120} - \ldots + \frac{(-1)^k(2k+1)x^{2k}}{(k+1)!} + \ldots\right] \tag{A46}$$

Integration yields

$$\varepsilon_0(x) = \frac{x^3}{\sqrt{\pi}} \cdot \left[\frac{1}{3} - \frac{3x^2}{2 \cdot 5} + \frac{5x^4}{6 \cdot 7} - \frac{7x^6}{24 \cdot 9} + \frac{9x^8}{120 \cdot 11} - \ldots + \frac{(-1)^k(2k+1)x^{2k}}{(2k+3)(k+1)!} + \ldots\right] \tag{A47}$$

and the following series for the error function then follows:

$$\begin{aligned} \operatorname{erf}(x) = & \frac{x}{\sqrt{\pi}} + \frac{x}{\sqrt{\pi}} \cdot e^{-x^2} + \\ & \frac{x^3}{\sqrt{\pi}} \cdot \left[\frac{1}{3} - \frac{3x^2}{2 \cdot 5} + \frac{5x^4}{6 \cdot 7} - \frac{7x^6}{24 \cdot 9} + \frac{9x^8}{120 \cdot 11} - \ldots + \frac{(-1)^k(2k+1)x^{2k}}{(2k+3)(k+1)!} + \ldots\right] \end{aligned} \tag{A48}$$

The series associated with first- and second-order approximations follow in an analogous manner.

Appendix G. Proof of Theorem 10

The filter output is given by the convolution integral:

$$\begin{aligned} y(t) = & \int_0^t \operatorname{erf}\left[\frac{\lambda}{\gamma}\right] \cdot \frac{(t-\lambda)e^{-[t-\lambda]/\tau}}{\tau^2} d\lambda \\ = & \frac{e^{-t/\tau}}{\tau^2} \cdot \left[t \int_0^t \operatorname{erf}\left[\frac{\lambda}{\gamma}\right] e^{\lambda/\tau} d\lambda - \int_0^t \operatorname{erf}\left[\frac{\lambda}{\gamma}\right] \lambda e^{\lambda/\tau} d\lambda\right]. \end{aligned} \tag{49}$$

Using the integral results, e.g., Equations (4.2.1) and (4.2.5) of [38]:

$$\int_0^t \operatorname{erf}(a\lambda)e^{b\lambda} d\lambda = \frac{1}{b}\operatorname{erf}(at)e^{bt} - \frac{1}{b}\exp\left[\frac{b^2}{4a^2}\right]\operatorname{erf}\left[at - \frac{b}{2a}\right] + \frac{1}{b}\exp\left[\frac{b^2}{4a^2}\right]\operatorname{erf}\left[-\frac{b}{2a}\right] \tag{A50}$$

$$\begin{aligned} \int_0^t \operatorname{erf}(a\lambda)\lambda e^{b\lambda} d\lambda = & \frac{1}{b}\left[t - \frac{1}{b}\right]\operatorname{erf}(at)e^{bt} - \frac{1}{b}\exp\left[\frac{b^2}{4a^2}\right]\left[\begin{array}{c} \left[\frac{b}{2a^2} - \frac{1}{b}\right]\operatorname{erf}\left[at - \frac{b}{2a}\right] - \\ \frac{1}{a\sqrt{\pi}}\exp\left[-\left[at - \frac{b}{2a}\right]^2\right] \end{array}\right] + \\ & \frac{1}{b}\exp\left[\frac{b^2}{4a^2}\right]\left[\left[\frac{b}{2a^2} - \frac{1}{b}\right]\operatorname{erf}\left[-\frac{b}{2a}\right] - \frac{1}{a\sqrt{\pi}}\exp\left[\frac{-b^2}{4a^2}\right]\right] \end{aligned} \tag{A51}$$

with $a = 1/\gamma$, $b = 1/\tau$, $b^2/4a^2 = \gamma^2/4\tau^2$, it then follows that

$$\begin{aligned} y(t) = & \frac{te^{-t/\tau}}{\tau^2}\left[\tau erf\left[\frac{t}{\gamma}\right]e^{t/\tau} - \tau\exp\left[\frac{\gamma^2}{4\tau^2}\right]erf\left[\frac{t}{\gamma} - \frac{\gamma}{2\tau}\right] + \tau\exp\left[\frac{\gamma^2}{4\tau^2}\right]erf\left[\frac{-\gamma}{2\tau}\right]\right] - \\ & \frac{e^{-t/\tau}}{\tau^2}\left[\begin{array}{c} \tau(t-\tau)erf\left[\frac{t}{\gamma}\right]e^{t/\tau} - \tau\exp\left[\frac{\gamma^2}{4\tau^2}\right]\left[\begin{array}{c} \left[\frac{\gamma^2}{2\tau} - \tau\right]\operatorname{erf}\left[\frac{t}{\gamma} - \frac{\gamma}{2\tau}\right] - \\ \frac{\gamma}{\sqrt{\pi}}\exp\left[-\left[\frac{t}{\gamma} - \frac{\gamma}{2\tau}\right]^2\right] \end{array}\right] + \\ \tau\exp\left[\frac{\gamma^2}{4\tau^2}\right]\left[\left[\frac{\gamma^2}{2\tau} - \tau\right]\operatorname{erf}\left[\frac{-\gamma}{2\tau}\right] - \frac{\gamma}{\sqrt{\pi}}\exp\left[\frac{-\gamma^2}{4\tau^2}\right]\right] \end{array}\right] \end{aligned} \tag{A52}$$

Simplifying, and using the fact that the error function is an odd function, yields the required result:

$$\begin{aligned} y(t) = & erf\left[\frac{t}{\gamma}\right] + \\ & \frac{e^{-t/\tau}}{\tau} \cdot \left[\begin{array}{c} \exp\left[\frac{\gamma^2}{4\tau^2}\right]\left[\frac{\gamma^2}{2\tau} - (t+\tau)\right]\left[\operatorname{erf}\left[\frac{\gamma}{2\tau}\right] - \operatorname{erf}\left[\frac{\gamma}{2\tau} - \frac{t}{\gamma}\right]\right] - \\ \frac{\gamma}{\sqrt{\pi}}\exp\left[\frac{\gamma^2}{4\tau^2}\right]\exp\left[-\left[\frac{t}{\gamma} - \frac{\gamma}{2\tau}\right]^2\right] + \frac{\gamma}{\sqrt{\pi}} \end{array}\right] \end{aligned} \tag{A53}$$

To prove convergence, consider

$$\lim_{n\to\infty} y_n(t) = \lim_{n\to\infty} \int_0^t f_n\left[\frac{\lambda}{\gamma}\right] h(t-\lambda)d\lambda = \int_0^t \operatorname{erf}\left[\frac{\lambda}{\gamma}\right] h(t-\lambda)d\lambda \tag{A54}$$

where $\lim_{n\to\infty} f_n(x) = \operatorname{erf}(x)$ and h is the impulse response of the second-order filter. The interchange of limit and integration is valid, consistent with Lemma 2, as the integrand comprises differentiable bounded functions.

References

1. Lebedev, N.N. *Special Functions and Their Applications*; Dover Publications: Dover, DE, USA, 1971.
2. Fujiwara, T. Wavelength response of harmonic distortion in AC-bias recording. *IEEE Trans. Magn.* **1980**, *16*, 501–506. [CrossRef]
3. Lee, J.; Woodring, D. Considerations of Nonlinear Effects in Phase-Modulation Systems. *IEEE Trans. Commun.* **1972**, *20*, 1063–1073. [CrossRef]
4. Klein, S.A. Measuring, estimating, and understanding the psychometric function: A commentary. *Percept. Psychophys.* **2001**, *63*, 1421–1455. [CrossRef] [PubMed]
5. Rinderknecht, M.D.; Lambercy, O.; Gassert, R. Performance metrics for an application-driven selection and optimization of psychophysical sampling procedures. *PLoS ONE* **2018**, *13*, e0207217. [CrossRef]
6. Shi, Q. OFDM in bandpass nonlinearity. *IEEE Trans. Consum. Electron.* **1996**, *42*, 253–258. [CrossRef]
7. Taggart, D.; Kumar, R.; Raghavan, S.; Goo, G.; Chen, J.; Krikorian, Y. Communication system performance—Detailed modeling of a power amplifier with two modulated input signals. In Proceedings of the 2005 IEEE Aerospace Conference, Bozeman, MT, USA, 4 December 2005; pp. 1398–1409. [CrossRef]
8. Li, W. Damage Models for Soft Tissues: A Survey. *J. Med. Biol. Eng.* **2016**, *36*, 285–307. [CrossRef]
9. Ogden, R.W.; Roxburgh, D.G. A pseudo-elastic model for the Mullins effect in filled rubber. *Proc. R. Soc. Lond. Ser. A Math. Phys. Eng. Sci.* **1999**, *455*, 2861–2877. [CrossRef]
10. Temme, N.M. Error Functions, Dawson's and Fresnel Integrals. In *NIST Handbook of Mathematical Functions*; Olver, F.W., Lozier, D.W., Boisvert, R.F., Clark, C.W., Eds.; National Institute of Standards and Technology and Cambridge University Press: Cambridge, MA, USA, 2010.
11. Schrerier, F. The Voigt and complex error function: A comparison of computational methods. *J. Quant. Spectrosc. Radiat. Transf.* **1992**, *48*, 743–762. [CrossRef]
12. Marsaglia, G. Evaluating the Normal Distribution. *J. Stat. Softw.* **2004**, *11*, 1–11. [CrossRef]
13. Sandoval-Hernandez, M.A.; Vazquez-Leal, H.; Filobello-Nino, U.; Hernandez-Martinez, L. New handy and accurate approximation for the Gaussian integrals with applications to science and engineering. *Open Math.* **2019**, *17*, 1774–1793. [CrossRef]
14. Menzel, R. Approximate closed form solution to the error function. *Am. J. Phys.* **1975**, *43*, 366–367, Erratum in *Am. J. Phys.* **1975**, *43*, 923–923 . [CrossRef]
15. Matíc, I.; Radoičić, R.; Stefanica, D. A sharp Pólya-based approximation to the normal CDF. *Appl. Math. Comput.* **2018**, *322*, 111–122. [CrossRef]
16. Chevillard, S. The functions erf and erfc computed with arbitrary precision and explicit error bounds. *Inf. Comput.* **2012**, *216*, 72–95. [CrossRef]
17. De Schrijver, S.K.; Aghezzaf, E.-H.; Vanmaele, H. Double precision rational approximation algorithms for the standard normal first and second order loss functions. *Appl. Math. Comput.* **2012**, *219*, 2320–2330. [CrossRef]
18. Cody, W.J. Rational Chebyshev approximations for the error function. *Math. Comput.* **1969**, *23*, 631–637. [CrossRef]
19. Howard, R.M. Dual Taylor Series, Spline Based Function and Integral Approximation and Applications. *Math. Comput. Appl.* **2019**, *24*, 35. [CrossRef]
20. Howard, R.M. Arbitrarily Accurate Spline Based Approximations for the Hyperbolic Tangent Function and Applications. *Int. J. Appl. Comput. Math.* **2021**, *7*, 1–59. [CrossRef]
21. Abramowitz, M.; Stegun, I.A. *Handbook of Mathematical Functions with Formulas, Graphs and Mathematical Tables*; US Department of Commerce: Dover, DE, USA, 1964.
22. Nandagopal, M.; Sen, S.; Rawat, A. A Note on the Error Function. *Comput. Sci. Eng.* **2010**, *12*, 84–88. [CrossRef]
23. Schöpf, H.M.; Supancic, P.H. On Bürmann's theorem and its application to problems of linear and nonlinear heat transfer and diffusion: Expanding a function in powers of its derivative. *Math. J.* **2014**, *16*, 1–44.
24. Winitzki, S. A Handy Approximation for the Error Function and Its Inverse. 2008. Available online: https://scholar.google.com/citations?user=Q9U40gUAAAAJ&hl=en&oi=sra (accessed on 10 November 2020).
25. Soranzo, A.; Epure, E. Simply explicitly invertible approximations to 4 decimals of error function and normal cumulative distribution function. *arXiv* **2012**, arXiv:1201.1320v1.
26. Vedder, J.D. Simple approximations for the error function and its inverse. *Am. J. Phys.* **1987**, *55*, 762–763. [CrossRef]

27. Vazquez-Leal, H.; Castaneda-Sheissa, R.; Filobello-Nino, U.; Sarmiento-Reyes, A.; Orea, J.S. High Accurate Simple Approximation of Normal Distribution Integral. *Math. Probl. Eng.* **2012**, *2012*, 1–22. [CrossRef]
28. Abrarov, S.M.; Quine, B.M. A rapid and highly accurate approximation for the error function of complex argument. *arXiv* **2013**, arXiv:1308.3399.
29. Champeney, D.C. *A Handbook of Fourier Theorems*; Cambridge University Press: Cambridge, UK, 1987.
30. Chiani, M.; Dardari, D. Improved exponential bounds and approximation for the Q-function with application to average error probability computation. In Proceedings of the Global Telecommunications Conference, GLOBECOM'02, IEEE, Taipei, Taiwan, 17–21 November 2002; Volume 2, pp. 1399–1402. [CrossRef]
31. Alzer, H. Error function inequalities. *Adv. Comput. Math.* **2010**, *33*, 349–379. [CrossRef]
32. Chu, J.T. On Bounds for the Normal Integral. *Biometrika* **1955**, *42*, 263. [CrossRef]
33. Neuman, E. Inequalities and Bounds for the Incomplete Gamma Function. *Results Math.* **2013**, *63*, 1209–1214. [CrossRef]
34. Yang, Z.-H.; Chu, Y.-M. On approximating the error function. *J. Inequalities Appl.* **2016**, *2016*, 311. [CrossRef]
35. Yang, Z.-H.; Qian, W.-M.; Chu, Y.; Zhang, E. On approximating the error function. *Math. Inequalities Appl.* **2018**, *21*, 469–479. [CrossRef]
36. Abuelma'atti, M.T. A note on the harmonic distortion in AC-bias recording. *IEEE Trans. Magn.* **1988**, *24*, 3259–3260. [CrossRef]
37. Salzer, H.E. Formulas for calculating the Error function of a complex variable. *Math. Tables Other Aids Comput.* **1951**, *5*, 67–70. [CrossRef]
38. Ng, E.W.; Geller, M. A table of integrals of the error functions. *J. Res. Nat. Bur. Stand. B Math. Sci.* **1969**, *73B*, 1–20. [CrossRef]

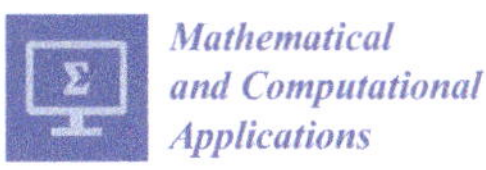

Article

Stochastic Neural Networks for Automatic Cell Tracking in Microscopy Image Sequences of Bacterial Colonies

Sorena Sarmadi [1], James J. Winkle [1], Razan N. Alnahhas [2], Matthew R. Bennett [3], Krešimir Josić [1,4], Andreas Mang [1] and Robert Azencott [1,*]

[1] Department of Mathematics, University of Houston, Houston, TX 77204, USA; sorena.sarmadi@gmail.com (S.S.); wink@utexas.edu (J.J.W.); kresimir.josic@gmail.com (K.J.); andreas@math.uh.edu (A.M.)
[2] Department of Biomedical Engineering, Boston University, Boston, MA 02215, USA; rnalnahhas@gmail.com
[3] Departments of Biosciences and Bioengineering, Rice University, Houston, TX 77005, USA; matthew.bennett@rice.edu
[4] Department of Biology and Biochemistry, University of Houston, Houston, TX 77204, USA
* Correspondence: razencot@math.uh.edu; Tel.: +1-713-743-3489

Abstract: Our work targets automated analysis to quantify the growth dynamics of a population of bacilliform bacteria. We propose an innovative approach to frame-sequence tracking of deformable-cell motion by the automated minimization of a new, specific cost functional. This minimization is implemented by dedicated Boltzmann machines (stochastic recurrent neural networks). Automated detection of cell divisions is handled similarly by successive minimizations of two cost functions, alternating the identification of children pairs and parent identification. We validate the proposed automatic cell tracking algorithm using (i) recordings of simulated cell colonies that closely mimic the growth dynamics of *E. coli* in microfluidic traps and (ii) real data. On a batch of 1100 simulated image frames, cell registration accuracies per frame ranged from 94.5% to 100%, with a high average. Our initial tests using experimental image sequences (i.e., real data) of *E. coli* colonies also yield convincing results, with a registration accuracy ranging from 90% to 100%.

Keywords: stochastic neural networks; cell tracking; microscopy image analysis; detection-and-association methods

MSC: 62H35; 62M45

Citation: Sarmadi, S.; Winkle, J.J.; Alnahhas, R.N.; Bennett, M.R.; Josić, K.; Mang, A.; Azencott, R. Stochastic Neural Networks for Automatic Cell Tracking in Microscopy Image Sequences of Bacterial Colonies. *Math. Comput. Appl.* **2022**, *27*, 22. https://doi.org/10.3390/mca27020022

Academic Editor: Leonardo Trujillo

Received: 16 December 2021
Accepted: 23 February 2022
Published: 2 March 2022

1. Introduction

Technology advances have led to increasing magnitudes of data generation with increasing levels of precision [1,2]. However, data generation presently far outpaces data analysis and drives the requirement for analyzing such large-scale data sets with automated tools [3–5]. The main goal of the present work is to develop computational methods for an automated analysis of microscopy image sequences of colonies of *E. coli* growing in a single layer. Such recordings can be obtained from colonies growing in microfluidic devices, and they provide a detailed view of individual cell-growth dynamics as well as population-level, inter-cellular mechanical and chemical interactions [6–8].

However, to understand both variability and lineage-based correlations in cellular response to environmental factors and signals from other cells requires the tracking of large numbers of individual cells across many generations. This can be challenging, as large cell numbers tightly packed in microfluidic devices can compromise spatial resolution, and toxicity effects can place limits on the temporal resolution of the recordings [9,10]. One approach to better understand and control the behavior of these bacterial colonies is to develop computational methods that capture the dynamics of gene networks within single cells [6,11,12]. For these methods to have a practical impact, one ultimately has to fit the models to the data, which allows us to infer hidden parameters (i.e., characteristics of the behavior of cells that cannot be measured directly). Image analysis and pattern recognition techniques for biological imaging data [13–15], like the methods developed in the present work, can be used to track lineages and thus automatically infer how gene expression varies over time. These methods can serve as an indispensable tool to extract information to fit and validate both coarse and detailed models of bacterial population, thus allowing us to infer model parameters from recordings.

Here, we describe an algorithm that provides *quantitative* information about the population dynamics, including the life cycle and lineage of cells within a population from recordings of cells growing in a mono-layer. A typical sequence of frames of cells growing in a microfluidic trap is shown in Figure 1. We describe the design and validation of algorithms for tracking individual cells in sequences of such images [11,16,17]. After segmentation of individual image frames to identify each cell, tracking individual cells from frame to frame is a combinatorial problem. To solve this problem we take into account the unknown cell growth, cell motion, and cell divisions that occur between frames. Segmentation and tracking are complicated by imaging noise and artifacts, overlap of bacteria, similarity of important cell characteristics across the population (shape; length; and diameter), tight packing of bacteria, and large interframe durations resulting in significant cell motion, and up to a 30% increase in individual cell volume.

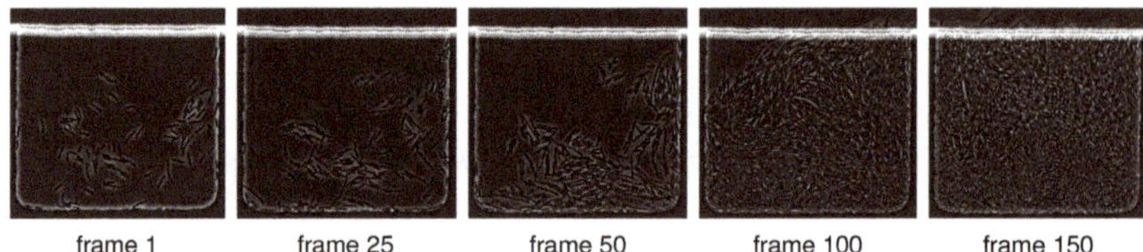

Figure 1. Typical microscopy image sequence. We show five frames out of a total of 150 frames of an image sequence showing the dynamics of *E. coli* in a microfluidic device [18] (real laboratory image data). These cells are are about 1 µm in diameter and on average 3 µm in length, and they divide about every 30 min. The original images exported from the microscope are 0.11 µm/pixel. We report results for these real datasets in Section 4.

1.1. Related Work

The present work focuses on tracking *E. coli* in a time series of images. A comparison of different cell-tracking algorithms can be found in [19,20]. Multi-object tracking in video sequences and object recognition in time series of images is a challenging task that arises in numerous applications in computer vision [21,22]. In (biomedical) image processing, motion tracking is often referred to as *"image registration"* [19,23–26] or *"optical flow"* [27–30]. Typical solutions used in the defense industry, for instance, track *small numbers* of fast moving targets by image sequence analysis at pixel levels and use sophisticated reconstruction of the optical flow, combined with real time segmentation, and quick combinatoric exploration at each image frame. Initially, we did implement several well-known algorithms for reconstruction of the optical flow, but the results we obtained were not satisfactory due to long interframe times and high noise levels. Moreover, we are not interested in tracking individual pixels but rather cells (i.e., rod-shaped, deformable shapes), while recognizing events of cell division and recording cell lineage. Consequently, we decided to first segment each image frame to isolate each cell, and then to match cells between successive frames.

As for the problem at hand, one approach proposed in prior work to simplify the tracking task is to make the experimental setup more rigid by confining individual cell lineages to small tubes; the associated microfluidic device is called a *"mother machine"* [31–36]. The microfluidic devices we consider here yield more complicated data as cells are allowed to move and multiply freely in two dimensions (constrained to a mono-layer). We refer to Figure 1 for a typical sequence of experimental images considered in the present work.

Turning to methods that work on more complex biological cell imaging data, we can distinguish different classes of tracking methods. *"Model-based evolution methods"* operate on the image intensities directly. They rely on particle filters [37–39] or active contour models [40–44]. These methods work well if the cells are not tightly packed. However, they may lead to erroneous results if the cells are close together, the inter-cellular boundaries are blurry, or the cells move significantly. Our work belongs to another class—the so called *"detection-and-association methods"* [45–55], which first detect cells in each frame and then solve the tracking problem/association task across successive frames. (We note that the segmentation and tracking of cells does not necessarily need to be implemented in two distinct steps. In many image sequence analyses, implementing these two steps jointly can be beneficial [37,49,54,56–58]. However, for the clarity of exposition and easier implementation of our new tracking technique, we present these steps separately.) Doing so necessitates the segmentation of cells within individual frames. We refer to [59] for an overview of cell segmentation approaches. Deep learning strategies have been widely used for this task [5,50,54,58,60–65]. We consider a framework based on convolutional neural networks (CNNs). Others have also used CNNs for cell segmentation [62,64,66–68]. We omit a detailed discussion of our segmentation approach within the main body of this paper, as we do not view it as our main contribution (see Section 1.2). However, the interested reader is referred to Appendix D for some insights. To solve the tracking problem after the cell detection, many of the methods cited above use hand-crafted association scores based on the proximity of the cells and shape similarity measures [46,48,51,54]. We follow this approach here. We note that we not only consider local association scores between cells but also include measures for the integrity of a cell's neighborhood (i.e., *"context information"*).

Our method is tailored for tracking cells in tightly packed colonies of rod-shaped *E. coli* bacteria. This problem has been considered previously [5,45,49,52]. However, we are not aware of any large-scale datasets that provide ground truth tracking data for these types of recordings, but note that there are community efforts for providing a framework for testing cell tracking algorithms [20,69] (see, e.g., [37,70]). (Cell tracking challenge: http://celltrackingchallenge.net (accessed on 15 December 2021).) Works that consider these data are for example [37,54,57,58,62,67]. The cells in this dataset have significantly different characteristics compared to those considered in the present work. As we describe below, our approach is based on distinct characteristics of the bacteria cells and, consequently, does not directly apply to these data. Therefore, we have developed our own validation and calibration framework (see Section 2.1).

Standard graph matching algorithms (see, e.g., [71]) do not directly apply to our problem. Indeed, a fundamental complication is that cells can divide between successive images. Hence, each assignment from one frame to its successor is not a one-to-one mapping but a one-to-many correspondence. More advanced graph matching strategies are described in [72,73]. Graph-based matching strategies for cell-tracking that are somewhat related to our approach are described in [70,74–77]. Like the methods mentioned above, they consider various association scores for tracking. Individual cells are represented as nodes, and neighbors are connected through edges. Our approach also introduces cost terms for structural matching of local neighborhoods by specific scoring for single nodes, pairs of nodes, and triplets of nodes, after a (modified) Delaunay triangulation. By using a graph-like structure, cell divisions can be identified by detecting changes in the topology of the graph [75,76]. We tested a similar strategy, but came to the conclusion that we cannot reliably construct neighborhood networks between frames for which topology changes only occur due to cell division; the main issue we observed is that the significant motion of cells between frames can introduce additional topology changes in our neighborhood structure. Consequently, we decided to relax these assumptions.

Refs. [71,78–80] implement multi-target tracking in videos by stochastic models based on random finite set densities and variants thereof. The fit to the data are based on Gibbs sampling to maximize the posterior likelihood. A key challenge of these approaches is the estimation of an adequate finite number of Gibbs sampling iterations when one computes posterior distributions. Most Gibbs samplers are ergodic Markov chains on a finite but huge state spaces, so that their natural exponential rate of convergence is not a practically reassuring feature.

As mentioned above, some recent works jointly solve the tracking and segmentation problem [37,49,54,56–58]. Contrary to observations we have made in our data, these approaches rely (with the exception of [49]) on the fact that the tracking problem is inherent to the segmentation problem (*"tracking-by-detection methods"* [54]; see also [5]). That is, the key assumption made by many of these algorithms is that cells belonging to the same lineage overlap across frames (see also [47]). In this case, cell-overlap can serve as a good proxy for cell-tracking [54]. We note that in our data we cannot guarantee that the frame rate is sufficiently high for this assumption to hold. Refs. [56,57,67] exploited machine learning techniques for segmentation *and* motion tracking. One key challenge here is to provide adequate training data for these methods to be successful. Here, we describe simulation-based techniques that can be extended to produce training data, which we use for parameter tuning [12,81].

The works that are most similar to ours are [45,49,52]. They perform a local search to identify the best cell-tracking candidates across frames. One key difference across these works are the matching criteria. Moreover, Refs. [45,49] employ a local greedy-search, whereas we consider stochastic neural network dynamics for optimization. Ref. [52] constructs score matrices within a score based neighborhood tracking method; an integer programming method is used to generate frame-to-frame correspondences between cells and the lineage map. Other approaches that consider linear programming to maximize an association score function for cell tracking can be found in [47,54,82].

As we have mentioned in the abstract, we obtain a tracking accuracy that ranges from 90% to 100%, respectively. Overall, our method is competitive with existing approaches: For example, Ref. [45] reports a tracking accuracy of up to 97% for data that are similar to ours, while Ref. [74] reports a tracking accuracy (spatial, temporal, and cell division detection) at the order of 95% (between about 93% and 98%, respectively). The second group also reports results for their prior approach [75], with an accuracy at the order of 90% (ranging from about 87% to 92%, respectively). Accuracies reported in [77] range from about 92% to 97%, respectively. This work also includes a comparison to one of their earlier approaches [76] with an accuracy of up to 85% and 89% if the datasets are pre-aligned. We note that the data considered in [74–77] are quite different from ours. Refs. [37,54,57,58,70] consider the data from the cell tracking challenge [20,69] to evaluate the performance of their methods. As in the previously mentioned work, these data are again quite different from ours. To evaluate the performance of the methodology, the so-called *acyclic oriented graph matching measure* [83] is considered. We refer to the webpage of the cell tracking challenge for details on the evaluation metrics (see http://celltrackingchallenge.net/evaluation-methodology, accessed on 15 December 2021). Based on these, the reported tracking scores are between 0.873 and 0.902 [37], 0.901 and 1.00 [70], 0.950 and 0.987 [54], 0.788 and 0.982 [58] and 0.765 and 0.915 [57] depending on the considered data set, respectively.

1.2. Contributions

For image segmentation, we first apply two well-known, powerful variational segmentation algorithms to generate a large training set of correctly delineated single cells. We can then train a CNN dedicated to segmenting out each single cell. Using a CNN significantly reduces the runtime of our computational framework for cell identification. The frame-to-frame tracking of individual cells in tightly packed colonies is a significantly more challenging task, and is hence the main topic discussed in the present work. We develop a set of innovative automatic cell tracking algorithms based on the successive minimization of three dedicated cost functionals. For each pair of successive image frames, minimizing these cost functionals over all potential cell registration mappings poses significant computational and mathematical challenges. Standard gradient descent algorithms are inefficient for these discrete and highly combinatorial minimization problems. Instead, we implement the stochastic neural network dynamics of BMs, with architectures and energy functions tailored to effectively solve our combinatorial tracking problem. Our major contributions are: (i) The design of a multi-stage cell tracking algorithm that starts with a parent–children pairing step, followed by removal of identified parent–children triplets, and concludes with a cell-to-cell registration step. (ii) The design of dedicated BM architectures, with several energy functions, respectively, minimized by true parent–children pairing and by true cell-to-cell registration. Energy minimizations are then implemented by simulation of BM stochastic dynamics. (iii) The development of automatic algorithms for the estimation of unknown weight parameters of our BM energy functions, using convex-concave programming tools [84–86]. (iv) The evaluation of our methodology on synthetic and real image sequences of cell colonies. The massive effort involved in human expert annotation of cell colony recordings limits the availability of "*ground truth tracking*" data for dense bacterial colonies. Therefore, we first validated the accuracy of our cell tracking algorithms on recordings of simulated cell colonies, generated by the dedicated cell colony simulation software [12,81]. This provided us with ground truth frame-by-frame registration for cell lineages, enabling us to validate our methodology.

1.3. Outline

In Section 2, we describe the synthetic image sequence (see Section 2.1) and experimental data (see Section 2.2) of cell colonies considered as benchmarks for our cell tracking algorithms. In Section 2.3, we describe key cell characteristics considered in our tracking methodology to define metrics that enter our cost functionals. Our tracking approach is developed in greater detail in Section 3. We define valid cell registration mappings between successive image frames in Section 3.1. We outline how to automatically calibrate the weights of our various penalty terms in Section 3.2. Our algorithms for pairing parent cells with their children and for cell-to-cell registration are developed in Sections 3.3–3.9. We present our main validation results on long image sequences (time series of images) in Section 4 and conclude with Section 5.

2. Datasets

Below, we introduce the datasets used to evaluate the performance of the proposed methodology. The synthetic data are described in Section 2.1. The experimental data (real imaging data) are described in Section 2.2.

2.1. Synthetic Videos of Simulated Cell Colonies

To validate our cell tracking algorithms, we consider simulated image sequences of dense cell populations. We refer to [12,81] for a detailed description of this mathematical model and its implementation. (The code for generating the synthetic data has been released at https://github.com/jwinkle/eQ, accessed on 15 December 2021) The simulated cell colony dynamics are driven by an agent based model [12,81], which emulates live colonies of growing, moving, and dividing rod-like *E. coli* cells in a 2D microfluidic trap environment. Between two successive frames J, J_+, cells are allowed to move until they nearly bump into each other, and to grow at multiplicative rate denoted $g.rate$ with an average value of 1.05 per minute.

The cells are modeled as 2D spherocylinders of constant 1 µm width. Each cell grew exponentially in length with a doubling time of 20 min. To prevent division synchronization across the population when a mother cell of length L_{div} divides, the two daughter cells are assigned random birth lengths $L_0(b_1) = L_1 = \delta L_{\text{div}}$ and $L_0(b_2) =: L_2 = (1-\delta)L_{\text{div}}$, where $\delta > 0$ is a random number sampled independently at each division from a uniform distribution on $[0.45, 0.55]$. Consequently, a bacterial cell b of length L_{div} divides into two cells b_1 and b_2, their lengths L_1, L_2 satisfy $L_1 + L_2 = L_{\text{div}}$ and L_i/L_{div}, $i = 1, 2$, is a random number. The cells have a length of approximately 2 µm after division and 4 µm right before division. We refer to [81] for additional details. The simulation keeps track of cell lineage, cell size, and cell location (among other parameters). The main output of each such simulation considered here is a binary image sequence of the cell colony with a fixed interframe duration. Each such synthetic image sequence is used as the sole input to our cell tracking algorithm. The remaining meta-data generated by the simulations are only used as ground truth to evaluate the performance of our tracking algorithms.

We consider several benchmark datasets of *synthetic* image sequences of simulated cell colonies of different complexity. We refer to these benchmarks as BENCH1 (500 frames), BENCH2 (300 frames), BENCH3 (300 frames), and BENCH6 (100 frames), with an interframe duration of 1, 2, 3 and 6 min, respectively. Notice that there is no explicit noise on the growth rate. However, due to the crowding of cells, the growth rate will vary from cell-to-cell. The generated binary images are of size 600×600 pixels. We summarize these benchmarks in Table 1. The associated image sequences involve between 100 up to 500 frames, respectively. In Figure 2, we display an example of two simulated consecutive frames separated by 1 min. To simplify our presentation and validation tests, we control our simulations to make sure that cells will not exit the region of interest from one frame to the next, and we exclude cells that are only partially visible in the current frames.

Table 1. Benchmark datasets. To test the tracking software, we consider simulated data. We have generated data of varying complexity with different interframe durations. We note that we also consider these data to train our algorithms for tracking cells. We report the label for each dataset, the interframe duration, as well as the number of frames generated. We set the *cell growth factor* to *g.rate* = 1.05 per min. We refer to the text for details about how these data have been generated.

Label	Interframe Duration	Number of Frames
BENCH1	1 min	500
BENCH2	2 min	300
BENCH3	3 min	300
BENCH6	6 min	100

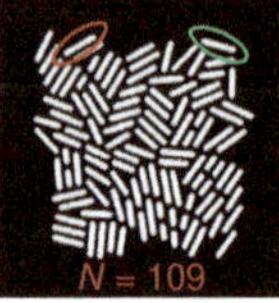

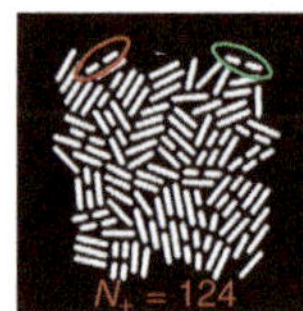

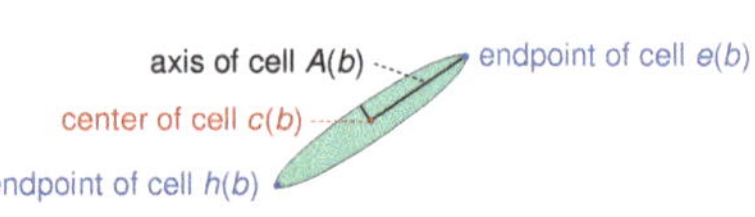

image J at time t image J_+ at time $t + \Delta t$

Figure 2. Simulated data and cell characteristics considered in the proposed algorithm. (**Left**): Two successive images generated by dynamic simulation for a colony of rod-shaped bacteria. The left image J displays $N = 109$ cells at time t. At time $t + \Delta t$ with $\Delta t = 1$ min, cells have moved and grown, and some have divided. These cells are displayed in image J_+, which contains $N_+ = 124$ cells. We highlight two cells that have undergone a division between the frames (red and green ellipses). (**Right**): Geometry of a rod shaped bacterium. We consider different quantities of interest in the proposed algorithm. These include the center $c(b)$ of a cell, the two end points $e(b)$ and $h(b)$, and the long axis $A(b)$, respectively.

2.2. Laboratory Image Sequences (Real Biological Data)

We also verify the performance of our approach on real datasets of *E. coli* bacteria. These bacteria are about 1 µm in diameter and on average 3 µm in length, and they divide about every 30 min. The original images exported from the microscope are 0.11 µm/pixel. The microscopy experimental data were obtained using JS006 [87] (BW25113 Δ*araC* Δ*lacI*) *E. coli* strains containing a plasmid constitutively expressing yellow or cyan fluorescent protein (*sfyfp* or *sfcfp*) for identification. The plasmid also contains an ampicillin resistance gene and p15A origin. These cells were grown overnight in LB medium with 100 µg/mL ampicillin for 18 h. These cultures were diluted in the morning into 1/1000 into 50 mL fresh LB with 100 µg/mL ampicillin and grown for 3 h until they reached an OD600 of about

0.3. The cells were then concentrated by centrifuging 30 mL of culture at 2000× g for 5 min and then resuspending in 10 mL of fresh LB. The concentrated culture was loaded into a hallway microfluidic device prewarmed and flushed with 0.1% (v/v) Tween-20 [88]. In the microfluidic device, the cells were provided with continuous fresh LB with 100 µg/mL ampicillin and 0.075% (v/v) Tween-20. The microfluidic device was placed onto an 60× oil objective and imaged every 6 min at phase contrast, YFP, and CFP filter settings using an inverted fluorescence microscope. We show a representative dataset in Figure 1.

2.3. Cell Characteristics

Next, we discuss characteristics of the *E. coli* bacteria important for our tracking algorithm.

Cell Geometry. In accordance with the dynamics of bacterial colonies in microfluidic traps, the dynamic simulation software generates colonies of rod-shaped bacteria. Cell shapes can be approximated by long and thin ellipsoids, which are geometrically well identified by their center, their long axis, and the two endpoints of this long axis. The center $c(b)$ is the centroid of all pixels belonging to cell b. The long axis $A(b)$ of cell b is computed by principal component analysis (PCA). The endpoints $e(b)$ and $h(b)$ of cell b are the first and last cell pixels closest to $A(b)$; see Figure 2 (right) for a schematic illustration.

Cell Neighbors. For each image frame J, denote $B = B(J)$ as the set of fully visible cells in J, and by $N = N(J) = \text{card}(B)$ the number of these cells. Let V be the set of all cell centers $c(b)$ with $b \in B$. Denote $delV$ the Delaunay triangulation [89] of the finite planar set V with N vertices. We say that two cells b_1, b_2 in B are *neighbors* if they verify the following three conditions: (i) (b_1, b_2) are connected by the edge *edg* of one triangle in $delV$. (ii) The edge *edg* does not intersect any other cell in B. (iii) Their centers verify $\|c(b_1) - c(b_2)\| \leq \rho$, where $\rho > 0$ is a user defined parameter.

For the synthetic images of size 600 × 600 that we considered (see Section 2.1), we take $\rho = 80$ pixels. We write $b_1 {\sim} b_2$ for short, whenever b_1, b_2 are neighbors (i.e, satisfy the three conditions identified above).

Cell Motion. Let J, J_+ denote two successive images (i.e., frames). Denote $B = B(J)$, $B_+ = B(J_+)$ as the associated sets of cells. Superpose temporarily the images J and J_+ so that they then have the same center pixel. Any cell $b \in B$, which does not divide in the interframe $J \to J_+$, becomes a cell b_+ in image J_+. The "*motion vector*" of cell b from frame J to J_+ is then defined by $v(b) = c(b_+) - c(b)$. If the cell b does divide between J and J_+, denote b_{div} as the last position reached by cell b at the time of cell division, and define similarly the motion $v(b) = c(b_{\text{div}}) - c(b)$. In our experimental recordings of real bacterial colonies with interframe duration 6 min, there is a *fixed number* $w > 0$ such that $\|v(b)\| \leq w/2$ for all cells $b \in B(J)$ for all pairs J, J_+. In particular, we observed that, for real image sequences, $w = 100$ pixels is an adequate choice. Consequently, we select $w = 100$ pixels for all simulated image sequences of BENCH6. For BENCH1, we select $w = 45$ pixels, again based on a comparison with real experimental recordings. Overall, the meta-parameter w is assumed to be a fixed number and to be known, since $w/2$ is an observable upper bound for the cell motion norm for a particular image sequence of a lab experiment.

Target Window. Recall that J, $J+$ are temporarily superposed. Let $U(b) \subset J_+$ be a square window of width w, with the same center as cell b. The *target window* $W(b)$ is the set of all cells in B_+ having their centers in $U(b)$. Since $\|v(b)\| \leq w/2$, the cell b_+ must belong to the target window $W(b) \subset B_+$.

3. Methodology

3.1. Registration Mappings

Next, we discuss our assumptions on a valid registration mapping that establishes cell-to-cell correspondences between two frames. Let J, J_+ denote two successive images, with cell sets B and B_+, respectively. As above, we let $N = \text{card}(B)$, and $N_+ = \text{card}(B_+)$. Our goal is to track each cell from J to J_+. For each cell $b \in B$, there exist three possible evolutions between J and J_+:

Case 1: Cell $b \in B$ did not divide in the interframe $J \to J_+$, and has become a cell $f(b) \in B_+$; that is, $f(b)$ has grown and moved during the interframe time interval.

Case 2: Cell $b \in B$ divided between J and J_+, and generated two children cells $b_1, b_2 \in B_+$; we then denote $f(b) = (b_1, b_2) \in B_+ \times B_+$.

Case 3: Cell $b \in B$ disappeared in the interframe $J \to J_+$, so that $f(b)$ is not defined.

To simplify our exposition, we *ignore Case 3*. We discuss Case 3 in greater detail in the conclusions in Section 5. Consequently, a valid (true) registration mapping f will take values in the set $\{B_+\} \cup \{B_+ \times B_+\}$.

3.2. Calibration of Cost Function Weights

With the notation we introduced, fix any two finite sets A, A_+. Let $G := \{g : A \to A_+\}$ be the set of all mappings g: $A \to A_+$. Fix m penalty functions $\text{pen}_k(g) \geq 0$, $k = 1, \ldots, m$. Let $g^* \in G$ be the ground truth mapping we want to discover through minimization in g of some given cost function $\text{COST}(g)$ defined by the linear combination of the penalty functions $\text{pen}_k(g)$, the contributions of which are controlled by the cost function weights $\lambda_k > 0$. In this section, we present a generic *weight calibration algorithm*, extending a technique introduced and applied in [90,91] for Markov random fields based image analysis.

The cost function must perform well (with the same weights) for hundreds of pairs of (synthetic) images J, J^+. We consider one such synthetic pair for which the ground truth registration mapping $f \in G$ is known, and use it to compute an adequate set of weights, which will then be used on all other synthetic pairs J, J^+. Notice that, for experimental recordings of real cell colonies, no ground truth registration mappings f are available. In this case, f should be replaced by a set of user constructed, correct *partial* mappings defined on small subsets of A. The proposed weight calibration algorithm will also work in those situations.

We now show how knowing one ground truth mapping f can be used to derive the best feasible weights ensuring that f should be a plausible minimizer of the cost functional $\text{COST}(g)$ over $g \in G$. Let $\text{PEN}(g) = [\text{pen}_1(g), \ldots, \text{pen}_m(g)]$ be the vector of m penalties for any mapping $g \in G$. Let $\Lambda = [\lambda_1, \ldots, \lambda_m]$ be the weight vectors. Then, $\text{COST}(g) = \langle \Lambda, \text{PEN}(g) \rangle$. Replacing g by another mapping $h \neq g$ induces the penalty changes $\Delta\,\text{PEN}_{g,h} = \text{PEN}(h) - \text{PEN}(g)$ and the cost change $\Delta\,\text{COST}(g,h) = \langle \Lambda, \Delta\,\text{PEN}_{g,h} \rangle$. Now, fix any known ground truth mapping $f \in G$. We want f to be a minimizer of COST, so we should have $\Delta\,\text{COST}(f, f') \geq 0$ for all modifications $f \to f' \in G$.

For each $a \in A$, select an arbitrary $s(a) \in W(a)$ (where $W(a)$ is the target window for cell a; see Section 2.3), to define a new mapping $f' = f'_a$ from A to A_+ by $f'_a(a) = s(a)$, and $f'_a(x) \equiv f(x)$ for all $x \neq a$. Since f is a minimizer of COST, this single point modification $f \to f'_a$ must generate the following cost increase

$$\langle \Lambda, \Delta\,\text{PEN}(f, f'_a) \rangle = \Delta\,\text{COST}(f, f'_a) \geq 0.$$

Denote $V_a \in \mathbb{R}^m$ the vector $V_a = \Delta\,\text{PEN}(f, f'_a)$. Then, the positive vector $\Lambda \in \mathbb{R}^m$, $\Lambda \succeq 0$, should verify the set of linear constraints $\langle \Lambda, V_a \rangle \geq 0$ for all $a \in A$. There may be too many such linear constraints. Consequently, we *relax* these constraints by introducing a vector $y = [y(a)] \in \mathbb{R}^{\text{card}(A)}$, $y \succeq 0$, of slack variables $y(a) \geq 0$ indexed by all the $a \in A$. (In optimization, slack variables are introduced as additional unknowns to transform inequality constraints to an equality constraint and a non-negativity constraint on the slack variables.) We require the unknown positive vector Λ and the slack variables vector y to verify the system of linear constraints:

$$\begin{aligned} \langle \Lambda, V_a \rangle + y(a) &= 0 \qquad \text{for all } a \in A \\ \Lambda \succeq 0, \;\; y &\succeq 0 \\ \langle \Lambda, Z \rangle &\leq 1000 \end{aligned} \tag{1}$$

where $Z = [1, \ldots, 1] \in \mathbb{R}^m$. The normalizing constant 1000 can be arbitrarily changed by rescaling. We seek high positive values for $\Delta\,\text{COST}(f, f'_a)$ and small L_1-norm for the slack variable vector y. Thus, we will seek two vectors $\Lambda \in \mathbb{R}^m$ and $y \in \mathbb{R}^{\text{card}(A)}$ solving the following *convex-concave* minimization problem, where $\gamma > 0$ is a user selected (large) meta parameter:

$$\underset{\Lambda, y}{\text{minimize}} \;\; \gamma \|y\|_{L1} - \sum_{a \in A} [\langle \Lambda, V_a \rangle]^+ \tag{2}$$

subject to (1), where we denote $[x]^+ := \max(x, 0)$ for arbitrary x. To numerically solve the constrained minimization problem (2), we use the libraries CVXPY and DCCP (disciplined convex-concave programming) [84–86]. DCCP is a package for convex-concave programming designed to solve non-convex problems. (DCCP can be downloaded at https://github.com/cvxgrp/dccp (last accessed on 20 January 2022)) It can handle objective functions and constraints with any known curvature as defined by the rules of disciplined convex programming [92]. We give examples of numerically computed weight vectors Λ below. The computing time was less than 30 s for the data that we have prepared. For simplicity, we just considered one step changes in our computations, which make the overlap penalty weak. To increase the accuracy of the model, it is possible to consider a larger number of samples (i.e., multi-step changes). Note that the solutions Λ of (2) are of course not unique, even after normalization by rescaling.

3.3. *Cell Divisions and Parent–Children Short Lineages*

Next, we discuss how we tackle the assignment problem when cells divide.

3.3.1. Cell Divisions

We now outline a cost function based methodology to detect cell divisions. The first step will be to seek the most likely parent for each potential pair of children cells. Fix two successive synthetic image frames J, J_+ with short interframe time equal to 1 minute. Their cell sets B, B_+ have cardinality N and N_+, respectively. For our synthetic image sequences, all cells $b \in B$ still exist in B_+—either as whole cells or after dividing into two children cells, and no new cell enters the field of view during the interframe $J \to J_+$. This forces $N_+ \geq N$, and implies that the number DIV of cell divisions occurring in this interframe verifies $DIV = N_+ - N$. Each children pair $(b_1, b_2) \in B_+ \times B_+$ is born from a single parent $b \in B$. Thus, the set *trueCH* of all such *true children pairs* must then verify

$$\text{card}(trueCH) = DIV = N_+ - N. \tag{3}$$

For our video recordings of actual cell populations, during any interframe, we may have n_{out} cells exiting the field of view and n_{in} cells entering it, so that $|\,\text{card}(trueCH) - DIV|$ may be of the order of $n_{in} + n_{out}$. To take this into account, we *relax* the constraint in (3) as follows:

$$|\,\text{card}(trueCH) - DIV| \leq REL, \tag{4}$$

where REL is a fixed bound estimated from our experiments. For simplicity, we have restricted our methodology to the situation where n_{in} and n_{out} are always 0. However, even in that case, there was a computational advantage to using the slightly relaxed constraint (4) with $REL = 1$.

3.3.2. Most Likely Parent Cell for a Given Children Pair

For successive images J, J_+ with 1 min interframe, define the set PCH of *plausible children pairs* by

$$PCH = \{(b_1, b_2) \in B_+ \times B_+ \text{ with centers } c_1, c_2 \text{ verifying } \|c_1 - c_2\| < \tau\}, \tag{5}$$

where the threshold $\tau > 0$ is user selected and fixed for the whole benchmark set BENCH1 of synthetic image sequences.

To evaluate if a pair of cells $(b_1, b_2) \in PCH$ can qualify as a pair of children generated by division of a parent cell $b \in B$, we now quantify the *"geometric distortion"* between b and (b_1, b_2). Cell division of b into $b_1, b_2 \in B_+$ occurs with small motions of b_1, b_2. During the short interframe duration, the initial centers c_1, c_2 of b_1, b_2 in image J move by at most $w/2$ pixels each (see Section 2.3), and their initial distance to the center c of b is roughly at most $\|A(b)\|/4$, where $A(b)$ is the long axis of cell b. Hence, the centers c, c_1, c_2 of b, b_1, b_2 should verify the constraint

$$\max\{\|c_1 - c\|, \|c_2 - c\|\} \leq w + \|A\|/4. \tag{6}$$

Define the set *SHLIN* of potential *short lineages* as the set all triplets (b, b_1, b_2) with $b \in B$, $(b_1, b_2) \in PCH$, verifying the preceding constraint (6). For each potential lineage $(b, b_1, b_2) \in SHLIN$, define three terms penalizing the geometric distortions between a parent $b \in B$ and a pair of children $(b_1, b_2) \in PCH$ by the following formulas, where we denote c, c_1, c_2, the centers of cells b, b_1, b_2 and A, A_1, A_2 their long axes, respectively: (*i*) center distortion $\mathrm{cen}(b, b_1, b_2) = \|c - (c_1 + c_2)/2\|$, (*ii*) size distortion $\mathrm{siz}(b, b_1, b_2) = |\|A\| - (\|A_1\| + \|A_2\|)|$, and (*iii*) angle distortion

$$\mathrm{ang}(b, b_1, b_2) = \mathrm{angle}(A, A_1) + \mathrm{angle}(A, A_2) + \mathrm{angle}(A, c_2 - c_1).$$

Here, angle denotes *"angles between non-oriented straight lines,"* with a range from 0 to $\pi/2$. Introduce three positive weights $\lambda_{\mathrm{cen}}, \lambda_{\mathrm{siz}}, \lambda_{\mathrm{ang}}$ (to be estimated), and for every short lineage $(b, b_1, b_2) \in SHLIN$ define its *distortion cost* by

$$\mathrm{dist}(b, b_1, b_2) = \lambda_{\mathrm{cen}}\, \mathrm{cen}(b, b_1, b_2) + \lambda_{\mathrm{siz}}\, \mathrm{siz}(b, b_1, b_2) + \lambda_{\mathrm{ang}}\, \mathrm{ang}(b, b_1, b_2). \tag{7}$$

For each plausible pair of children $(b_1, b_2) \in PCH$, we will compute the most likely *parent cell* $b^* = \mathrm{parent}(b_1, b_2)$ as the cell $b^* \in B$ minimizing $\mathrm{dist}(b, b_1, b_2)$ in (7) over all $b \in B$, as summarized by the formula

$$b^* = \mathrm{parent}(b_1, b_2) = \underset{\{b \in B | (b, b_1, b_2) \in SHLIN\}}{\mathrm{argmin}} \mathrm{dist}(b, b_1, b_2). \tag{8}$$

To force this minimization to yield a reliable estimate of $b^* = \mathrm{parent}(b_1, b_2)$ for most true pairs of children (b_1, b_2), we calibrate the weights λ_j, $j \in \{\mathrm{cen}, \mathrm{siz}, \mathrm{ang}\}$ by the algorithm outlined in Section 3.2, using as *"ground truth set"* a fairly small set of visually identified true short lineages (parent, children). For fixed (b_1, b_2), the set of potential parent cells $b \in B$ has very *small size* due to the constraint (6). Hence, brute force minimization of the functional $\mathrm{dist}(b, b_1, b_2)$ in (7) over all $b \in B$ such that $(b, b_1, b_2) \in SHLIN$, is a *fast computation* for each (b_1, b_2) in *PCH*. The distortion minimizing $b = b^*$ yields the most likely parent cell $\mathrm{parent}(b_1, b_2) = b^*$. The brute force minimization in b of $\mathrm{dist}(b, b_1, b_2)$ is still a greedy minimization in the sense that other soft constraints introduced further on are not taken into consideration during this preliminary fast computation of b^*.

3.3.3. Penalties to Enforce Adequate Parent–Children Links

Any true pair of children cells $pch = (b_1, b_2)$ should belong to *PCH*, but must also verify lineage and geometric constraints which we now enforce via several penalties. Note that the new penalties introduced here are fully distinct from the three penalties specified above to define $\mathrm{dist}(b, b_1, b_2)$.

***"Lineage"* Penalty.** Valid children pairs $(b_1, b_2) \in PCH$ should be correctly matchable with their most likely parent cell $b^* = \mathrm{parent}(b_1, b_2)$ (see (8)). Thus, we define the *"lineage"* penalty $\mathrm{lin}(b_1, b_2) = \mathrm{dist}(b^*, b_1, b_2)$ by

$$\mathrm{lin}(b_1, b_2) = \underset{\{b \in b | (b, b_1, b_2) \in shlin\}}{\mathrm{argmin}} \mathrm{dist}(b, b_1, b_2) = \mathrm{dist}(\mathrm{parent}(b_1, b_2), b_1, b_2).$$

Notice that the computation of $\text{lin}(b_1, b_2)$ is quite fast.

***"Gap"* Penalty.** Denote $tips(b)$ as the set of two endpoints of any cell b. For any pair $pch = (b_1, b_2) \in PCH$, define endpoints $x_1 \in tips(b_1), x_2 \in tips(b_2)$ and the *gap* penalty $\text{gap}(b_1, b_2)$ by

$$\text{gap}(b_1, b_2) = \|x_1 - x_2\| = \min\{\|x - y\| \text{ for } (x, y) \in TIPS\} \tag{9}$$

with $TIPS = tips(b_1) \times tips(b_2)$.

***"Dev"* Penalty.** For rod-shaped bacteria, a true pair $(b_1, b_2) \in PCH$ of just born children must have a small $\text{gap}(b_1, b_2) = \|x_1 - x_2\|$ and roughly aligned cells b_1 and b_2. For $(b_1, b_2) \in PCH$, we quantify the *deviation from alignment* $\text{dev}(b_1, b_2)$ as follows. Let x_1, x_2 be the closest endpoints of b_1, b_2 (see (9)). Let str_{12} be the straight line linking the centers c_1, c_2 of b_1, b_2. Let d_1, d_2 be the distances from x_1, x_2 to the line str_{12}. Then, set

$$\text{dev}(b_1, b_2) = \frac{d_1 + d_2}{\|c_2 - c_1\|}.$$

***"Ratio"* Penalty.** True children pairs must have nearly equal lengths. Thus, for $(b_1, b_2) \in PCH$ with lengths L_1, L_2, we define the length *ratio penalty* by

$$\text{ratio}(b_1, b_2) = |(L_1/L_2) + (L_2/L_1) - 2|.$$

***"Rank"* Penalty.** Let $L_{\min}$ be the minimum cell length over all cells in B_+. In B_+, children pairs (b_1, b_2) just born during interframe $J \to J_+$ must have lengths L_1, L_2 close to $L_{\min}$. Thus, for $(b_1, b_2) \in PCH$, we define the *rank* penalty by

$$\text{rank}(b_1, b_2) = |(L_1/L_{\min}) - 1| + |(L_2/L_{\min}) - 1|.$$

Given two successive images J, J_+, we seek the set $X = trueCH$ of true children pairs in $B_+ \times B_+$, which is an unknown subset of PCH. In Section 3.5 below, we replace X by its indicator function z and we build a cost function $E(z)$ which should be nearly minimized when z is close to the indicator of *trueCH*. A key term of $E(z)$ will be a weighted linear combination of the penalty functions $\{\text{lin}, \text{gap}, \text{dev}, \text{ratio}, \text{rank}\}$. Since these penalties are different from those introduced in Section 3.3.2, we estimate their weights in the cost function $E(z)$ by the algorithm outlined in Section 3.2. The minimization of $E(z)$ will be implemented by simulations of a BM with energy function $E(z)$. We present these stochastic neural networks in the next section.

3.4. Generic Boltzmann Machines (BMs)

Minimization of our main cost functionals is a heavily combinatorial task, since the unknown variable is a mapping between two finite sets of sizes ranging from 80 to 120. To handle these minimizations, we use BMs originally introduced by Hinton et al. (see [93,94]). Indeed, these recurrent stochastic neural networks can efficiently emulate some forms of simulated annealing.

Each BM implemented here is a network $BM = \{U_1, \ldots, U_N\}$ of N *stochastic neurons* U_j. In the BM context, the time $t = 0, 1, 2, \ldots$ is discretized and represents the number of steps in a Markov chain, where the successive updates $Z(t) \to Z(t+1)$ of the BM configuration $Z(t)$ are analogous to the steps of a Gibbs sampler. The *configuration* $Z(t) = \{Z_1(t), \ldots, Z_N(t)\}$ of the whole network BM at time t is defined by the random states $Z_j(t)$ of all neuron U_j. Each $Z_j(t)$ belongs to a fixed finite set $W(j)$. Hence, $Z(t)$ belongs to the *configurations set* $CONF = W(1) \times \cdots \times W(N)$.

Neuron interactivity is specified by a finite set CLQ of *cliques*. Each clique K is a subset of $S = \{1, \ldots, N\}$. During configuration updates $Z(t) \to Z(t+1)$, neurons may interact only if they are in the same clique. Here, all cliques K are of small sizes 1, or 2, or 3.

For each clique K, one specifies an energy function $J_K(z)$ defined for all $z \in CONF$, with $J_K(z)$ depending only on the z_j such that $j \in K$. The full energy $E(z)$ of configuration z is then defined by

$$E(z) = \Sigma_{K \in CLQ} J_K(z).$$

The BM stochastic dynamics $Z(t) \to Z(t+1)$ is driven by the energy function $E(z)$, and by a fixed decreasing sequence of *virtual temperatures* $Temp(t) > 0$, tending slowly to 0 as $t \to \infty$. Here, we use standard temperature schemes of the form $Temp(t) \equiv c\eta^t$ with fixed $c > 0$ and slow *decay rate* $0.99 < \eta < 1$.

We have implemented the classical *"asynchronous"* BM dynamics. At each time t, *only one* random neuron U_j may modify its state, after reading the states of all neurons belonging to cliques containing U_j. A much faster alternative, implementable on GPUs, is the *"synchronous"* BM dynamics, where at each time t roughly 50% of all neurons may simultaneously modify their states (see [95–97]). The detailed BM dynamics are presented in the appendix (see Appendix A).

When the virtual temperatures $Temp(t)$ decrease slowly enough to 0, the energy $E(Z(t))$ converges in probability to a local minimum of the BM energy $E(z)$ over all configurations $z \in CONF$.

3.5. Optimized Set of Parent–Children Triplets

Next, we formulate the search for bona fide parent–children triplets as an optimization problem. For brevity, this outline is restricted to situations where (3) holds, as is the case for our synthetic image data. Simple modifications extend this approach to the relaxed constraint (4), which we used for lab videos of live cell populations. Fix successive images J, J_+ with a positive number of cell divisions $DIV = N_+ - N$. Denote $PCH = \{pch_1, pch_2, \ldots, pch_m\}$ the set of m plausible children pairs (b_1, b_2) in B_+. The penalties lin, gap, dev, ratio, and rank defined above for all pairs $(b_1, b_2) \in PCH$ determine five numerical vectors $LIN, GAP, DEV, RAT, RANK$ in $\mathbb{R}^m$ with coordinates $LIN_j = \text{lin}(pch_j)$, $GAP_j = \text{gap}(pch_j)$, $DEV_j = \text{dev}(pch_j)$, $RAT_j = \text{ratio}(pch_j)$, $RANK_j = \text{rank}(pch_j)$.

We now define a *binary* BM constituted by m *binary* stochastic neurons U_j, $j = 1 \ldots m$. At time $t = 0, 1, 2, \ldots$, each U_j has a random *binary valued state* $Z_j(t) = 1$ or 0. The random configuration $Z(t) = [Z_1(t), \ldots, Z_m(t)]$ of this BM belongs to the configuration space $CONF = \{0,1\}^m$ of all binary vectors $z = [z_1, \ldots, z_m]$. Let SUB be the set of all subsets of PCH. Each configuration $z \in CONF$ is the indicator function of a subset $sub(z)$ of PCH. We view each $sub(z) \in SUB$ as a possible estimate for the unknown set $trueCH \subset B_+ \times B_+$ of true children pairs (b_1, b_2). For each potential estimate $sub(z)$ of $trueCH$, the *"lack of quality"* of the estimate $sub(z)$ will be penalized by the *energy function* $E(z) \geq 0$ of our binary BM. We now specify the energy $E(z)$ for all $z \in CONF$ by combining the penalty terms introduced above. Note that the penalty terms introduced in Section 3.3.2 are quite different from those introduced in Section 3.3.3. No cell in B_+ can be assigned to more than one parent in b. To enforce this constraint, define the symmetric $m \times m$ binary matrix $[Q_{j,k}]$ by (i) $Q_{j,k} = 1$ if $j \neq k$ and the two pairs pch_j, pch_k have one cell in common, (ii) $Q_{j,k} = 0$ if $j \neq k$ and the two pairs pch_j, pch_k have no cell in common, (iii) $Q_{j,j} = 0$ for all j.

The quadratic penalty $z \mapsto \langle z, Qz \rangle$ is non-negative for $z \in CONF$, and must be zero if $sub(z) = trueCH$. Introduce six positive weight parameters to be selected further on λ_j, $j \in \{\text{lin}, \text{gap}, \text{dev}, \text{rat}, \text{rank}, Q\}$. Define the vector $V \in \mathbb{R}^m$ as a weighted linear combination of the penalty vectors $LIN, GAP, DEV, RAT, RANK$

$$V = \lambda_{\text{lin}} LIN + \lambda_{\text{gap}} GAP + \lambda_{\text{dev}} DEV + \lambda_{\text{rat}} RAT + \lambda_{\text{rank}} RANK.$$

For any configuration $z \in CONF$, the BM energy $E(z)$ is defined by the *quadratic function*

$$E(z) = \langle V, z \rangle + \lambda_Q \langle z, Qz \rangle. \tag{10}$$

We already know that the unknown set *trueCH* of true children pairs must have cardinal $DIV = N_+ - N$. Thus, we seek a configuration $z^* \in CONF$ minimizing the energy $E(z)$ under the rigid constraint $\text{card}\{sub(z)\} = DIV$. Let $ONE \in \mathbb{R}^m$ be the vector with all its coordinates equal to 1. The constraint on z can be reformulated as $\langle ONE, z\rangle = DIV$. We want the unknown *trueCH* to be close to the solution z^* of the constrained minimization problem

$$z^* = \underset{z \in CONF}{\text{argmin}}\, E(z) \quad \text{subject to} \quad \langle ONE, z\rangle = DIV.$$

To force this minimization to yield a reliable estimate of *trueCH*, we calibrate the six weights

$$\lambda_j, j \in \{\text{lin}, \text{gap}, \text{dev}, \text{rat}, \text{rank}, Q\}$$

by the algorithm in Section 3.2. Denote $CONF_1$ the set of all $z \in CONF$ such that $\langle ONE, z\rangle = DIV$. To minimize $E(z)$ under the constraint $z \in CONF_1$, fix a slowly decreasing temperature scheme $Temp(t)$ as in Section 3.4. We need to force the BM stochastic configurations $Z(t)$ to remain in $CONF_1$. Then, for large time step t, the $Z(t)$ will converge in probability to a configuration $z^* \in CONF_1$ approximately minimizing $E(z)$ under the constraint $z \in CONF_1$.

Start with any $Z(0) \in CONF_1$. Assume that, for $0 \leq s \leq t$, one has already dynamically generated BM configurations $Z(s) \in CONF_1$. Then, randomly select two sites j, k such that $Z_j(t) = 1$ and $Z_k(t) = 0$. Compute a virtual configuration Y by setting $Y_j = 0$, $Y_k = 1$, and $Y_i \equiv Z_i$ for all sites i different from j and k. Compute the energy change $\Delta E = E(Y) - E(Z(t))$, and the probability $p(t) = \exp(-D/Temp(t))$, where $D = \max\{0, \Delta E\}$. Then, randomly select $Z(t+1) = Y$ or $Z(t+1) = Z(t)$ with respective probabilities $p(t)$ and $(1 - p(t))$. Clearly, this forces $Z(t+1) \in CONF_1$.

3.6. *Performance of Automatic Children Pairing on Synthetic Videos*

In the following subsections, we provide experimental results for pairing children and parent cells.

3.6.1. Children Pairing: Fast BM Simulations

For $m = \text{card}(PCH) \leq 1000$, one can reduce the computational cost for BM dynamics simulations by pre-computing and storing the $m \times m$ symmetric binary matrix Q, as well as the m-dimensional vectors LIN, GAP, DEV, RAT, $RANK$ and their linear combination V. A priori reduction of m significantly reduces the computing times, and can be implemented by trimming away the pairs $pch_j \in PCH$ for which the penalties LIN_j, GAP_j, DEV_j, RAT_j, and $RANK_j$ are all larger than predetermined empirical thresholds. We performed a study on 100 successive (synthetic) images. We show scatter plots for the most informative penalty terms in Figure 3. These plots allow us to determine adequate thresholds for the penalty terms. We observed that, for the synthetic and real data, we considered the trimming of DEV, GAP, and $RANK$ reduced the percentage of invalid children pairs by 95%, therefore drastically reducing the combinatorial complexity of the problem.

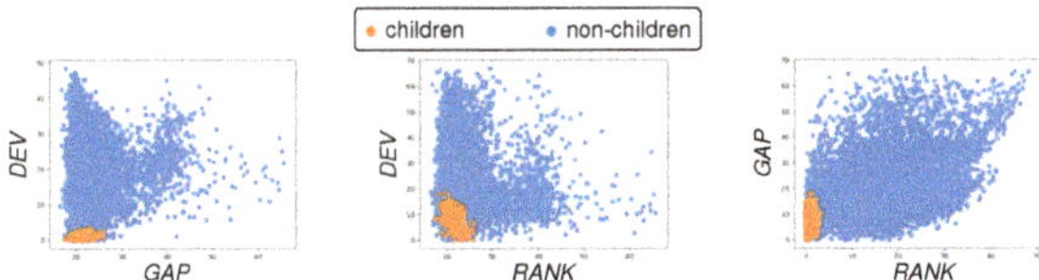

Figure 3. Scatter plots for tandems of the penalty terms DEV, GAP, and $RANK$. We mark in orange the true children pairs and in blue invalid children pairs. These plots allow us to identify appropriate empirical thresholds to trim the (considered synthetic) data in order to reduce the computational complexity of the parent–children pairing.

The quadratic energy function $E(z)$ is the sum of clique energies $J_K(z)$ involving only cliques of cardinality 1 and 2. For any clique $K = \{j\}$ of cardinality 1, with $1 \leq j \leq m$, one has $J_K(z) = V_j z_j$. For any clique $K = \{j, k\}$ of cardinality 2, with $1 \leq j < k \leq m$, one has $J_K(z) = 2Q_{j,k} z_j z_k$. A key computational step when generating $Z(t+1)$ is to evaluate the energy change ΔE when one flips the binary values $Z_j(t) = 1$ and $Z_k(t) = 0$ by the new value $(1 - z_i)$ for a fixed single site i. This step is quite fast since it uses only the numbers V_j, V_k, and $\langle q(j), Z(t) \rangle$, $\langle q(k), Z(t) \rangle$, where $q(i)$ is the ith row of the matrix Q.

3.6.2. Children Pairing: Implementation on Synthetic Videos

We have implemented our children pairing algorithms on synthetic image sequences having 100 to 500 image frames with 1 min interframe (benchmark set BENCH1; see Section 2.1). The cell motion bound $w/2$ per interframe was defined by $w = 20$ pixels. The parameter τ that defines the sets PCH of plausible children pairs (see (5)) was set at $\tau = 45$ pixels.

The known true cell registrations indicated that, in our typical BENCH1 image sequence, the successive sets PCH had average cardinalities of 120, while the number of true children pairs per PCH roughly ranged from 2 to 6 with a median of 4. The size of the reduced configuration space $CONF1$ per image frame thus ranged from 10^4 to $120^6/6! = 4.2 \times 10^9$ with a median of 9×10^6.

Our weights estimation technique introduced in Section 3.2 yields the weights

$$[\lambda_{\text{cen}}, \lambda_{\text{siz}}, \lambda_{\text{ang}}] = [0.255, 0.05, 0.05]$$

and

$$[\lambda_{\text{gap}}, \lambda_{\text{dev}}, \lambda_{\text{rat}}, \lambda_{\text{rank}}] = [0.01, 1, 0.0001, 0.05]$$

for the penalties introduced in Section 4. To reduce the computing time for hundreds of BM energy minimizations on the BENCH1 image sequences, we excluded obviously invalid children pairs in each PCH set, by simultaneously thresholding of the penalty terms. The BM temperature scheme was $Temp(t) = 1000\,(0.995)^t$, with the number of epochs capped at 5000. The average CPU time for BM energy minimization dedicated to optimized children pairing was about 30 seconds per frame. (We provide hardware specifications in Appendix B).

3.6.3. Parent–Children Matching: Accuracy on Synthetic Videos

For each successive image pair J, J_+, with cells B, B_+ of cardinality $N < N_+$, our parent–children matching algorithm computes a set SHL of short lineages (b, b_1, b_2), where the cell $b \in B$ is expected to be the parent of cells $b_1, b_2 \in B_+$. Recall that $DIV = N_+ - N$ provides the number of cell divisions during the interframe $J \to J_+$. The number VAL of correctly reconstructed short lineages $(b, b_1, b_2) \in SHL$ is obtained by direct comparison to the known ground truth registration $J \to J_+$. For each frame J, we define the *pcp-accuracy* of our parent–children pairing algorithm as the ratio VAL/DIV.

We have tested our parent–children matching algorithm on three long synthetic image sequences BENCH1 (500 frames), BENCH2 (300 frames), and BENCH3 (300 frames), with respective interframes of 1, 2, and 3 min. For each frame J_k, we computed the pcp-accuracy between J_k and J_{k+1}.

We report the accuracies of our parent–children pairing algorithms in Table 2. For BENCH1, all 500 pcp-accuracies reached 100%. For BENCH2, pcp-accuracies reach 100% for 298 frames out of 300, and for the remaining two frames, accuracies were still high at 93% and 96%. For BENCH3, where interframe duration was longest (3 min), the 300 pcp-accuracies decreased slightly but still averaged 99%, and never fell below 90%.

Table 2. Accuracies of parent–children pairing algorithm. We applied our parent–children pairing algorithm to three long synthetic image sequences BENCH1 (500 frames), BENCH2 (300 frames), and BENCH3 (300 frames), with interframe intervals of 1, 2, 3 min, respectively. The table summarizes the resulting pcp-accuracies. Note that pcp-accuracies are practically always at 100%. For BENCH2, pcp-accuracies are 100% for 298 frames out of 300, and for the remaining two frames, accuracies were still high at 93% and 96%. For BENCH3, the average pcp-accuracy for the 3 min interframe is 99%.

Sequence	Pcp-Accuracy	Frames
BENCH1	$acc = 100\%$	500 out of 500
BENCH2	$acc = 100\%$	298 out of 300
BENCH2	$99\% \geq acc \geq 93\%$	2 out of 300
BENCH3	$acc = 100\%$	271 out of 300
BENCH3	$99\% \geq acc \geq 95\%$	17 out of 300
BENCH3	$94\% \geq acc \geq 90\%$	12 out of 300

3.7. Reduction to Registrations with No Cell Division

Fix successive frames J, J_+ and their cell sets B, B_+. We seek the unknown registration mapping $f : B \to \{B_+ \cup (B_+ \times B_+)\}$, where $f(b) \in B_+$ iff cell b did not divide during the interframe $J \to J_+$ and $f(b) = (b_1, b_2) \in B_+ \times B_+$ iff cell b divided into (b_1, b_2) during the interframe.

If $\text{card}(B) = N < N_+ = \text{card}(B_+)$, we know that the number of cell divisions during the interframe $J \to J_+$ should be $DIV = DIV(B, B_+) = N_+ - N > 0$. We then apply the parent–children matching algorithm outlined above to compute a set $SHL = SHL(B, B_+)$ of short lineages (b, b_1, b_2) with $b \in B$, $b_1, b_2 \in B_+$ and $\text{card}(SHL) = DIV$. For each $(b, b_1, b_2) \in SHL$, the cell b is computed by $b = \text{parent}(b_1, b_2)$ as the parent cell of the two children cells $b_1, b_2 \in B_+$.

For each $(b, b_1, b_2) \in SHL$, eliminate from B the parent cell, b, and eliminate from B_+ the two children cells b_1, b_2. We are left with two residual sets, $resB \subset B$ and $resB_+ \subset B_+$, having the same cardinality, $N - DIV = N_+ - 2DIV$. Assuming that our set SHC of short lineages is correctly determined, the cells $b \in redB$ should not divide in the interframe $J \to J_+$, and hence have a single (still unknown) registration $f(b) \in redB_+$. Thus, the still unknown part of the registration f is a bijection from $redB$ to $redB_+$.

Let $divB = B - redB$ and $divB_+ = B_+ - redB_+$. For each $b \in divB$, the cell b divides into the unique pair of cells, $(b_1, b_2) \in divB_+ \times divB_+$, such that $(b, b_1, b_2) \in SHL$. Hence, we can set $f(b) = (b_1, b_2)$ for all $b \in divB$. Thus, the remaining problem to solve is to compute the bijective registration $f : redB \to redB_+$. We have reduced the registration discovery to a new problem, where *no cell divisions occur* in the interframe duration. In what follows, we present our algorithm to solve this registration problem.

3.8. Automatic Cell Registration after Reduction to Cases with No Cell Division

As indicated above, we can *explicitly reduce* the generic cell tracking problem to a problem where there is *no cell division*. We consider images J, J_+ with associated cell sets B, B_+ such that $N = \text{card}(B) = \text{card}(B_+)$. Hence, there are no cell divisions in the interframe $J \to J_+$ and the map f of this reduced problem is (in principle) a bijection $f : B \to B_+$ with $\text{card}(B) = \text{card}(B_+)$. In Figure 4, we show two typical successive images we use for testing with no cell division generated by the simulation software [12,81] (see Section 2.1).

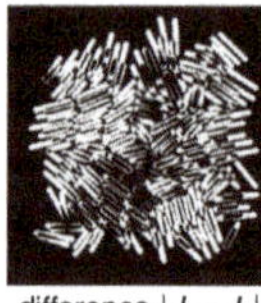

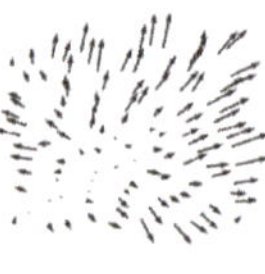

Figure 4. Simulated cell dynamics. From left to right, two successive simulated images J and J_+ with an interframe time of six minutes and no cell division, their image difference $|J - J_+|$, and the associated motion vectors. For the image J and J_+ we color four pairs of cells in $B \times B_+$, which should be matched by the true cell registration mapping. Notice that the motion for an interframe time of six minutes is significant. We can observe that, even without considering cell division, we can no longer assume that corresponding cells in frame J and J_+ overlap.

3.8.1. The Set *MAP* of Many-to-One Cell Registrations

We have reduced the registration search to a situation where, during the interframe $J \to J_+$, no cell has divided, no cell has disappeared, and no cell has suddenly emerged in B_+ without originating from B. The unknown registration $f : B \to B_+$ should then in principle be injective and onto. However, for computational efficiency, we will temporarily relax the bijectivity constraint on f. We will seek f in the set *MAP* of all *many-to-one mappings* $f : B \to B_+$ such that for each $b \in B$, the cell $f(b)$ is in the target window $W(b) \subset B_+$ (see Section 2.3).

3.8.2. Registration Cost Functional

To design a cost functional $\text{cost}(f)$, which should be roughly minimized when $f \in MAP$ is very close to the true registration from B to B_+, we linearly combine penalties $\text{match}(f)$, $\text{over}(f)$, $\text{stab}(f)$, $\text{flip}(f)$ weighted by unknown positive weights $\lambda_{\text{match}}, \lambda_{\text{over}}, \lambda_{\text{stab}}, \lambda_{\text{flip}}$, to write, for all registrations $f \in MAP$,

$$\text{cost}(f) = \lambda_{\text{match}}\,\text{match}(f) + \lambda_{\text{over}}\,\text{over}(f) + \lambda_{\text{stab}}\,\text{stab}(f) + \lambda_{\text{flip}}\,\text{flip}(f). \quad (11)$$

We specify the individual terms that appear in (11) below. Ideally, the minimizer of $\text{cost}(f)$ over all $f \in MAP$ is close to the unknown true registration mapping $f : B \to B_+$. To enforce a good approximation of this situation, we first estimate efficient positive weights by applying our calibration algorithm (see Section 3.2). The actual minimization of $\text{cost}(f)$ over all $f \in MAP$ is then implemented by a BM described in Section 3.9.

Cell Matching Likelihood: $\text{match}(f)$**.** Here, we extend a pseudo likelihood approach used to estimate parameters in Markov random fields modeling by Gibbs distributions (see [98]). Recall that *g.rate* is the *known* average cell growth rate. For any cells $b \in B$, $b_+ \in B_+$, the geometric quality of the matching $b \mapsto b_+$ relies on three main characteristics: (*i*) motion $c(b_+) - c(b)$ of the cell center $c(b)$, (*ii*) angle between the long axes $A(b)$ and $A(b_+)$, (*iii*) cell length ratio $\|A(b_+)\|/\|A(b)\|$. Thus, for all $b \in B$ and b_+ in the target window $W(b)$, define (i) Kinetic energy: $\text{kin}(b, b_+) = \|c(b) - c(b_+)\|^2$. (ii) Distortion of cell length: $\text{dis}(b, b_+) = |\log(\|A(b_+)\|/\|A(b)\|) - \log g.rate|^2$. (iii) Rotation angle: $0 \le \text{rot}(b, b_+) \le \pi/2$ is the geometric angle between the straight lines carrying $A(b)$ and $A(b_+)$.

Fix $b \in B$, and let b' run through the whole target window $W(b)$. The finite set of values thus reached by the kinetic penalties $\text{kin}(b, b')$ has two smallest values $kin_1(b)$, $kin_2(b)$. Define $list.kin = \bigcup_{b \in B}\{\text{kin}_1(b), \text{kin}_2(b)\}$, which is a list of $2N$ *"low"* kinetic penalty values. Repeat this procedure for the penalties $\text{dis}(b, b')$ and $\text{rot}(b, b')$ to similarly define a *list.dis* of $2N$ *"low"* distortion penalty values, and a *list.rot* of $2N$ *"low"* rotation penalty values.

The three sets *list.kin*, *list.dis*, *list.rot* can be viewed as three random samples of size $2N$, respectively, generated by three unknown probability distributions $P_{\text{kin}}, P_{\text{dis}}, P_{\text{rot}}$. We approximate these three probabilities by their *empirical* cumulative distribution functions

CDF_{kin}, CDF_{dis}, CDF_{rot}, which can be readily computed. We now use the right tails of these three CDFs to compute separate probabilistic evaluations of how *likely* the matching of cell $b \in B$ with cell $b_+ \in W(b)$ is. For any fixed mapping $f \in MAP$, and any $b \in B$, set $b_+ = f(b)$. Compute the three penalties $vkin = \text{kin}(b, b_+)$, $vdis = \text{dis}(b, b_+)$, $vrot = \text{rot}(b, b_+)$, and define three associated *"likelihoods"* for the matching $b \to b_+ = f(b)$:

$$\begin{aligned}
\text{LIK}_{\text{kin}}(b, b_+) &= 1 - \text{CDF}_{\text{kin}}(vkin),\\
\text{LIK}_{\text{dis}}(b, b_+) &= 1 - \text{CDF}_{\text{dis}}(vdis),\\
\text{LIK}_{\text{rot}}(b, b_+) &= 1 - \text{CDF}_{\text{rot}}(vrot).
\end{aligned}$$

High values of the penalties *vkin, vdis, vrot* thus will yield three small likelihoods for the matching $b \to b_+ = f(b)$. With this, we can define a *"joint likelihood"* $0 \leq \text{LIK}(b, b_+) \leq 1$ evaluating how likely is the matching $b \to b_+ = f(b)$:

$$\text{LIK}(b, b_+) = \prod_{j \in \{\text{kin,dis,rot}\}} \text{LIK}_j(b, b_+). \tag{12}$$

Note that higher values of $\text{LIK}(b, b_+)$ correspond to a better geometric quality for the matching of b with $b_+ = f(b)$. To avoid vanishingly small likelihoods, whenever $\text{LIK}(b, b_+) < 10^{-6}$, we replace it by 10^{-6}. Then, for any mapping $f \in MAP$, we define its *likelihood* $\text{lik}(f)$ by the finite product

$$\text{lik}(f) = \prod_{b \in B} \text{LIK}(b, f(b)).$$

The product of these N likelihoods is typically very small, since $N = \text{card}(B)$ can be large. Thus, we evaluate the geometric matching quality $\text{match}(f)$ of the mapping f via the averaged *log-likelihood of* f, namely,

$$\text{match}(f) = -\frac{1}{N} \log \text{lik}(f) = -\frac{1}{N} \sum_{b \in B} \log \text{LIK}(b, f(b)).$$

Good registrations $f \in MAP$ should yield small values for the criterion $\text{match}(f)$.

Overlap: $\text{over}(f)$. We expect *bona fide* cell registrations $f \in MAP$ to be bijections. Consequently, we want to penalize mappings f which are many-to-one. We say that two distinct cells $(b, b') \in B \times B$ do *overlap* for the mapping $f \in MAP$ if $f(b) = f(b')$. The total number of overlapping pairs (b, b') for f defines the *overlap penalty*:

$$\text{over}(f) = \frac{1}{\text{card}(B)} \sum_{b \in B} \sum_{b' \in B} 1_{f(b) = f(b')}.$$

Neighbor Stability: $\text{stab}(f)$. Let $B = \{b_1, \dots, b_N\}$. Denote G_i as the set of all neighbors for cell b_i in B (i.e., $b_j \sim b_i \iff b_j \in G_i$; see Section 2.3). For *bona fide* registrations $f \in MAP$, and for most pairs of neighbors $b_i \sim b_j$ in B, we expect $f(b_i)$ and $f(b_j)$ to remain neighbors in B_+. Consequently, we penalize the lack of *"neighbors stability"* for f by

$$\text{stab}(f) = \sum_i \sum_{j \neq i} \frac{1}{N|G_i||G_j|} 1_{b_i \sim b_j} 1_{f(b_i) \nsim f(b_j)}.$$

Neighbor Flip: $\text{flip}(f)$. Fix any mapping $f \in MAP$, any cell $b \in B$ and any two neighbors b', b'' of b in B. Let $z = f(b)$, $z' = f(b')$, $z'' = f(b'')$. Let c, c', c'' and d, d', d'' be the centers of cells b, b', b'' and z, z', z''. Let α be the oriented angle between $c' - c$ and $c'' - c$, and let α_f be the angle between $d' - d$ and $d'' - d$, respectively. We say that the mapping f has *flipped* cells b', b'' around b, and we set $\text{FLIP}(f, b, b', b'') = 1$ if z', z'' are both neighbors of z, and the two angles α, α_f have *opposite signs*. In all other cases, we set $\text{FLIP}(f, b, b', b'') = 0$.

For any registration $f \in MAP$, define the *flip penalty* for f by

$$\text{flip}(f) = \sum_{b \in B} \sum_{b' \in B} \sum_{b'' \in B} \frac{1}{N|G(b)|^2} \text{FLIP}(f, b, b', b''),$$

where $G(b)$ is the neighborhood of cell b in B. In Figure 5, we illustrate an example of an unwanted cell flip.

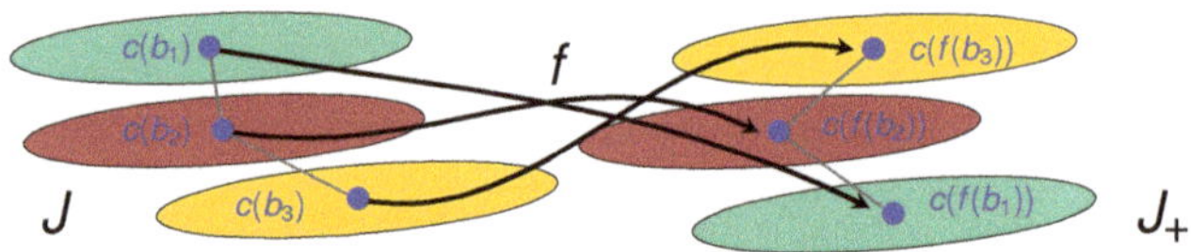

Figure 5. Illustration of an undesirable flip for the mapping f. The cells b_1 and b_3 are neighbors of b_2, and mapped by f on neighbors $z_1 = f(b_1), z_3 = f(b_3)$ of $z_2 = f(b_2)$, as should be expected for bona fide cell registrations. However, for this mapping f, we have z_3 above z_2 above z_1, whereas, for the original cells, we had b_1 above b_2 above b_3. Our cost function penalizes flips of this nature.

3.9. BM Minimization of Registration Cost Function

In what follows, we define the optimization problem for the registration of cells from one frame to another (i.e., cell tracking), as well as associated methodology and parameter estimates.

3.9.1. BM Minimization of Cost(f) over $f \in MAP$

Let B, B_+ be two successive sets of cells. As outlined above, we have reduced the problem to one in which we can assume that $N = \text{card}(B) = \text{card}(B_+)$, so that there is no cell division during the interframe. Write $B = \{b_1, \ldots, b_N\}$. For short, denote $W(j) \subset B_+$ instead of $W(b_j)$ the target window of cell b_j. We seek to minimize cost(f) over all registrations $f \in MAP$. Let *BM* be a BM with sites $S = \{1, \ldots, N\}$ and stochastic neurons $\{U_1, \ldots, U_N\}$. At time t, the random state $Z_j(t)$ of U_j will be some cell z_j belonging to the target window $W(j)$ and the random configuration $Z(t) = \{Z_1(t), \ldots, Z_N(t)\}$ of the whole *BM* belongs to the configurations set $CONF = W(1) \times \ldots \times W(N)$.

To any configuration $z = \{z_1, \ldots, z_N\} \in CONF$, we associate a unique cell registration $f \in MAP$ defined by $f(b_j) = z_j$ for all j, denoted by $f = \text{map}(z)$. This determines a bijection $z \mapsto f = \text{map}(z)$ from *CONF* onto *MAP*. The inverse of map : $CONF \to MAP$ will be called range : $MAP \to CONF$, and is defined by $z = \text{range}(f)$, when $z_j = f(b_j)$ for all j.

3.9.2. BM Energy Function $E(z)$

We now define the energy function $E(z) \geq 0$ of our BM for all $z \in CONF$. Denote $E^* = \text{minimize}_{z \in CONF} E(z)$. Since $f \mapsto z = \text{range}(f)$ is a bijection from *MAP* to *CONF*, we must have

$$E^* = \underset{z \in CONF}{\text{minimize}}\, E(z) = \underset{f \in MAP}{\text{minimize}}\, E(\text{range}(f)).$$

Our goal is to minimize cost(f), and we know that BM simulations should roughly minimize $E(z)$ over all $z \in CONF$. Thus, we define the BM energy function $E(z)$ by forcing

$$\text{cost}(f) = E(\text{range}(f)) \tag{13}$$

for any registration mapping $f \in MAP$, which—due to the preceding subsection—is equivalent to

$$E(z) = \text{cost}(\text{map}(z)) \tag{14}$$

for all configurations $z \in CONF$. The next subsection will explicitly express the energy $E(z)$ in terms of *cliques* of neurons. Due to (13) and (14), we have

$$E^* = \underset{f \in MAP}{\text{minimize}} \ \text{cost}(f) = \underset{z \in CONF}{\text{minimize}} \ E(z).$$

For large time t, the BM stochastic configuration $Z(t)$ tends with high probability to concentrate on configurations $z \in CONF$, which roughly minimize $E(z)$. The random registration $F^t = \text{map}(Z(t))$ will belong to MAP and verify $Z(t) = \text{range}(F^t)$, so that $E(Z(t)) = E(\text{range}(F^t)) = \text{cost}(F^t)$. Consequently, for large t—with high probability—the random mapping $F^t = \text{map}(Z(t))$ will have a value of the cost functional $\text{cost}(F^t)$ close to $\text{minimize}_{f \in MAP} \text{cost}(f)$.

3.9.3. Cliques of Interactive Neurons

The BM energy function $E(z)$ just defined turns out to involve only three sets of small cliques: (i) CL_1 is the set of all singletons $K = \{i\}$, with $i = 1 \dots N$. (ii) CL_2 is the set of all pairs $K = \{i, j\}$ such that cells b_i and b_j are *neighbors* in B. (iii) CL_3 is the set of all triplets $K = \{i, j, k\}$ such that cells b_j and b_k are both *neighbors* of b_i in B. Denote $CLQ = CL_1 \cup CL_2 \cup CL_3$ as the set of all cliques for our BM.

Cliques in CL_1. For each clique $K = \{i\}$ in CL_1, and each $z \in CONF$, define its energy $J_{\text{match},K}(z) = J_{\text{match},K}(z_i)$ by

$$J_{\text{match},K}(z) = -\frac{1}{N} \log \text{LIK}(b_i, z_i) \text{ for all } z \in ZW,$$

where LIK is given by (12). Set $J_{\text{match},K} \equiv 0$ for K in $CL_2 \cup CL_3$. For all $z \in CONF$, define the energy $\text{E}_{\text{match}}(z)$ by

$$E_{\text{match}}(z) = \sum_{K \in CLQ} J_{\text{match},K}(z) = \sum_{K \in CL_1} J_{\text{match},K}(z),$$

which implies that the registration $f = \text{map}(z)$ verifies $\text{match}(f) = E_{\text{match}}(z)$.

Cliques in CL_2. For all $z \in CONF$, all cliques $K = \{i, j\}$ in CL_2, define the clique energies $J_{\text{over},K}(z) = J_{\text{over},K}(z_i, z_j)$ and $J_{\text{stab},K}(z) = J_{\text{stab},K}(z_i, z_j)$ by $J_{\text{over},K}(z) = 1_{z_i = z_j}/N$ and

$$J_{\text{stab},K}(z) = \frac{1}{N|G_i||G_j|} 1_{b_j \sim b_i} 1_{z_j \not\sim z_i},$$

where $|G_i|$ and $|G_j|$ are the numbers of neighbors in B for cells z_i and z_j, respectively. Set $J_{\text{over},K} = J_{\text{stab},K} \equiv 0$ for K in $CL_1 \cup CL_3$. Define the two energy functions

$$E_{\text{over}}(z) = \sum_{K \in CLQ} J_{\text{over},K}(z) = \sum_{K \in CL_2} J_{\text{over},K}(z),$$
$$E_{\text{stab}}(z) = \sum_{K \in CLQ} J_{\text{stab},K}(z) = \sum_{K \in CL_2} J_{\text{stab},K}(z),$$

which implies that $f = \text{map}(z)$ verifies $\text{over}(f) = E_{\text{over}}(z)$ and $\text{stab}(f) = E_{\text{stab}}(z)$.

Cliques in CL_3. For each clique $K = \{i, j, k\}$ in CL_3, define the clique energy $J_{\text{flip},K}$ by

$$J_{\text{flip},K}(z) = J_{\text{flip}}^{i,j,k}(z) = \frac{1}{N|G_i|^2} \ \text{FLIP}(f^{i,j,k}, b_i, b_j, b_k),$$

where $f^{i,j,k}$ is any registration mapping b_i, b_j, b_k onto z_i, z_j, z_k. The indicator FLIP was defined in Section 3.8.2. Set $J_{\text{flip},K} \equiv 0$ for K in $CL_1 \cup CL_2$. Define the energy

$$E_{\text{flip}}(z) = \sum_{K \in CLQ} J_{\text{flip},K}(z) = \sum_{K \in CL_3} J_{\text{flip},K}(z),$$

which implies that $f = F(z)$ verifies $\text{flip}(f) = E_{\text{flip}}(z)$.

Finally, define the clique energy J_K for all $K \in CLQ$ by the linear combination

$$J_K = \lambda_{\text{match}} J_{\text{match},K} + \lambda_{\text{over}} J_{\text{over},K} + \lambda_{\text{stab}} J_{\text{stab},K} + \lambda_{\text{flip}} J_{\text{flip},K}.$$

Summing this relation over all $K \in CLQ$ yields

$$\sum_{K \in CLQ} J_K = \lambda_{\text{match}} E_{\text{match}} + \lambda_{\text{over}} E_{\text{over}} + \lambda_{\text{stab}} E_{\text{stab}} + \lambda_{\text{flip}} E_{\text{flip}}. \tag{15}$$

Define then the final BM energy function $z \mapsto E(z)$ by

$$E(z) = \sum_{K \in CLQ} J_K(z) \text{ for all } z \text{ in } CONF. \tag{16}$$

For any $z \in CONF$, the associated registration $f = \text{map}(z)$ verifies $\text{match}(f) = E_{\text{match}}(z), \text{over}(f) = E_{\text{over}}(z)$, $\text{stab}(f) = E_{\text{stab}}(z)$, $\text{flip}(f) = E_{\text{flip}}(z)$. By weighted linear combination of these equalities, and, due to (15), we obtain for all configurations $z \in CONF$, $E(z) = \text{cost}(f)$ when $f = \text{map}(z)$ or, equivalently, when $z = \text{range}(f)$.

3.9.4. Test Set of 100 Synthetic Image Pairs

As shown above, the minimization of $\text{cost}(f)$ over all registrations $f \in MAP$ is equivalent to seeking BM configurations $z \in CONF$ with minimal energy $E(z)$. We have implemented this minimization of $E(z)$ by the long-term asynchronous dynamics of the BM just defined. This algorithm was designed for the registration of image pairs exhibiting no cell division, and was, therefore, implemented after the automatic reduction of the generic registration problem, as indicated earlier. We have tested this specialized registration algorithm on a set of 100 pairs of successive images of simulated cell colonies exhibiting no cell divisions. These 100 image pairs were extracted from the benchmark set BENCH6 of synthetic image sequence described in Section 2.1. The 100 pairs of cell sets B, B_+ had sizes $N = \text{card}(B) = \text{card}(B_+)$ ranging from 80 to 100 cells. For each test pair B, B_+, each target window $W(j)$ typically contained 30 to 40 cells. The set $CONF$ of configurations had huge cardinality ranging from 10^{130} to 10^{160}. However, the average number of neighbors of a cell was around 4 to 5.

3.9.5. Implementation of BM Minimization for Cost(f)

The numbers clq_1, clq_2, clq_3 of cliques in CL_1, CL_2, CL_3 have the following rough ranges $80 \leq clq1 \leq 100$, $160 \leq clq_2 \leq 250$, and $450 \leq clq_3 \leq 600$. For $k = 1,2,3$, denote $val(k)$ the numbers of non-zero values for $J_K(z)$ when z runs through $CONF$ and K runs through all cliques of cardinality k. One easily checks the rough upper bounds $val(1) < 4000$; $val(2) < 200{,}000$; $val(3) < 300{,}000$. Hence, to automatically register B to B_+, one could pre-compute and store all the possible values of $J_K(z)$ for all cliques $K \in CL_1 \cup CL_2 \cup CL_3$ and all the configurations $z \in CONF$. This accelerates the key computing steps of the asynchronous BM dynamics, namely, for the evaluation of energy change $\Delta E = E(z') - E(z)$, when configurations z and z' differ at only one site $j \in S$. Indeed, the single site modification $z_j \to z'_j$ affects only the energy values $J_K(z)$ for the very small number $r(j)$ of cliques K, which contain the site j. In our benchmark sets of synthetic images, one had $r(j) < 24$ for all $j \in S$. Hence, the computation of ΔE was fast since it requires retrieving at most 24 pairs of pre-computed $J_K(z)$, $J_K(z')$, and evaluating the 24 differences $J_K(z') - J_K(z)$. Another practical acceleration step is to replace the ubiquitous computations of probabilities $p(t) = \exp(-D/Temp(t))$ by simply testing the value $-D/Temp(t)$ against 100 precomputed logarithmic thresholds.

In our implementation of ABM dynamics, we used virtual temperature schemes such as $Temp(t) = 50 \cdot \rho^t$ with $0.995 \leq \rho \leq 0.999$. The BM simulation was stopped when the stochastic energy $E(Z(t))$ had remained roughly stable during the last N steps. Since all target windows $W(j)$ had cardinality smaller 40, the initial configuration $Z(0) = x$ was computed via

$$x_j = \underset{y \in W(j)}{\text{argmax}} \, \text{LIK}(b_j, y) \quad \text{for } j = 1, \ldots, N,$$

where the likelihoods LIK were defined by (12).

3.9.6. Weight Calibration

For the pair of successive *synthetic* images J, J_+ displayed in Figure 4, we have $N = \text{card}(B) = \text{card}(B_+) = 513$ cells. The ground truth registration f is known by construction; we used it to apply the weight calibration described in Section 3.2. We set the meta-parameter γ to 10^{10} and obtained the vector of weights

$$\Lambda^* = [\lambda^*_{\text{match}}, \lambda^*_{\text{over}}, \lambda^*_{\text{stab}}, \lambda^*_{\text{flip}}] = [110, 300, 300, 290]. \quad (17)$$

These weights are *kept fixed for all the 100 pairs* of images taken from the set BENCH6. The determined weights are used in the cost function $\text{cost}(f)$ defined above. This correctly parametrized the BM energy function $E(z)$. We then simulated the BM stochastic dynamics to minimize the BM energy $E(Z(t))$.

3.9.7. BM Simulations

We launched 100 simulations of the asynchronous BM dynamics, one for each pair of successive images in our test set of 100 images taken from BENCH6. For each such pair, the ground truth mapping $f : B \to B_+$ was known by construction and the stochastic minimization of the BM energy generated an estimated cells registration $f' : B \to B_+$. For each pair B, B_+ in the considered set of 100 images, the accuracy of this automatically computed registration f' was evaluated by the percentage of cells $b \in B$ such that $f'(b) = f(b)$. When $\text{card}(B) = N$, our BM has N stochastic neurons, and the asynchronous BM dynamics proceeds by successive *epochs*. Each epoch is a sequence of N single site updates of the BM configuration. For each one of our 100 simulations of BM asynchronous dynamics, the number of epochs ranged from 250 to 450.

The average computing time was about eight minutes per epoch, which entailed a computing time ranging from 30 to 50 min for each one of our 100 automatic registrations $f' : B \to B_+$ reported here. (We specify the hardware used to carry out these computations in Appendix B). Each image contains about 100 to 150 cells. Consequently, the runtime for the algorithm is approximately 20 s per cell for our prototype implementation. We note that this is only a rough estimate. The runtime depends on several factors, such as the number of cells in an image; the number of mother and daughter cells (i.e., how many cells divide); the size of the neighborhood of each individual cell (window size); the weights used in the cost function (which affects the number of epochs), etc. We note that the temperature scheme had not been optimized yet, so that these computing times are upper bounds. Earlier SBM studies [99–102] indicate that the same energy minimizations on GPUs could provide a computational speedup by a factor ranging between 30 and 50. We report registration accuracies in Table 3. For each pair of images in the considered set of 100 images, the accuracy of automatic registration was larger than 94.5%. The overall average registration accuracy was quite high at 99%.

Table 3. Registration accuracy for synthetic image sequence $BENCH_{100}$. We consider 100 pairs of consecutive synthetic images taken from the benchmark dataset BENCH6. Automatic registration was implemented by BM minimization of the cost function cost(f), which was parametrized by the vector of optimized weights Λ^* in (17). The average registration accuracy was 99%.

Registration Accuracy	Number of Frames
$acc = 100\%$	55 frames out of 100
$99\% \geq acc > 97\%$	40 frames out of 100
$96\% \geq acc > 94.5\%$	5 frames out of 100

4. Results

In this section, we report results for the registration for cell dynamics involving growth, motion, and cell divisions.

4.1. Tests of Cell Registration Algorithms on Synthetic Data

We now consider more generic long synthetic image sequences of simulated cell colonies, with a small interframe duration of one minute. We still impose the mild constraint that no cell is lost between two successive images. The main difference with the earlier benchmark of 100 images from BENCH6 is that cells are *allowed to freely divide* during interframes, as well as to grow and to move. For the full implementation on 100 pairs of successive images, we first execute the parent–children pairing, and remove the identified parent–children triplets; we can then apply our cell registration algorithmic on the reduced sets cells. Our image sequence contained 760 true parent–children triplets, which we automatically identified with an accuracy of 100%. As outlined earlier, we removed all these identified cell triplets and then applied our tracking algorithm. This left us with a total of 12,631 cells (spread over 100 frames). Full automatic registration was then implemented with an accuracy higher than 99.5%.

4.2. Tests of Cell Registration Algorithms on Laboratory Image Sequences

To test our cell tracking algorithm on pairs of consecutive images extracted from recorded image sequences of bacterial colonies (real data), we had to automatically delineate all individual cells in each image. Representative frames of these data are shown in Figure 1. We describe these data in more detail in Section 2.2. We will only briefly outline the overall segmentation approach to not distract from our main contribution—the cell tracking algorithm. We use the watershed algorithm [103] (also used, e.g., in [76]) to segment each frame into individual image segments containing one single cell each. Consequently, these regions represent over segmentations of the individual cells; we only know that each region will contain a bacteria cell b. To segment individual cells, an additional step is necessary. We then apply ad hoc nonlinear filters to remove minor segmentation artifacts. In a second step, we then identified the contour of each single cell b by applying the Mumford–Shah algorithm [104] within the image segment containing a cell b. Since this procedure is quite time-consuming for large images, we have implemented it to produce a training set of delineated individual cells to train a CNN for image segmentation. After automatic training, this CNN substantially reduces the runtime of the cell segmentation/delineation procedure. We show the resulting segmentations in Figure 6. We provide additional information regarding our approach for the segmentation of individual bacteria cells in the appendix (see Appendix D).

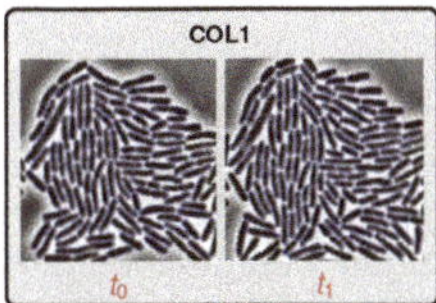

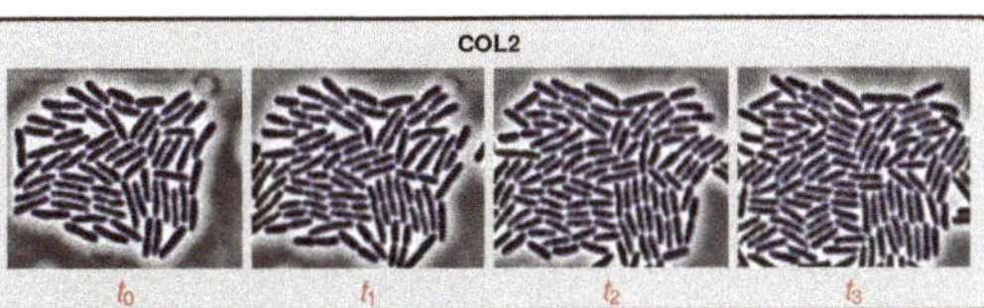

Figure 6. Segmentation results for experimental recordings of live cell colonies. We show two short image sequences extracts COL1 (**left**) and COL2 (**right**). The interframe duration is six minutes. The image sequence extract COL1 has only two successive image frames. The image sequence extract COL2 has four successive image frames. We are going to automatically compute four cell registrations, one for each pair of successive images in COL1 and COL2.

After each cell has been identified (i.e., segmented out) in each pair J, J_+ of successive images, we transform J, J_+ into binary images, where cells appear in white on a black background. For each resulting pair B, B_+ of successive sets of cells, we apply the parent–children pairing algorithm outlined in Section 3.3 to identify all the short lineages. For the two successive images in COL1, the discovered short lineages are shown in Figure 7 (left pair of images). Here, color designates the cell triplet algorithmically identified: parent cell in image J and its two children in image J_+. We then remove each identified *"parent"* from B and its two children from B_+. This yields the reduced cell sets $redB$ and $redB_+$. We can then apply our tracking algorithm (see Section 3.7) dedicated to situations where cells do not divide during the interframe.

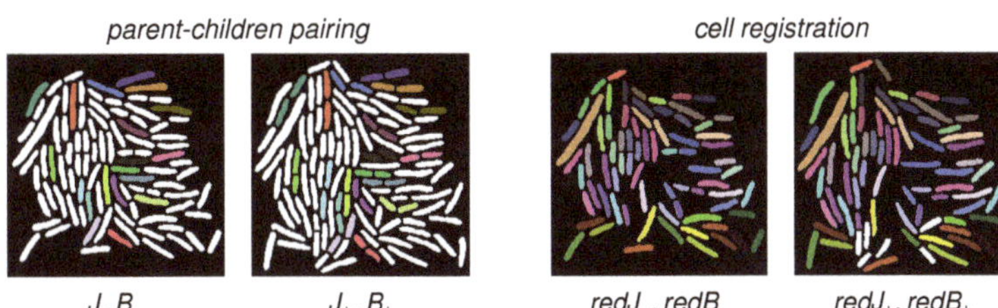

Figure 7. Cell tracking results for the pair COL1 of successive images J, J_+ shown in Figure 6. The interframe duration is six minutes. (**Left**): Results for parent–children pairing on COL1. Automatically detected parent–children triplets are displayed in the same color. (**Right**): Computed registration. The removal of the automatically detected parent–children triplets (see **left column**) generates the reduced cell sets $redB$ and $redB_+$. Automatic registration of $redB$ and $redB_+$ is again displayed via identical color for the registered cell pairs (b, b_+). Mismatches are mostly due to previous errors in parent–children pairing (see Figure 8 for a more detailed assessment).

For image sequences of live cell colonies, we had to re-calibrate most of our weight parameters. The weight parameters used for these image sequences are summarized in Table 4.

The BM temperature scheme was $Temp(t) = 2000\,(0.995)^t$, with the number of epochs capped at 5000. We illustrate our COL1 automatic registration results in Figure 7 (right pair of images). Here, if cell $b \in redB$ has been automatically registered onto cell $b_+ \in redB_+$, b, b_+ share the same color. The cells colored in white in $redB_+$ are cells which the registration algorithm did not succeed in matching to some cell in $redB$. These errors can essentially be attributed to errors in the parent–children pairing step. By visual inspection, we have determined that there are 14 true parent–children triplets in the successive images of COL1. Our parent–children pairing algorithm did correctly identify 11 of these 14 triplets. To check further the performance of our registration algorithm on live images, we also report automatic registration results for *"manually prepared"* true versions of $redB$ and $redB_+$, obtained by removing *"manually"* the true parent–children triplets determined by visual inspection. For the short image sequence COL2, results are displayed in Figure 8.

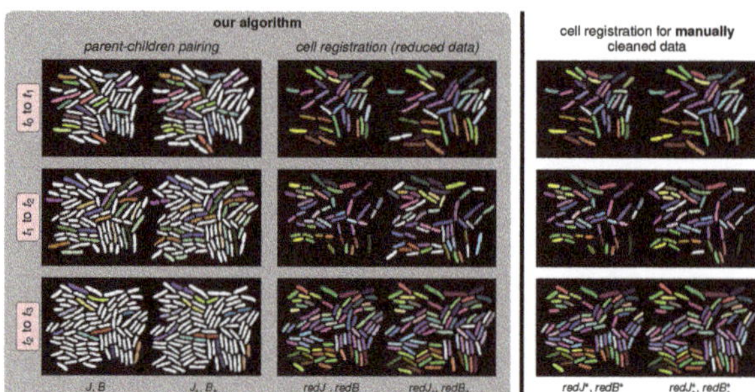

Figure 8. Cell tracking results for the short image sequence COL2 in Figure 6. The interframe duration for COL2 is six minutes. COL2 involves four successive images $J(t_i), i = 0,1,2,3$. In our figure, each one of the three rows displays the automatic cell registration results between images $J(t_i)$ and $J(t_{i+1})$ for $i = 0,1,2$. We report the accuracies of parent–children pairing and of the registration in Table 5. (**Left column**): Results for parent–children pairing. Each parent–children triplet is identified by the same color for each parent cell and its two children. (**Middle column**): Display of the automatically computed registration after removing the parent–children triplets already identified in order to generate two reduced sets *redB* and $redB_+$ of cells. Again, the same color is used for each pair of automatically registered cells. The white cells in $redB_+$ are cells which could not be registered to some cell in *redB*. (**Right column**): To differentiate between errors induced during automatic identification of and errors generated by automatic registration between *redB* and $redB_+$, we manually removed all *"true"* parent–children triplets and then applied our registration algorithm to this *"cleaned"* (reduced) cell sets $redB^*$ and $redB^*_+$.

Table 4. *Cost function weights* for parent–children pairing in the COL1 images displayed in Figure 6.

Weights	λ_{cen}	λ_{siz}	λ_{ang}	λ_{gap}	λ_{dev}	λ_{rat}	λ_{rank}	λ_{over}
Value	3	7	100	0.8	4	0.01	0.01	600

The display setup is the same: The left column shows the results of automatic parent–children pairing. The middle column illustrates the computed registration after automatic removal of the computer identified parent–children triplets. The third column displays the computed registration after *"manually"* removing the true parent–children triplets determined by visual inspection. Note that the overall matching accuracy can be improved if we reduce errors in the parent–children pairing. We report quantitative accuracies in Table 5. For parent–children pairing, accuracy ranges between 70% and 78%. For pure registration after correct parent–children pairing, accuracy ranges between 90% and 100%.

Table 5. Cell tracking accuracy for the short image sequence COL2 in Figure 6 with an interframe of six minutes. We report the ratio of correctly predicted cell matches over the total number of true cell matches and the associated percentages. The accuracy results quantify four distinct percentages of correct detections (i) for parent cells in image J, (ii) for children cells in image J_+, (iii) for parent–children triplets, and (iv) for registered pairs of cells $(b, b_+) \in redB \times redB_+$.

Task	Accuracy					
		$\{t_0, t_1\}$		$\{t_1, t_2\}$		$\{t_2, t_3\}$
correctly detected parents	15/19	79%	20/21	95%	7/10	70%
correctly detected children	35/38	92%	32/42	76%	14/20	70%
correct parent–children triplets	15/19	78%	16/21	76%	7/10	70%
correctly registered cell pairs	36/36	100%	44/49	90%	76/80	95%

5. Conclusions and Future Work

We have developed a methodology for automatic cell tracking in recordings of dense bacterial colonies growing in a mono-layer. We have also validated our approach using synthetic data from agent based simulations, as well as experimental recordings of *E. coli* colonies growing in microfluidic traps. Our next goal is to streamline our implementation for systematic cell registration on experimentally acquired recordings of such cell colonies, to enable automated quantitative analysis and modeling of cell population dynamics and lineages.

There are a number of challenges for our cell tracking algorithm: Inherent imaging artifacts such as noise or intensity drifts, cell overlaps, similarity of cell shape characteristics across the population, tight packing of cells, somewhat large interframe times, cell growth combined with cell motion, and cell divisions represent just a few of these challenges. Overall, the cell tracking problem has combinatorial complexity, and for large frames is beyond the concrete patience of human experts. We tackle these challenges by developing a two-stage algorithm that first identifies parent–children triplets and subsequently computes cell registration from one frame to the next, after reducing the two original cell sets by automatic removal of the identified parent–children triplets. Our algorithms specify innovative cost functions dedicated to these registration challenges. These cost functions have combinatorial complexity. To discover good registrations, we minimize these cost functions numerically by intensive stochastic simulations of specifically structured BMs. We have validated the potential of our approach by reporting promising results obtained on long synthetic image sequences of simulated cell colonies (which naturally provide a ground truth for cell registration from one frame to the next). We have also successfully tested our algorithms on experimental recordings of live bacterial colonies.

The choice of adequate cost functions to drive each major cost optimization step in our multi-step cell tracking algorithms is essential for obtaining good tracking. Selecting the proper formulation had a strong impact on actual tracking accuracy. Our cost functions are fundamentally nonlinear, which entails additional complications. We introduced a set of meta-parameters for each cost function, and proposed an original learning algorithm to automatically identify good ranges for these meta-parameters.

Our BMs are focused on stochastic minimization of dedicated cost functions. An interesting feature of BMs we will explore in future work is the simplicity of their natural massive parallelization for fast stochastic minimization [90]. This allows us to mitigate the slow convergence typically observed for Gibbs samplers on discrete state spaces with high cardinality. Parallelized BMs implement a form of massively parallel simulated annealing. Sequential simulated annealing has been explored by physicists [105–108] seeking to minimize spin–glasses energies. For these clique-based energies, reaching global minima requires unfeasible CPU times, and much faster parallel simulated annealing yields only good local minima, via a sophisticated but still greedy stochastic search. Parallel stochasticity favors ending in rather stable local minima, which in turn enforces low sensitivity to small changes in energy parameters. Robustness to small changes in the coefficients of our cost functions is a desirable feature, since our algorithmic calibration of cost coefficients focuses on computing good ranges for these meta-parameters. We do not aim to seek global minima, generally a very elusive search because computing speed and scalability are important features in our problem. Recall the established results of Huber [109] showing that optimal estimators of the mean for a Gaussian distribution lose efficiency very quickly when the Gaussian data are slightly perturbed.

In future work, we will further improve the stability and accuracy of our cell registration algorithms by exploring natural modifications of our cost functions. In the present work, we have not yet explicitly considered the case of cells vanishing between successive frames. This is a critical issue that can occur due to cells exiting or entering the field of view as well as due to errors in cell segmentation. The problem is somewhat controlled and/or mitigated in our experimental setup, where we expect cells to enter or vanish close to a precisely positioned trap edge and/or near frame boundaries. Since we intend to track lineages, each frame-to-frame error of this type may be problematic, and it will be instrumental for our future work to address these issues.

Linking parents to children involves an optimization distinct from the final optimization of frame-to-frame registrations. This did reduce computing time without reducing the quality for our benchmark results. However, in future work, one could attempt to iterate this sequence of two optimizations in order to reach a better minimum.

We note that our algorithm does work for experimental setups in which the frame rate of the video recordings is not fixed. This will require an adaptive parameter selection that depends on the frame rate. This can be implemented based on a trivial rescaling procedure. However, note that, for larger interframe times, more errors will impact tracking results. Indeed, large interframe durations intensify fluctuations in key parameters of cell dynamics, and increase the range of cell displacement, imposing searches in larger cell neighborhoods for cell pairing, as well as increased combinatorial complexity.

We have considered synthetic data to evaluate the performance of our method. One clear practical issue is that some of the parameters of our tracking algorithms may change when applied to laboratory image sequences acquired from colonies of different cells, with various image acquisition setups. One can design a computational framework to automatically fit the parameters of the simulation model to the imaging data acquired on specific live cell colonies, using specific camera hardware and setup. In future work, we will attempt to implement this type of fitting for our simulation model, before launching intensive model simulations to calibrate the parameters of our new tracking algorithms. We have not yet removed physical scales in the implementation of our tracking algorithm. Implementing such a non-dimensionalization will allow us to reduce the sensitivity of our methodology with respect to new datasets.

Identification of full lineages is an interesting concrete goal for cell tracking. Evaluating the accuracy of lineage identification on real cell colonies is quite challenging since it requires inheritable biological tagging of cells. This is probably feasible for populations mixing two or three cell types, but not for individualized tagging in populations of moderate size. However even partial tagging of sub-populations would provide some control on lineage identification accuracies.

Author Contributions: Conceptualization, R.A., M.R.B., K.J. and A.M.; methodology, R.A., A.M., S.S.; software, S.S. and J.J.W.; validation, S.S. and J.J.W.; formal analysis, R.A. and A.M.; investigation, R.A., M.R.B., K.J., A.M. and S.S.; resources, R.A., M.R.B., K.J. and A.M.; data curation, R.N.A. and M.R.B.; writing—original draft preparation, R.N.A., R.A., M.R.B., K.J., A.M., S.S. and J.J.W.; writing—review and editing, R.N.A., R.A., M.R.B., K.J., A.M., S.S. and J.J.W.; visualization, S.S., J.J.W. and A.M.; supervision, R.A. and A.M.; project administration, R.A., M.R.B., K.J. and A.M.; funding acquisition, R.A., M.R.B., K.J. and A.M. All authors have read and agreed to the published version of the manuscript.

Funding: This research was partly supported by the National Science Foundation (NSF) through the grants GRFP 1842494 (R.N.A.), DMS-1854853 (R.A. and A.M.), DMS-2009923 (R.A. and A.M.), 1662305 (K.J.), MCB-1936770 (K.J.), and DMS-2012825 (A.M.); the joint NSF-National Institutes of General Medical Sciences Mathematical Biology Program grant DMS-1662290 (M.R.B.); and the Welch Foundation grant C-1729 (M.R.B.). Any opinions, findings, and conclusions or recommendations expressed herein are those of the authors and do not necessarily reflect the views of the NSF or the Welch Foundation.

Acknowledgments: This work was completed in part with resources provided by the Research Computing Data Core at the University of Houston.

Conflicts of Interest: The authors declare no conflict of interest.

Appendix A. Stochastic Dynamics of BMs

Notations and terminology refer to Section 3.4. Consider a BM network of N stochastic neurons U_j, with finite configuration set $CONF = W(1) \times \ldots \times W(N)$. At time t, let $Z_j(t) \in W(j)$ be the random state of neuron U_j, and the BM configuration $Z(t) \in CONF$

is then $Z(t) = \{Z_1(t), \ldots, Z_N(t)\}$. Fix as in Section 3.4 a sequence *Temp*(t) of virtual temperatures slowly decreasing to 0 for large t.

There are two main options to implement the Markov chain dynamics $Z(t) \to Z(t+1)$ (see [95]).

Appendix A.1. Asynchronous BM Dynamics

Generate a long random sequence of sites $m(t) \in S = \{1, \ldots, N\}$, for instance by concatenating successive random permutations of the set S. At time t, the only neuron that may modify its current state is $U_{m(t)}$. For brevity, write $M = m(t)$. The neuron U_M will compute its new random state $Z_M(t+1) \in W(M)$ by the following *updating procedure*: (*i*) For each y in $W(M)$, define a new configuration $Y \in CONF$ by $Y_M(t) = y$, and $Y_j(t) = Z_j(t)$ for all $j \neq M$. Let $\Delta(y) = E(Y) - E(Z(t))$ be the corresponding BM energy change. (*ii*) In the finite set $W(M)$, select any z such that $\Delta(z) = \min_{y \in W(M)} \Delta(y)$, and set $D = \max\{0, \Delta(z)\}$. (*iii*) Compute the probability $p = \exp(-D/Temp(t))$. (*iv*) The new random state $Z_M(t+1)$ of neuron U_M will be equal to z with probability p and equal to the current state $Z_M(t)$ with probability $1-p$. (*v*) For all $j \neq M$, the new state $Z_j(t+1)$ of neuron U_j remains equal to its current state $U_j(t)$.

Appendix A.2. Synchronous BM Dynamics

Fix a *synchrony* parameter $0 < \alpha < 1$, usually around 50%. At each time t, *all* neurons U_j synchronously, but independently compute their own random *binary tag* $tag_j(t)$, equal to 1 with probability α, and to 0 with probability $(1-\alpha)$. Let $SYN(t)$ be the set of all neurons. All the neurons U_j such that $tag_j(t) = 1$ then synchronously and independently compute their new random states $Z_j(t+1) \in W(j)$ by applying the updating procedure given above. In addition, for all j such that $tag_j(t) = 0$, the new state $Z_j(t+1)$ of Uj remains equal to $Z_j(t)$.

Appendix A.3. Comparing Asynchronous and Synchronous BM Dynamics

As t becomes large, and for temperatures *Temp*(t) slowly decreasing to 0, both BM dynamics generate with high probability configurations $Z(t)$ which provide deep local minima $E(Z(t))$ of the BM energy function. The asynchronous dynamics can be fairly slow. However, the synchronous dynamics are much faster since they emulate efficient forms of *parallelel simulated annealing* (see [90,110]) and are directly implementable on GPUs.

Appendix B. Computer Hardware

The computations were carried out on a dedicated server at the Department of Mathematics of the University of Houston. The hardware specifications are 64 Intel(R) Xeon(R) Gold 6142 CPU cores at 2.60 GHz with 128 GB of memory.

Appendix C. Parameters for Simulation Software

Our tracking module is a collection of `python` functions and has been released to the public at https://github.com/scopagroup/BacTrak (accessed on 15 December 2021). We refer to [12,81] for a detailed description of this mathematical model and its implementation. The code for generating the synthetic data has been released at https://github.com/jwinkle/eQ (accessed on 15 December 2021). We note that detailed installation instructions for the software can be found on this page. The parameters for this agent-based simulation software are as follows: Cells were modeled as 2D spherocylinders of constant, 1 µm width. The computational framework takes into account mechanical constraints that can impact cell growth and influence other aspects of cell behavior. The growth rate of the cells is exponential and is controlled by the doubling time. The time until cells double is set to 20 min (default setting; resulting in a growth rate of $g.rate = 1.05$). The cells have a length of approximately 2 µm after division and 4 µm right before division (minimum division length of 4 µm; subject to some random perturbation). In our data set of simulated videos, there is no "trap wall" (as opposed to the simulations carried out in [12,81]). The "trap" encompassing all cells on a given frame has a size of 30 μm× 30 μm subdivided into 400 × 400 pixels of size 0.075 μm × 0.075 μm. The size of the resulting binary image used in our tracking algorithm is 600 × 600 pixels. (We add a boundary of 100 pixels on each side). Bacteria are moving, growing and dividing within the trap. However, at this stage of our study, we consider only video segments where no cell disappears and where cells do not enter the trap from outside so that the trap is a confined environment. Cells move only due to soft shocks' interactions with other neighboring cells. The time interval between any two successive image frames ranges from one minute to six minutes (see Table 1). All other simulation parameters remain unchanged; i.e., we use the default parameters specified in the simulation software.

Appendix D. Cell Segmentation

In the next couple of sections, we outline the framework we have developed to segment individual cells from real world laboratory imaging data. In a first step, we consider traditional segmentation algorithms—a watershed algorithm [103,111,112] in combination with a variational contour based model—to generate a sufficiently large dataset to train a neuronal network. The actual segmentations on real data can subsequently be carried out efficiently using segmentation predictions generated by the trained neuronal network. Note that the proposed segmentation algorithm is only included for completeness. We do not view this as a major contribution of the present work.

Appendix D.1. Watershed Algorithm

We consider a *watershed algorithm* based on immersion that compares high intensity values to local intensity minima for cell segmentation [103,111,112].

We consider Matlab's implementation of the watershed algorithm in the present work. This version of the watershed algorithm is unseeded and yields n regions $R = \{R_1, R_2, \ldots, R_n\}$. To identify these regions, we perform a statistical analysis of each image histogram to compute adaptive rough thresholds for interiors and exterior of cells. This leads to watershed results which identify each cell by a segment slightly larger than the cell itself. The very small percentage of oversegmented cells is automatically detected by cell length and width computations through PCA analysis of each cell shape viewed as a cloud of planar points. Since our segments are slightly too wide, we reduce each segment to the exact outer cell contour by applying a Mumford–Shah algorithm to each segment computed by the watershed algorithm. In an ideal case, after applying the watershed algorithm, each individual bacteria cell b_i, $i = 1, \ldots, n$, will be located in a single region $R_i \subset \mathbb{R}^2$. However, we observed several segmentation errors after applying the watershed algorithm to the considered data. A common error is that a line segment that defines the boundary of a region crosses through a cell. That is, two regions contain parts of one bacterium cell. In what follows, we devise strategies to correct these errors. For this processing step, we have normalized the intensities of the data to $[0, 1]$.

Appendix D.2. Segmentation Errors: Correction Steps

We define the boundary segment $B_{i,j}$ as a non-empty intersection of two region's boundaries, i.e., $B_{i,j} = \partial S_i \cap \partial S_j$. Moreover, we denote the area of a region R_i as $\text{area}(R_i)$. We know that the interior of a bacteria cell b_i has a lower intensity than the exterior region of a cell. More precisely, the interior of a cell tends to have intensity values of zero, whereas the exterior of a cell (i.e., the background) tends to have an intensity that is close or equal to one. For this reason, we define a function for the intensity of the boundary. To remove outliers, we consider the average intensity value of the pixels located along a boundary segment. We denote this mean intensity value along a boundary $B_{i,j}$ by $\text{mint}(B_{i,j})$ and the average intensity of a region R_i by $\text{mint}(R_i)$. One difficulty is that we cannot assume that the intensity of the pixels on the interior of each cell corresponds to the same value (i.e., there exist intensity and contrast drifts depending on location). We hypothesize that, if

$\text{mint}(B_{i,j})$ of a boundary segment is close to the average intensity of the regions on both sides of the boundary segment $B_{i,j}$, this boundary segment does not separate two bacteria cells; it is erroneous. Conversely, if the difference between the mean intensity along a boundary segment and the mean intensity of the interior regions it separates is high, we consider that the boundary segment represents a good segmentation (i.e., represents a segment that does separate two cells). To quantify this notion, we define the height of a boundary segment as $H_{i,j} = \text{mint}(B_{i,j}) - (\text{mint}(R_i) + \text{mint}(R_j))/2$.

In Table A1, we report some statistics associated with the quantities of interest introduced above. There are several key observations we can draw from this table which confirm our qualitative (i.e., visual) assessment of the segmentation results. Most notably, we can observe that there seem to exist outliers in terms of cell size. Moreover, we can observe that, in some cases, we obtain a height of the boundary segment that is negative, and by that nonsensical. These observations allow us to develop some heuristic rules to remove erroneous segmentations.

Table A1. Statistics of some quantities of interest related to the intensity of boundary segments and regions. These quantities allow us to define heuristics to identify erroneous segmentations computed by the watershed algorithm. We state the characteristic and report the minimum, maximum 5% quantile, mean, and standard deviation for the reported quantities of interest.

Characteristic	5% Quantile	Min	Max	Mean
Watershed area	56.00	43.00	984.00	211.00 ± 138.00
Mean intensity of area	0.34	0.00	0.57	0.41 ± 0.06
Mean intensity of boundary segment	0.46	0.30	0.99	0.74 ± 0.14
Height of boundary segment	0.05	−0.09	0.62	0.33 ± 0.14

We introduce the following post-processing steps: (i) We connect small regions to their neighbors (i.e., regions that are too small in area to realistically contain any cells). We select the threshold for the area to be 65. This threshold is selected in accordance with the scale of the image and the expected size of bacteria cells observed in the image data. We merge each small region with one of its neighboring regions by removing the segment that separates the two. To select an appropriate region for merging, we choose the region that gave the lowest height $H_{i,j}$ from all available candidate regions that share the same boundary segment. (ii) We remove all boundary segments B_{ij} with a height H_{ij} that is below the 5% quantile of all heights. (iii) We remove all incomplete regions from our segmentation. We define a region as incomplete, if the region or the associated boundary segments touch an edge of the image. This step is necessary since we cannot guarantee that the regions close to the boundary contain an entire cell or only parts of a cell. Consequently, we decided to remove them to prevent any issues with our post-analysis.

Appendix D.3. Cell Boundary Detection

The next step is to identify the boundaries of individual cells contained within a subregion defined by the watershed algorithm. To identify the boundaries of the cells (and by that segment the individual cells), we use the Mumford–Shah algorithm [104]. Notice that we can execute the Mumford–Shah algorithm for each region R_i separately making this an embarrassingly parallelizable problem. Denote the cell in each R_i region by b_i. We divide each of these regions into three different zones. The first zone is the interior of the cell b_i denoted by $\text{in}(b_i)$. The second zone is exterior of the cell (i.e., the background) contained in the region and denoted by $\text{out}(b_i)$. The third zone is the boundary of the cell b_i, denoted by ∂b_i. The Mumford–Shah algorithm represents a variational approach that allows us to segment cartoon like images. Mathematically speaking, we model information contained in each region R_i as piecewise-smooth functions. In our model, the associated regions we seek to identify are given by the zones defined above—the interior and the exterior of the cell b_i. Let $u_{\text{int}}(b_i)$ denote the mean intensity for the interior of the cell b_i and

$u_{\text{ext}}(b_i)$ denote the mean intensity for the exterior of the cell b_i. With this definition, we obtain the cost functional

$$\text{cost}_{\text{MS}}(\text{int}(b_i), \text{ext}(b_i)) = \sum_{x \in \text{ext}(b_i)} (u(x) - u_{\text{ext}}(b_i))^2) + \sum_{x \in \text{int}(b_i)} (u(x) - u_{\text{int}}(b_i))^2 + \nu \, \text{bl}(b_i),$$

where the first two terms measure the discrepancy between the piecewise smooth function u_{ext} and u_{int} and the image intensities u and the third term is a penalty that measures the length of the boundary of a particular cell b_i with parameter $\nu > 0$. Notice that our formulation slightly deviates from the traditional definition of the Mumford–Shah cost functional; we drop the penalty for the smoothness of the function u. The minimizer of the cost function cost_{MS} defined above provides the sought after segmentation: the boundary, interior, and exterior of a cell. We have implemented the minimization of the cost function formula for each cell separately.

Appendix D.4. Convolutional Neural Networks (CNNs)

Next, we introduce our actual method for cell segmentation that can be efficiently applied to a large dataset (as opposed to the prototype method described above to generate the underlying training data). The biggest issue with the methodology outlined above is that our prototype implementation is computationally costly. While we envision that an improved implementation as well as the use of parallel computing can significantly reduce the time to solution, we decided not to further pursue a reduction in runtime but extend our methodology by taking advantage of existing machine learning algorithms. Replacing the approach outlined above by CNNs allowed us to reduce the runtime by factor of 60 to less than 3 min, without any significant loss in accuracy.

Training and Testing Data. In the absence of any ground truth data set for the classification of rod-shape bacteria cells from movies of cell populations, we consider the output of the Mumford–Shah algorithm introduced above as ground truth classification for training and testing our machine learning methodology. Above, we introduced three different zones: The interior $\text{in}(b_i)$, the exterior $\text{ext}(b_i)$, and the boundary ∂b_i of a cell b_i. We reduce these three regions to two zones—the interior and exterior of a cell b_i. We assign pixels that belong to $\text{int}(b_i)$ the label 0 and pixels that belong to $\text{ext}(b_i)$ and ∂b_i the label of 1. For an image of size 200×200, we obtain 40,000 binary labels. We limit the training of the CNN to a subregion of size 200×200 in the center of each preprocessed image to avoid issues associated with mislabeled training data of cells located at the boundary of our data. We consider X as the set of features and Y as the set of labels. We want to assign to each pixel a label of either 0 or 1. For pixel p, we define X_p to be a 7×7 square window with center p located in the original image. The corresponding label Y_p is denoted by $C(p)$, which corresponds to the class of the pixel p in the binarized image.

CNN Algorithm. The considered CNN algorithm consists of two parts, (i) the convolutional auto-encoder and (ii) a fully connected multilayer perceptron (MLP). The input for the auto-encoder is a window of 7×7 pixels. In the first layer of the encoder, we have a $5 \times 5 \times 4$ convolution layer Conv1 with 3×3 kernel. We feed Conv1 to a max-pooling layer MPool2 with one stride and pooling window 2×2. The output of MPool2 is the input of a $3 \times 3 \times 8$ convolution layer Conv3. For decoding, we have almost the same structure in reverse order: We feed Conv3 to a $5 \times 5 \times 4$ deconvolution with 3×3 kernel. Subsequently, we feed the output of this layer to a $7 \times 7 \times 1$ deconvolution with 3×3 kernel. The decoder's output is a window of 7×7 pixels. We compare this output with the input window (since it is an auto-encoder, features and labels are the same) by using the mean square error as a cost function. We train the auto-encoder for all training sets using a mini-batch gradient descent. When the training is finished, we freeze the weights for Conv1 and Conv3.

After training the auto-encoder and freezing the weights, we feed X as the input to Conv1 and obtain the output of Conv3 denoted by $\hat{X}$. In the next step, we train an MLP with features $\hat{X}$ and labels Y. We flatten $\hat{X}$, which is a $3 \times 3 \times 8$ matrix to a vector of size

72×1, called FCL4. FCL4 is fully connected to the hidden layer HID5 with 10 nodes. We use ReLu as a nonlinear function for HID5. We connect HID5 to the output layer OUT6, which possess two nodes for the two classes 0 and 1. We use a softmax function to find two probabilistic outputs p_0 and $p_1 = 1 - p_0$ for related classes. We use maximum-entropy as a cost function. We train the MLP for training set of $(\hat{X}, Y)$ with mini-batch gradient descent.

We have trained the model with two images of size 200×200 pixels; the training set is 80,000 7×7 images. We train the model for 100 epochs. The accuracy of the model for the image is 93%. The confusion matrix is shown in Table A2. Based on this confusion matrix, we can observe that the proposed methodology can predict the pixels located in the interior of a cell quite well. However, we can also observe that there is a slightly lower accuracy for the pixels outside the cells. This can be probably explained by the fact that the data sets are tightly packed with cells so that we have available more observations of foreground pixels (interior of cells) than pixels that belong to the background.

Table A2. Confusion matrix for the CNN.

	0	1
0	0.97	0.03
1	0.11	0.89

References

1. Butts-Wilmsmeyer, C.J.; Rapp, S.; Guthrie, B. The technological advancements that enabled the age of big data in the environmental sciences: A history and future directions. *Curr. Opin. Environ. Sci. Health* **2020**, *18*, 63–69. [CrossRef]
2. Sivarajah, U.; Kamal, M.M.; Irani, Z.; Weerakkody, V. Critical analysis of Big Data challenges and analytical methods. *J. Bus. Res.* **2017**, *70*, 263–286. [CrossRef]
3. Balomenos, A.D.; Tsakanikas, P.; Aspridou, Z.; Tampakaki, A.P.; Koutsoumanis, K.P.; Manolakos, E.S. Image analysis driven single-cell analytics for systems microbiology. *BMC Syst. Biol.* **2017**, *11*, 1–21. [CrossRef]
4. Klein, J.; Leupold, S.; Biegler, I.; Biedendieck, R.; Münch, R.; Jahn, D. TLM-Tracker: Software for cell segmentation, tracking and lineage analysis in time-lapse microscopy movies. *Bioinformatics* **2012**, *28*, 2276–2277. [CrossRef]
5. Stylianidou, S.; Brennan, C.; Nissen, S.B.; Kuwada, N.J.; Wiggins, P.A. SuperSegger: Robust image segmentation, analysis and lineage tracking of bacterial cells. *Mol. Microbiol.* **2016**, *102*, 690–700. [CrossRef] [PubMed]
6. Bennett, M.R.; Hasty, J. Microfluidic devices for measuring gene network dynamics in single cells. *Nat. Rev. Genet.* **2009**, *10*, 628–638. [CrossRef]
7. Danino, T.; Mondragón-Palomino, O.; Tsimring, L.; Hasty, J. A synchronized quorum of genetic clocks. *Nature* **2010**, *463*, 326–330. Available online: http://xxx.lanl.gov/abs/15334406 (accessed on 15 December 2021). [CrossRef]
8. Mather, W.; Mondragon-Palomino, O.; Danino, T.; Hasty, J.; Tsimring, L.S. Streaming instability in growing cell populations. *Phys. Rev. Lett.* **2010**, *104*, 208101. [CrossRef] [PubMed]
9. El Najjar, N.; Van Teeseling, M.C.; Mayer, B.; Hermann, S.; Thanbichler, M.; Graumann, P.L. Bacterial cell growth is arrested by violet and blue, but not yellow light excitation during fluorescence microscopy. *BMC Mol. Cell Biol.* **2020**, *21*, 35. [CrossRef]
10. Icha, J.; Weber, M.; Waters, J.C.; Norden, C. Phototoxicity in live fluorescence microscopy, and how to avoid it. *BioEssays* **2017**, *39*, 1700003. [CrossRef] [PubMed]
11. Kim, J.K.; Chen, Y.; Hirning, A.J.; Alnahhas, R.N.; Josić, K.; Bennett, M.R. Long-range spatio-temporal coordination of gene expression in synthetic microbial consortia. *Nat. Chem. Biol.* **2019**, *15*, 1102–1109. [CrossRef]
12. Winkle, J.; Igoshin, O.A.; Bennett, M.R.; Josic, K.; Ott, W. Modeling mechanical interactions in growing populations of rod-shaped bacteria. *Phys. Biol.* **2017**, *14*, 055001. [CrossRef] [PubMed]
13. Carpenter, A.E.; Jones, T.R.; Lamprecht, M.R.; Clarke, C.; Kang, I.H.; Friman, O.; Guertin, D.A.; Chang, J.H.; Lindquist, R.A.; Moffat, J.; et al. CellProfiler: Image analysis software for identifying and quantifying cell phenotypes. *Genome Biol.* **2006**, *7*, R100. [CrossRef] [PubMed]
14. Kamentsky, L.; Jones, T.R.; Fraser, A.; Bray, M.; Logan, D.; Madden, K.; Ljosa, V.; Rueden, C.; Harris, G.B.; Eliceiri, K.; et al. Improved structure, function, and compatibility for CellProfiler: modular high-throughput image analysis software. *Bioinformatics* **2011**, *27*, 1179–1180. [CrossRef]
15. McQuin, C.; Goodman, A.; Chernyshev, V.; Kamentsky, L.; Cimini, B.A.; Karhohs, K.W.; Doan, M.; Ding, L.; Rafelski, S.M.; Thirstrup, D.; et al. CellProfiler 3.0: Next,-generation image processing for biology. *PLoS Biol.* **2018**, *16*, e2005970. [CrossRef]
16. Alnahhas, R.N.; Sadeghpour, M.; Chen, Y.; Frey, A.A.; Ott, W.; Josić, K.; Bennett, M.R. Majority sensing in synthetic microbial consortia. *Nat. Commun.* **2020**, *11*, 1–10. [CrossRef] [PubMed]
17. Locke, J.C.W.; Elowitz, M.B. Using movies to analyse gene circuit dynamics in single cells. *Nat. Rev. Microbiol.* **2009**, *7*, 383–392. [CrossRef]

18. Alnahhas, R.N.; Winkle, J.J.; Hirning, A.J.; Karamched, B.; Ott, W.; Josić, K.; Bennett, M.R. Spatiotemporal Dynamics of Synthetic Microbial Consortia in Microfluidic Devices. *ACS Synth. Biol.* **2019**, *8*, 2051–2058. [CrossRef] [PubMed]
19. Hand, A.J.; Sun, T.; Barber, D.C.; Hose, D.R.; MacNeil, S. Automated tracking of migrating cells in phase-contrast video microscopy sequences using image registration. *J. Microsc.* **2009**, *234*, 62–79. [CrossRef] [PubMed]
20. Ulman, V.; Maška, M.; Magnusson, K.E.G.; Ronneberger, O.; Haubold, C.; Harder, N.; Matula, P.; Matula, P.; Svoboda, D.; Radojevic, M.; et al. An objective comparison of cell-tracking algorithms. *Nat. Methods* **2017**, *14*, 1141–1152. [CrossRef] [PubMed]
21. Marvasti-Zadeh, S.M.; Cheng, L.; Ghanei-Yakhdan, H.; Kasaei, S. Deep learning for visual tracking: A comprehensive survey. *IEEE Trans. Intell. Transp. Syst.* **2021**, 1–26. [CrossRef]
22. Yilmaz, A.; Javed, O.; Shah, M. Object tracking: A survey. *ACM Comput. Surv. (CSUR)* **2006**, *38*, 13-es. [CrossRef]
23. Lucas, B.D.; Kanade, T. An iterative image registration technique with an application to stereo vision. In Proceedings of the International Conference on Artificial Intelligence, Vancouver, BC, Canada, 24–28 August 1981; pp. 674–679.
24. Mang, A.; Biros, G. An inexact Newton–Krylov algorithm for constrained diffeomorphic image registration. *SIAM J. Imaging Sci.* **2015**, *8*, 1030–1069. [CrossRef] [PubMed]
25. Mang, A.; Ruthotto, L. A Lagrangian Gauss–Newton–Krylov solver for mass- and intensity-preserving diffeomorphic image registration. *SIAM J. Sci. Comput.* **2017**, *39*, B860–B885. [CrossRef] [PubMed]
26. Mang, A.; Gholami, A.; Davatzikos, C.; Biros, G. CLAIRE: A distributed-memory solver for constrained large deformation diffeomorphic image registration. *SIAM J. Sci. Comput.* **2019**, *41*, C548–C584. [PubMed]
27. Borzi, A.; Ito, K.; Kunisch, K. An optimal control approach to optical flow computation. *Int. J. Numer. Methods Fluids* **2002**, *40*, 231–240. [CrossRef]
28. Horn, B.K.P.; Shunck, B.G. Determining optical flow. *Artif. Intell.* **1981**, *17*, 185–203. [CrossRef]
29. Delpiano, J.; Jara, J.; Scheer, J.; Ramírez, O.A.; Ruiz-del Solar, J.; Härtel, S. Performance of optical flow techniques for motion analysis of fluorescent point signals in confocal microscopy. *Mach. Vis. Appl.* **2012**, *23*, 675–689.
30. Madrigal, F.; Hayet, J.B.; Rivera, M. Motion priors for multiple target visual tracking. *Mach. Vis. Appl.* **2015**, *26*, 141–160. [CrossRef]
31. Banerjee, D.S.; Stephenson, G.; Das, S.G. Segmentation and analysis of mother machine data: SAM. *bioRxiv* **2020**. [CrossRef]
32. Jug, F.; Pietzsch, T.; Kainmüller, D.; Funke, J.; Kaiser, M.; van Nimwegen, E.; Rother, C.; Myers, G. Optimal Joint Segmentation and Tracking of Escherichia Coli in the Mother Machine. In *Bayesian and Graphical Models for Biomedical Imaging*; Springer: Cham, Switzerland, 2014; Volume LNCS 8677, pp. 25–36.
33. Lugagne, J.B.; Lin, H.; Dunlop, M.J. DeLTA: Automated cell segmentation, tracking, and lineage reconstruction using deep learning. *PLoS Comput. Biol.* **2020**, *16*, e1007673. [CrossRef] [PubMed]
34. Ollion, J.; Elez, M.; Robert, L. High-throughput detection and tracking of cells and intracellular spots in mother machine experiments. *Nat. Protoc.* **2019**, *14*, 3144–3161. [CrossRef]
35. Sauls, J.T.; Schroeder, J.W.; Brown, S.D.; Le Treut, G.; Si, F.; Li, D.; Wang, J.D.; Jun, S. Mother machine image analysis with MM3. *bioRxiv* **2019**, 810036. [CrossRef]
36. Smith, A.; Metz, J.; Pagliara, S. MMHelper: An automated framework for the analysis of microscopy images acquired with the mother machine. *Sci. Rep.* **2019**, *9*, 10123.
37. Arbelle, A.; Reyes, J.; Chen, J.Y.; Lahav, G.; Raviv, T.R. A probabilistic approach to joint cell tracking and segmentation in high-throughput microscopy videos. *Med. Image Anal.* **2018**, *47*, 140–152.
38. Okuma, K.; Taleghani, A.; De Freitas, N.; Little, J.J.; Lowe, D.G. A boosted particle filter: Multitarget detection and tracking. In Proceedings of the European Conference on Computer Vision, Prague, Czech Republic, 11–14 May 2004; Springer: Berlin/Heidelberg, Germany, 2004; pp. 28–39.
39. Smal, I.; Niessen, W.; Meijering, E. Bayesian tracking for fluorescence microscopic imaging. In Proceedings of the 3rd IEEE International Symposium on Biomedical Imaging: Nano to Macro, Arlington, WA, USA, 6–9 April 2006; pp. 550–553.
40. Kervrann, C.; Trubuil, A. Optimal level curves and global minimizers of cost functionals in image segmentation. *J. Math. Imaging Vis.* **2002**, *17*, 153–174. [CrossRef]
41. Li, K.; Miller, E.D.; Chen, M.; Kanade, T.; Weiss, L.E.; Campbell, P.G. Cell population tracking and lineage construction with spatiotemporal context. *Med. Image Anal.* **2008**, *12*, 546–566. [CrossRef] [PubMed]
42. Wang, X.; He, W.; Metaxas, D.; Mathew, R.; White, E. Cell segmentation and tracking using texture-adaptive snakes. In Proceedings of the IEEE International Symposium on Biomedical Imaging: From Nano to Macro, Arlington, VA, USA, 12–15 ApriL 2007; pp. 101–104.
43. Yang, F.; Mackey, M.A.; Ianzini, F.; Gallardo, G.; Sonka, M. Cell segmentation, tracking, and mitosis detection using temporal context. In Proceedings of the International Conference on Medical Image Computing and Computer-Assisted Intervention, Palm Springs, CA, USA, 26–29 October 2005; pp. 302–309.
44. Sethuraman, V.; French, A.; Wells, D.; Kenobi, K.; Pridmore, T. Tissue-level segmentation and tracking of cells in growing plant roots. *Mach. Vis. Appl.* **2012**, *23*, 639–658. [CrossRef]
45. Balomenos, A.D.; Tsakanikas, P.; Manolakos, E.S. Tracking single-cells in overcrowded bacterial colonies. In Proceedings of the Annual International Conference of the IEEE Engineering in Medicine and Biology Society (EMBC), Milano, Italy, 25–29 August 2015; IEEE: Piscataway, NJ, USA, 2015; pp. 6473–6476.

46. Bise, R.; Yin, Z.; Kanade, T. Reliable cell tracking by global data association. In Proceedings of the IEEE International Symposium on Biomedical Imaging: From Nano to Macro, Chicago, IL, USA, 30 March–2 April 2011; pp. 1004–1010.
47. Bise, R.; Li, K.; Eom, S.; Kanade, T. Reliably tracking partially overlapping neural stem cells in DIC microscopy image sequences. In Proceedings of the International Conference on Medical Image Computing and Computer-Assisted Intervention Workshop, London, UK, 20–24 September 2009; pp. 67–77.
48. Kanade, T.; Yin, Z.; Bise, R.; Huh, S.; Eom, S.; Sandbothe, M.F.; Chen, M. Cell image analysis: Algorithms, system and applications. In Proceedings of the 2011 IEEE Workshop on Applications of Computer Vision (WACV), Kona, HI, USA, 5–7 January 2011; IEEE: Piscataway, NJ, USA, 2011; pp. 374–381.
49. Primet, M.; Demarez, A.; Taddei, F.; Lindner, A.; Moisan, L. Tracking of cells in a sequence of images using a low-dimensional image representation. In Proceedings of the IEEE International Symposium on Biomedical Imaging, Paris, France, 14–17 May 2008; pp. 995–998.
50. Ronneberger, O.; Fischer, P.; Brox, T. U-Net: Convolutional Networks for Biomedical Image Segmentation. In Proceedings of the Medical Image Computing and Computer Assisted Intervention, Munich, Germany, 5–9 October 2015; Volume LNCS 9351, pp. 234–241.
51. Su, H.; Yin, Z.; Huh, S.; Kanade, T. Cell segmentation in phase contrast microscopy images via semi-supervised classification over optics-related features. *Med. Image Anal.* **2013**, *17*, 746–765. [CrossRef]
52. Wang, Q.; Niemi, J.; Tan, C.M.; You, L.; West, M. Image segmentation and dynamic lineage analysis in single-cell fluorescence microscopy. *Cytom. Part A J. Int. Soc. Adv. Cytom.* **2010**, *77*, 101–110. [CrossRef] [PubMed]
53. Jiuqing, W.; Xu, C.; Xianhang, Z. Cell tracking via structured prediction and learning. *Mach. Vis. Appl.* **2017**, *28*, 859–874. [CrossRef]
54. Zhou, Z.; Wang, F.; Xi, W.; Chen, H.; Gao, P.; He, C. Joint multi-frame detection and segmentation for multi-cell tracking. In Proceedings of the International Conference on Image and Graphics, Beijing, China, 23–25 August 2019; Volume LNCS 11902, pp. 435–446.
55. Sixta, T.; Cao, J.; Seebach, J.; Schnittler, H.; Flach, B. Coupling cell detection and tracking by temporal feedback. *Mach. Vis. Appl.* **2020**, *31*, 1–18.
56. Hayashida, J.; Nishimura, K.; Bise, R. MPM: Joint representation of motion and position map for cell tracking. In Proceedings of the IEEE/CVF Conference on Computer Vision and Pattern Recognition, Seattle, WA, USA, 14–19 June 2020; pp. 3823–3832.
57. Payer, C.; Stern, D.; Neff, T.; Bishof, H.; Urschler, M. Instance segmentation and tracking with cosine embeddings and recurrent hourglass networks. In Proceedings of the Medical Image Computing and Computer Assisted Intervention, Granada, Spain, 16–20 September 2018; Volume LNCS 11071, pp. 3–11.
58. Payer, C.; Štern, D.; Feiner, M.; Bischof, H.; Urschler, M. Segmenting and tracking cell instances with cosine embeddings and recurrent hourglass networks. *Med. Image Anal.* **2019**, *57*, 106–119. [PubMed]
59. Vicar, T.; Balvan, J.; Jaros, J.; Jug, F.; Kolar, R.; Masarik, M.; Gumulec, J. Cell segmentation methods for label-free contrast microscopy: Review and comprehensive comparison. *BMC Bioinform.* **2019**, *20*, 1–25.
60. Al-Kofahi, Y.; Zaltsman, A.; Graves, R.; Marshall, W.; Rusu, M. A deep learning-based algorithm for 2D cell segmentation in microscopy images. *BMC Bioinform.* **2018**, *19*, 1–11.
61. Falk, T.; Mai, D.; Bensch, R.; Çiçek, Ö.; Abdulkadir, A.; Marrakchi, Y.; Böhm, A.; Deubner, J.; Jäckel, Z.; Seiwald, K.; et al. U-Net: Deep learning for cell counting, detection, and morphometry. *Nat. Methods* **2019**, *16*, 67–70. [CrossRef]
62. Lux, F.; Matula, P. DIC image segmentation of dense cell populations by combining deep learning and watershed. In Proceedings of the 2019 IEEE 16th International Symposium on Biomedical Imaging (ISBI 2019), Venice, Italy, 8–11 April 2019; pp. 236–239.
63. Moen, E.; Bannon, D.; Kudo, T.; Graf, W.; Covert, M.; Van Valen, D. Deep learning for cellular image analysis. *Nat. Methods* **2019**, *16*, 1233–1246. [PubMed]
64. Rempfler, M.; Stierle, V.; Ditzel, K.; Kumar, S.; Paulitschke, P.; Andres, B.; Menze, B.H. Tracing cell lineages in videos of lens-free microscopy. *Med. Image Anal.* **2018**, *48*, 147–161.
65. Stringer, C.; Wang, T.; Michaelos, M.; Pachitariu, M. Cellpose: A generalist algorithm for cellular segmentation. *Nat. Methods* **2021**, *18*, 100–106. [CrossRef]
66. Akram, S.U.; Kannala, J.; Eklund, L.; Heikkilä, J. Joint cell segmentation and tracking using cell proposals. In Proceedings of the IEEE 13th International Symposium on Biomedical Imaging (ISBI), Prague, Czech Republic, 13–16 April 2016; pp. 920–924.
67. Nishimura, K.; Hayashida, J.; Wang, C.; Bise, R. Weakly-Supervised Cell Tracking via Backward-and-Forward Propagation. In Proceedings of the European Conference on Computer Vision, Seoul, Korea, 27 October–2 November 2019; pp. 104–121.
68. Rempfler, M.; Kumar, S.; Stierle, V.; Paulitschke, P.; Andres, B.; Menze, B.H. Cell lineage tracing in lens-free microscopy videos. In Proceedings of the International Conference on Medical Image Computing and Computer-Assisted Intervention, Quebec City, QC, Canada, 11–13 September 2017; pp. 3–11.
69. Maska, M.; Ulman, V.; Svoboda, D.; Matula, P.; Matula, P.; Ederra, C.; Urbiola, A.; Espana, T.; Venkatesan, S.; Balak, D.M.W.; et al. A benchmark for comparison of cell tracking algorithms. *Bioinformatics* **2014**, *30*, 1609–1617.
70. Löffler, K.; Scherr, T.; Mikut, R. A graph-based cell tracking algorithm with few manually tunable parameters and automated segmentation error correction. *bioRxiv* **2021**, *16*, e0249257.
71. Vo, B.T.; Vo, B.N.; Cantoni, A. The cardinality balanced multi-target multi-Bernoulli filter and its implementations. *IEEE Trans. Signal Process.* **2008**, *57*, 409–423.

72. Pierskalla, W.P. The multidimensional assignment problem. *Oper. Res.* **1968**, *16*, 422–431. [CrossRef]
73. Gilbert, K.C.; Hofstra, R.B. Multidimensional assignment problems. *Decis. Sci.* **1988**, *19*, 306–321. [CrossRef]
74. Chakraborty, A.; Roy-Chowdhury, A.K. Context aware spatio-temporal cell tracking in densely packed multilayer tissues. *Med. Image Anal.* **2015**, *19*, 149–163. [CrossRef] [PubMed]
75. Liu, M.; Yadav, R.K.; Roy-Chowdhury, A.; Reddy, G.V. Automated tracking of stem cell lineages of Arabidopsis shoot apex using local graph matching. *Plant J.* **2010**, *62*, 135–147. [CrossRef]
76. Liu, M.; Chakraborty, A.; Singh, D.; Yadav, R.K.; Meenakshisundaram, G.; Reddy, G.V.; Roy-Chowdhury, A. Adaptive cell segmentation and tracking for volumetric confocal microscopy images of a developing plant meristem. *Mol. Plant* **2011**, *4*, 922–931. [CrossRef]
77. Liu, M.; Li, J.; Qian, W. A multi-seed dynamic local graph matching model for tracking of densely packed cells across unregistered microscopy image sequences. *Mach. Vis. Appl.* **2018**, *29*, 1237–1247. [CrossRef]
78. Vo, B.N.; Vo, B.T. A multi-scan labeled random finite set model for multi-object state estimation. *IEEE Trans. Signal Process.* **2019**, *67*, 4948–4963. [CrossRef]
79. Punchihewa, Y.G.; Vo, B.T.; Vo, B.N.; Kim, D.Y. Multiple object tracking in unknown backgrounds with labeled random finite sets. *IEEE Trans. Signal Process.* **2018**, *66*, 3040–3055. [CrossRef]
80. Kim, D.Y.; Vo, B.N.; Thian, A.; Choi, Y.S. A generalized labeled multi-Bernoulli tracker for time lapse cell migration. In Proceedings of the 2017 International Conference on Control, Automation and Information Sciences, Jeju, Korea, 18–21 October 2017; pp. 20–25.
81. Winkle, J.J.; Karamched, B.R.; Bennett, M.R.; Ott, W.; Josić, K. Emergent spatiotemporal population dynamics with cell-length control of synthetic microbial consortia. *PLoS Comput. Biol.* **2021**, *17*, e1009381.
82. Bise, R.; Sato, Y. Cell detection from redundant candidate regions under non-overlapping constraints. *IEEE Trans. Med Imaging* **2015**, *34*, 1417–1427. [CrossRef] [PubMed]
83. Matula, P.; Maška, M.; Sorokin, D.V.; Matula, P.; Ortiz-de Solórzano, C.; Kozubek, M. Cell tracking accuracy measurement based on comparison of acyclic oriented graphs. *PLoS ONE* **2015**, *10*, e0144959.
84. Agrawal, A.; Verschueren, R.; Diamond, S.; Boyd, S. A rewriting system for convex optimization problems. *J. Control Decis.* **2018**, *5*, 42–60. [CrossRef]
85. Diamond, S.; Boyd, S. CVXPY: A Python-embedded modeling language for convex optimization. *J. Mach. Learn. Res.* **2016**, *17*, 1–5.
86. Shen, X.; Diamond, S.; Gu, Y.; Boyd, S. Disciplined convex-concave programming. In Proceedings of the 2016 IEEE 55th Conference on Decision and Control (CDC), Las Vegas, NV, USA, 12–14 December 2016; pp. 1009–1014.
87. Stricker, J.; Cookson, S.; Bennett, M.R.; Mather, W.H.; Tsimring, L.S.; Hasty, J. A fast, robust and tunable synthetic gene oscillator. *Nature* **2008**, *456*, 516–519. [CrossRef]
88. Chen, Y.; Kim, J.K.; Hirning, A.J.; Josić, K.; Bennett, M.R. Emergent genetic oscillations in a synthetic microbial consortium. *Science* **2015**, *349*, 986–989. [CrossRef]
89. Sloan, S.W. A fast algorithm for constructing Delauny triangulations in the plane. *Adv. Eng. Softw.* **1987**, *9*, 34–55. [CrossRef]
90. Azencott, R. *Simulated Annealing: Parallelization Techniques*; Wiley-Interscience: Hoboken, NJ, USA, 1992; Volume 27.
91. Azencott, R.; Chalmond, B.; Coldefy, F. Markov Image Fusion to Detect Intensity Valleys. *Int. J. Comput. Vis.* **1994**, *16*, 135–145. [CrossRef]
92. Boyd, S.; Vandenberghe, L. *Convex Optimization*; Campridge University Press: Campridge, UK, 2004.
93. Ackley, D.H.; Hinton, G.E.; Sejnowski, T.J. A learning algorithm for Boltzmann machines. *Cogn. Sci.* **1985**, *9*, 147–169. [CrossRef]
94. Hinton, G.E.; Sejnowski, T.J. Chapter Learning and Relearning in Boltzmann Machines. In *Parallel Distributed Processing: Explorations in the Microstructure of Cognition*; MIT Press: Cambridge, MA, USA, 1986; pp. 282–317.
95. Azencott, R. Synchronous Boltzmann machines and Gibbs fields: Learning algorithms. In *Neurocomputing*; Springer: Berlin/Heidelberg, Germany, 1990; pp. 51–63.
96. Azencott, R. Synchronous Boltzmann machines and artificial vision. *Neural Netw.* **1990**, 135–143. Available online: https://www.math.uh.edu/~razencot/MyWeb/Research/Selected_Reprints/1990SynchronousBoltzmanMachinesArtificialVision.pdf (accessed on 15 December 2021).
97. Azencott, R.; Graffigne, C.; Labourdette, C. Edge Detection and Segmentation of Textured Plane Images. In *Stochastic Models, Statistical Methods, and Algorithms in Image Analysis*; Springer-Verlag: New York, NY, USA, 1992; Volume 74, pp. 75–88.
98. Kong, A.; Azencott, R. Binary Markov Random Fields and Interpretable Mass Spectra Discrimination. *Stat. Appl. Genet. Mol. Biol.* **2017**, *16*, 13–30. [CrossRef] [PubMed]
99. Azencott, R.; Doutriaux, A.; Younes, L. Synchronous Boltzmann Machines and Curve Identification Tasks. *Netw. Comput. Neural Syst.* **1993**, *4*, 461–480.
100. Garda, P.; Belhaire, E. An Analog Circuit with Digital I/O for Synchronous Boltzmann Machines. In *VLSI for Artificial Intelligence and Neural Networks*; Springer: Berlin, Germany, 1991; pp. 245–254.
101. Lafargue, V.; Belhaire, E.; Pujol, H.; Berechet, I.; Garda, P. Programmable Mixed Implementation of the Boltzmann Machine. In *International Conference on Artificial Neural Networks*; Springer: Berlin, Germany, 1994; pp. 409–412.

102. Pujol, H.; Klein, J.-O.; Belhaire, E.; Garda, P. RA: An analog neurocomputer for the synchronous Boltzmann machine. In Proceedings of the Fourth International Conference on Microelectronics for Neural Networks and Fuzzy Systems, Turin, Italy, 26–28 September 1994; IEEE: Piscataway, NJ, USA, 1994; pp. 449–455.
103. Beucher, S.; Lantuejoul, C. Use of watersheds in contour detection. In *Workshop on Image Processing*; CCETT/IRISA: Rennes, France, 1979.
104. Mumford, D.B.; Shah, J. Optimal approximations by piecewise smooth functions and associated variational problems. *Commun. Pure Appl. Math.* **1989**, *42*, 577–685. [CrossRef]
105. Mézard, M.; Parisi, G.; Virasoro, M.A. *Spin Glass Theory and Beyond: An Introduction to the Replica Method and Its Applications*; World Scientific Publishing Company: Singapore, 1987; Volume 9.
106. Kirkpatrick, S.; Gelatt, C.D., Jr.; Vecchi, M.P. Optimization by simulated annealing. *Science* **1983**, *220*, 671–680. [CrossRef] [PubMed]
107. Roussel-Ragot, P.; Dreyfus, G. A problem independent parallel implementation of simulated annealing: Models and experiments. *IEEE Trans. Comput.-Aided Des. Integr. Circuits Syst.* **1990**, *9*, 827–835. [CrossRef]
108. Burda, Z.; Krzywicki, A.; Martin, O.C.; Tabor, Z. From simple to complex networks: Inherent structures, barriers, and valleys in the context of spin glasses. *Phys. Rev. E* **2006**, *73*, 036110. [CrossRef] [PubMed]
109. Huber, P.J. The 1972 Wald Lecture Robust Statistics: A Review. *Ann. Math. Stat.* **1972**, *43*, 1041–1067. [CrossRef]
110. Ram, D.J.; Sreenivas, T.; Subramaniam, K.G. Parallel simulated annealing algorithms. *J. Parallel Distrib. Comput.* **1996**, *37*, 207–212. [CrossRef]
111. Digabel, H.; Lantuejoul, C. Iterative Algorithms. In Proceedings of the 2nd European Symposium Quantitative Analysis of Microstructures in Material Science, Biology and Medicine, Caen, France, 4–7 October 1977; pp. 85–89.
112. Vincent, L.; Soille, P. Watersheds in digital spaces: An efficient algorithm based on immersion simulations. *IEEE Trans. Pattern Anal. Mach. Intell.* **1991**, *13*, 583–598. [CrossRef]

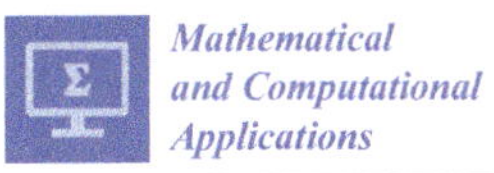
Mathematical and Computational Applications

Article

Image Segmentation with a Priori Conditions: Applications to Medical and Geophysical Imaging

Guzel Khayretdinova [1,2,†], Christian Gout [2,*,†], Théophile Chaumont-Frelet [3,†] and Sergei Kuksenko [1,†]

1 TUSUR CSR, 634050 Tomsk, Russia; guzel.khayretdinova@insa-rouen.fr (G.K.); ksergp@tu.tusur.ru (S.K.)
2 INSA Rouen, LMI, 76000 Rouen, France
3 Inria Sophia Antipolis-Méditerranée Research Centre, Université Côte d'Azur, 2004 Route des Lucioles, 06902 Valbonne, France; theophile.chaumont@inria.fr
* Correspondence: christian.gout@insa-rouen.fr
† These authors contributed equally to this work.

Abstract: In this paper, we propose a method for semi-supervised image segmentation based on geometric active contours. The main novelty of the proposed method is the initialization of the segmentation process, which is performed with a polynomial approximation of a user defined initialization (for instance, a set of points or a curve to be interpolated). This work is related to many potential applications: the geometric conditions can be useful to improve the quality the segmentation process in medicine and geophysics when it is required (weak contrast of the image, missing parts in the image, non-continuous contour...). We compare our method to other segmentation algorithms, and we give experimental results related to several medical and geophysical applications.

Keywords: image segmentation; a priori segmentation; level set method; geodesic active contour

Citation: Khayretdinova, G.; Gout, C.; Chaumont-Frelet, T.; Kuksenko, S. Image Segmentation with a Priori Conditions: Applications to Medical and Geophysical Imaging. *Math. Comput. Appl.* **2022**, *27*, 26. https://doi.org/10.3390/mca27020026

Academic Editor: Gianluigi Rozza

Received: 3 March 2021
Accepted: 7 March 2022
Published: 11 March 2022

1. Introduction

The problem of segmenting an image into its significant components has been studied for over 30 years in computers science, applied mathematics and more generally in computer vision.

Recently, from convolutional neural networks (see Fukushima [1], Waibel et al. [2], and LeCun et al. [3], among others...) to recurrent neural networks (Hochreiter and Schmidhuber [4]), encoder–decoders (Badrinarayanan et al. [5]), or generative adversarial networks (Goodfellow et al. [6]), deep learning based techniques have achieved huge successes in the field of artificial intelligence and image segmentation (see [7] for a recent survey). These approaches leads to excellent results, including in medical applications (see for example Fantazzini et al. [8]). However, in general, the performance heavily depends on labeled data, and this point is a main difficulty on several applications when lacking labeled data such as in many medical and geophysical applications. The data augmentation is possible [9], but it is often complicated, and requires more CPU/GPU time than other segmentation techniques based on energy minimization or variational approaches (Kass et al. [10], Chand and Vese [11], Mumford and Shah [12], Vese and Le Guyader [13]...) or geometrical ones (fast marching methods, see Sethian [14] or Forcadel et al. [15]).

In this work, we focus on a modeling based on a variational approach: it basically consists of evolving an initial contour subject to constraints towards the boundary of the object to be detected. This deformation is done by minimizing a functional depending on the curve and defined so that a local minimum is obtained at the boundary of the object. This energy-like functional minimization problem has led to many research works in the last 25 years. Caselles, Kimmel and Sapiro [16] have shown that by setting one of the regularization parameters to zero in the classical active contour model, one gets a problem equivalent to finding a geodesic curve in a Riemann space whose metric depends on the image contents. The issue was then no longer seen as an energy-like minimization problem but as a curve evolution one. Let us note that this approach is very suitable to our problems, and it could be possible to use deep learning (DL) based algorithm (as used by Le Guyader et al. [17]), but the a priori conditions are complicated to integrate in such DL models (and such DL models require a lot of hardware, much more that our proposed algorithm).

An edge can be viewed as the locus of connected points for which the image gradient varies abruptly. However, when data acquisition cannot be performed in an optimal manner (e.g., liver in medical imaging), this criterion can no longer be applied. In many applications, image data are missing or of poor quality, and occultation phenomena can often occur: two objects that are adjacent to one another can own intrinsic homogeneous textures so that it is hard to clearly identify the interface between them. It is then relevant to introduce geometric constraints in the modeling to help the segmentation process. In [18–21], the authors introduce the concept of a finite set of points to be interpolated during the segmentation process (like in Gout and Le Guyader [20]) or a variational approach to integrate higher-level shape priors into level set based segmentation method (see Cremers et al. [18]). However, in some applications (in geophysics or in several medical applications), this is not sufficient and adding new a priori conditions to be satisfied is required both to improve the segmentation process and also to simplify the initial condition (in [20], the method requires a priori conditions plus an initial guess to start the segmentation process, which constitutes a drawback of their algorithm).

In this work, we propose to consider a curve (or a set of curves) belonging to the searched contour of interest, integrated as an initial condition to be satisfied by the final segmentation contour obtained at the end of the segmentation process. This is an added value considering previous models (like [19]), where it was necessary to both give an initial condition and geometric conditions. In order to define this curve, it is possible to use D^m-spline functions (see Gout et al. [22]) obtained by minimizing an energy functional on a suitable Hilbert space (and so a set of points given by the user is sufficient to create such geometric conditions).

In this paper, we generalize and improve the approach of Gout and Le Guyader [20]: we combine the curve evolution approach developed by Caselles, Kimmel and Sapiro [16], with a geometrical approach consisting in a curve given by the user (a set of points can also be considered like in Gout and Le Guyader [20]), and the initial condition is then automatically generated (from the geometric conditions).

In Section 2, we give the modeling of our problem, based on a level set method (LSM) approach (Osher and Sethian [23], see also [24]). Using Euler-Lagrange variational principle, we obtain the PDE satisfied by the 3D LSM function Φ and the corresponding evolution problem is then given. The existence and uniqueness of the viscosity solution of the associated parabolic problem is also established: we give the main steps of the proof (based on Caselles et al. [16] and Gout and Le Guyader [20]). The discretization of the problem is tackled in Section 3, and is solved using an Additive Operator Splitting scheme (Weickert and Kühne [25]). Comparisons and experimental results with the proposed approach are given for several medical and geophysical imaging applications.

2. Mathematical Modelling

The well-known level set approach (see Osher and Sethian [23]) consists in considering the evolving active contour $\Gamma = \Gamma(t)$ as the zero level set of a function Φ, which is a Lipschitz continuous function defined by:

$$\left\{ \begin{array}{llll} \Phi : & \Omega \times [0, +\infty[& \longrightarrow & \mathbb{R}, \\ & (x,t) & \longmapsto & \Phi(x,t), \end{array} \right. \tag{1}$$

such that:

$$\Gamma(t) = \{x \in \Omega \mid \Phi(x,t) = 0\}, \tag{2}$$

and $\Phi(\cdot, t)$ takes opposite signs on each side of $\Gamma(t)$. This level set approach enables us to re-write the energy in terms of Φ:

$$E(\Phi) = \int_{\Omega} g(|\nabla I(x)|)|\nabla H(\Phi(x))|dx, \tag{3}$$

where H is the one-dimensional Heaviside function and where $g(s) = \dfrac{1}{1+s^2}$. We formulate the shape optimization problem as follows: "Find a shape such that the energy E is minimized":

$$\begin{cases} \text{Search for } \Phi \text{ such that:} \\ E(\Phi) = \min\limits_{\xi} E(\xi). \end{cases} \tag{4}$$

The functional E being not Gâteaux-differentiable, we regularize the problem by considering slightly regularized versions (C^1 or C^2 regularization) of the functions H and δ (the one dimensional Dirac measure) denoted, respectively, by H_ϵ and δ_ϵ, and considering that $\delta_\epsilon = H'_\epsilon$, the associated regularized functional E_ϵ becomes:

$$E_\epsilon(\Phi) = \int_{\Omega} g(|\nabla I(x)|)\delta_\epsilon(\Phi(x))|\nabla \Phi(x)|dx, \tag{5}$$

where:

$$\int_{\Omega} \delta_\epsilon(\Phi(x))|\nabla \Phi(x)|dx \tag{6}$$

is an approximation of the length of the zero level set of Φ.

2.1. A Priori Conditions

We now propose to include a priori condition in the process. We consider the a priori condition as a smooth arc curve $\mathcal{C}$ given by the user. The curve $\mathcal{C}$ is parametrised by $c \in C^1([0,1],\Omega)$. By construction, we impose that the curve $\mathcal{C}$ belongs to the level set $\{\Phi_0 = 0\}$. The first step consists in constructing the initial condition from the geometric conditions given by the user. We have chosen this option, because it permits to avoid the stage of giving an initial condition plus the a priori conditions.

The first step consists in closing the arc to get a closed curve. We introduce a point $a \in \Omega$ and a vector $v = (v_1, v_2) \in \mathbb{R}^2$. It is easy to construct two (unique) polynomials curves $\mathcal{C}_1$ and $\mathcal{C}_2 \subset \Omega$ that are parameterized using c_1 and $c_2 \in P_3([0,1],\Omega)$ such that:

$$\begin{array}{lll} c(1) = c_1(0), & c_1(1) = c_2(0) = a, & c_2(1) = c(0), \\ -c'(1) = c'_1(0), & c'_1(1) = -c'_2(0) = v, & c'_2(1) = -c'(0). \end{array} \tag{7}$$

We set $\Gamma_0 = \mathcal{C} \cup \mathcal{C}_1 \cup \mathcal{C}_2$, and we consider Φ_0 as the signed distance to Γ_0 :

$$\Phi_0(x) = \begin{cases} d(x,\Gamma_0) & \text{if } x \in Int\ (\Gamma_0), \\ -d(x,\Gamma_0) & \text{otherwise.} \end{cases} \tag{8}$$

To define the energy E_C, we also introduce the mask η: $\Omega \to \mathbb{R}$:

$$\eta(x) = \exp\left(-\frac{d(x,C)^2}{\sigma}\right) : \forall x \in \Omega, \tag{9}$$

where $\sigma \in \mathbb{R}^+$, is a positive parameter which controls the width of the mask. Then, we can define the following energy functional:

$$E_C(\Phi) = \int_{\Omega} \eta(x)(\Phi(x) - \Phi_0(x))^2 dx. \tag{10}$$

The goal is to minimize the L^2 norm between Φ and Φ_0: we impose the contour to be close to Γ_0. The mask permits to impose such influence in (and only in) a local neighborhood of $\mathcal{C}$.

2.2. Minimization Problem and Evolution Equation

The modeling of our problem consists in minimizing the energy:

$$E(\Phi) = E_\epsilon(\Phi) + \frac{\alpha}{2} E_C(\Phi) \tag{11}$$

with $\alpha > 0$.

We now give the evolution problem. We minimize E_ϵ with respect to Φ and deduce the associated Euler–Lagrange equation for Φ:

$$E'(\Phi) = \delta_\epsilon(\Phi) div\left(g(|\nabla I|)\frac{\nabla\Phi}{|\nabla\Phi|}\right) + \alpha\eta(\Phi - \Phi_0) \tag{12}$$

with the boundary conditions (homogeneous Neumann):

$$\frac{\delta_\epsilon(\Phi)}{|\nabla\Phi|}\frac{\partial\Phi}{\partial\nu}_{|\partial\Omega} = 0, \tag{13}$$

ν denoting the exterior normal to the boundary of Ω.

When a local minimum is reached, then the quantity $\frac{\partial\Phi}{\partial t}$ tends to 0, which means that the steady state is reached. A rescaling can be made so that the motion is applied to all level sets by replacing δ_ϵ by $|\nabla\Phi|$ (It makes the flow independent of the scaling of Φ).

Let us also note that the speed of convergence of the model can be increased by adding the component $\beta g(|\nabla I|)|\nabla\Phi|$ in the evolution equation, β being a constant. This component can be seen as an area constraint. An analogy with the Balloon model developed by Cohen [26] can be made: this constant motion term prevents the curve from stopping on a non-significative local minimum and is also of importance when initializing the process with a curve inside the object to be detected.

Thus, the proposed parabolic problem with the associated boundary conditions $\frac{\partial\Phi}{\partial\nu} = 0$ can be written:

$$\left\{\begin{array}{ll} \Phi(x,0) & = \Phi_0(x), \\ \frac{\partial\Phi}{\partial t} = & |\nabla\Phi|\left[div(g(|\nabla I|)\frac{\nabla\Phi}{|\nabla\Phi|})\right] + \alpha\eta(\Phi - \Phi_0) \\ & +\beta g(|\nabla I|)|\nabla\Phi|, \\ \frac{\partial\Phi}{\partial\nu} = & 0 \text{ on } \partial\Omega. \end{array}\right. \tag{14}$$

2.3. Existence, Uniqueness of the Solution

This section is a theoretical part showing the existence and uniqueness of the solution of our segmentation model. We can show the existence and uniqueness of a viscosity solution to the evolution problem (14). We rewrite our problem as:

$$\left\{\begin{array}{rcl} u_t + \tilde{F}(t,x,u,Du,D^2u) & = & 0 \quad in \quad]0,T[\times\Omega \\ B(x,Du) & = & 0 \quad in \quad]0,T[\times\partial\Omega, \end{array}\right. \tag{15}$$

where, for $(t,x,u,p,X) \in [0,T] \times \bar{\Omega} \times \mathbb{R}^2 \times \mathcal{S}^2$:

$$\tilde{F}(t,x,u,p,X) = F(t,x,p,X) + g(x,u), \quad B(x,p) = p\cdot\vec{n}, \tag{16}$$

and:

$$\begin{array}{rcl} F(t,x,p,X) & = & -trace\Big(g(|\nabla I(x)|)\Big(I - \frac{p\otimes p}{|p|^2}\Big)X\Big) \\ & & -\langle\nabla(g(|\nabla I(x)|)),p\rangle \\ g(x,u) & = & \alpha\eta(x)(u - \Phi_0(x)). \end{array} \tag{17}$$

We assume that Ω is a bounded domain in $\mathbb{R}^n$ with a C^1 boundary. Under the following conditions, the Equation (15) has a unique viscosity solution.

1. $\tilde{F} \in C([0,T] \times \bar{\Omega} \times \mathbb{R} \times (\mathbb{R}^n - \{0\}) \times S^n)$, where S^n denotes the space of $n \times n$ symmetric matrices equipped with the usual ordering.
2. There exists a constant $\gamma \in \mathbb{R}$ such that for each:

$$(t,x,p,X) \in [0,T] \times \bar{\Omega} \times (\mathbb{R}^n - \{0\}) \times S^n,$$

the function:

$$u \mapsto \tilde{F}(t,x,u,p,X) - \gamma u \tag{18}$$

is non-decreasing on $\mathbb{R}$.
3. For each $R > 0$, there exists a continuous function $w_R : [0,\infty[\longrightarrow [0,\infty[$ satisfying $w_R(0) = 0$ such that if $X, Y \in S^n$ and $\mu_1, \mu_2 \in [0,\infty[$ satisfy:

$$\begin{pmatrix} X & 0 \\ 0 & Y \end{pmatrix} \leq \mu_1 \begin{pmatrix} I & -I \\ -I & I \end{pmatrix} + \mu_2 \begin{pmatrix} I & 0 \\ 0 & I \end{pmatrix} \tag{19}$$

then:

$$\begin{aligned} &\tilde{F}(t,x,u,p,X) - \tilde{F}(t,y,u,q,-Y) \\ &\geq -w_R(\mu_1(|x-y|^2 + \rho(p,q)^2) \\ &+\mu_2 + |p-q| + |x-y|(1 + max(|p|,|q|))), \end{aligned} \tag{20}$$

for all $t \in [0,T], x,y \in \overline{\Omega}, u \in \mathbb{R}$ with $|u| \leq R$ and $p,q \in \mathbb{R}^n \backslash \{0\}$.
4. $B \in C(\mathbb{R}^n \times \mathbb{R}^n) \bigcap C^{1,1}(\mathbb{R}^n \times (\mathbb{R}^n \backslash \{0\}))$.
5. For each $x \in \mathbb{R}^n$, the function $p \mapsto B(x,p)$ is positively homogeneous of degree one in p, i.e., $B(x,\lambda p) = \lambda B(x,p), \forall \lambda \geq 0, p \in \mathbb{R}^n \backslash \{0\}$.
6. There exists a positive constant θ such that $< \nu(z), D_p B(z,p) > \geq \theta$ for all $z \in \partial\Omega$ and $p \in \mathbb{R}^n - \{0\}$. Here, $\nu(z)$ denotes the unit outer normal vector of Ω at $z \in \partial\Omega$.

All those conditions are clearly satisfied for $\tilde{F} = F$, and B (see for instance [20]). We here have to extend the proof to $\tilde{F} = F + g$.

1. Thanks to [20], we already know that $F \in C([0,T] \times \bar{\Omega} \times (\mathbb{R}^n \setminus \{0\}) \times \mathcal{S}^n)$. Since, $\eta, \Phi_0 \in C(\bar{\Omega})$, it is clear that $g \in C(\bar{\Omega} \times \mathbb{R})$ and therefore $\tilde{F} \in C([0,T] \times \bar{\Omega} \times \mathbb{R} \times (\mathbb{R}^n \setminus \{0\}) \times \mathcal{S}^n$.
2. Note that since F does not depend on u. It suffices to show that $u \rightarrow g(x,u) - \gamma u$ is non-decreasing on $\mathbb{R}$. On can easily see that the condition is satisfied for all $x \in \bar{\Omega}$ as soon as $\gamma < 0$.
3. Since g does not depend on X, it suffices to check the condition on F. We refer the reader to Gout and Le Guyader [20] for the proof.

The proof for conditions 4–6 are basic, and can be found for instance in [20]. Therefore, all the conditions are satisfied and the evolution problem (14) has a unique viscosity solution.

3. Experimental Results

For the discretization scheme, we use the additive operator splitting scheme (AOS) proposed by Weickert and Kuhne ([25], see also [20]). This corresponding algorithm requires a computational cost linear in $M \times N$ (size of the unknown vector Φ) at each step. This scheme is well-suited to our problem that differs only in the introduction of the function $\eta(x)$.

We use the method proposed in this paper knowing that:

- Following the considered application, the initial condition can be a set of points, or a curve (then can be constructed from a set points using a basic spline function (see Gout et al. [22])).
- The stop criterion can be either a preset number of iterations or a check that the solution is stationary.
- The distance is normalized in order to have the same weight between *a priori* information of the image and geometrical constraints.

- The discretization is made using finite differences as done in Chan and Vese [11].
- In the numerical examples, we take $\delta t = 0.1$, the regularization term is equal to 0.8.

To illustrate the interest of our approach, we start with a simple test case, using a three-band image. Then, we run the method on blood vessel image. We obtain the following results. Without a priori condition, the segmentation process will of course give the entire white band as a result. This is the interest of adding such a priori condition to enforce the contour to arbitrary stop following the choice of the user.

We can see that the final contour correctly takes into account the curve given by the user imposing a constraint independent of the pixel values of the images, and correctly segment the white zone.

Of course, our approach allows us (to help) the segmentation process on "complicated" images, when ground truth informations are required like in Figure 1 or in many other cases, especially in geophysical or medical imaging. In Section 3.1, we illustrate the added value brought by our approach considering the initial guess choice improvement. Two examples are given to show that, compared to Gout et al. method [19], the approach given in this paper is much faster to get the same result since the initial guess is (strongly) optimized.

In Section 3.2, we propose to compare our approach with well-known algorithms (Chan-Vese [11], U-Net [27]).

In Sections 3.3 and 3.4, we give experimental results in medicine and geophysics.

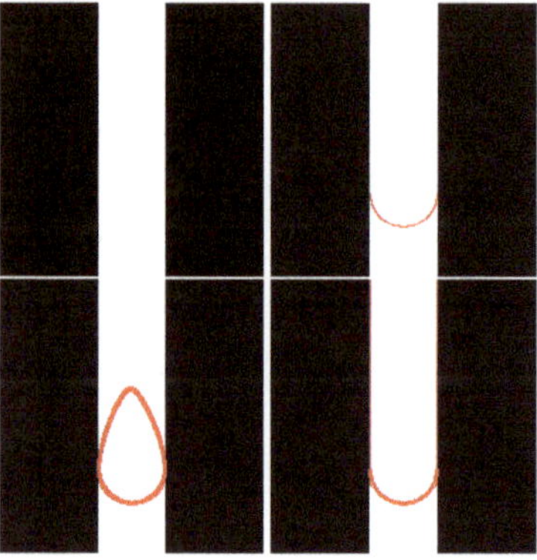

Figure 1. Three-band image (**top left**), user-defined constraint (**top right**), initial contour automatically defined by our approach (**bottom left**) and final contour (**bottom right**).

3.1. Impact of the Initial Guess on the Segmentation Process

In this part, we focus on the added value brought by the automatic initial guess we propose, obtained from the geometrical condition given by the user. We compare our approach to the one introduced by Gout, Le Guyader and Vese [19], we take the same geometrical conditions, and then compare the number iteration to get the same final contour. The comparison is relevant because the segmentation algorithm is analogous in the two considered methods. In [19], note that we have to add a closed contour to the geometric conditions to start the segmentation process, while our proposed approach leads to an automatic initial guess (obtained from the geometric conditions given by the user). In the following two numerical examples, we take exactly the same interpolation conditions (3 points).

In Figures 2–4, we can see that our approach requires 5 times less iterations.

We repeat the same process in the following example (Figures 5–7). The initial contour for Gout et al. [19] algorithm is here chosen outside the region of interest. Of course, we take exactly the same interpolation conditions (3 points) in both cases.

In the example given on Figures 5–7, we can see that our approach requires 30 times less iterations.

In this subsection, we illustrate the gain of the automatic generated initial guess we propose from the geometric conditions (see Table 1).

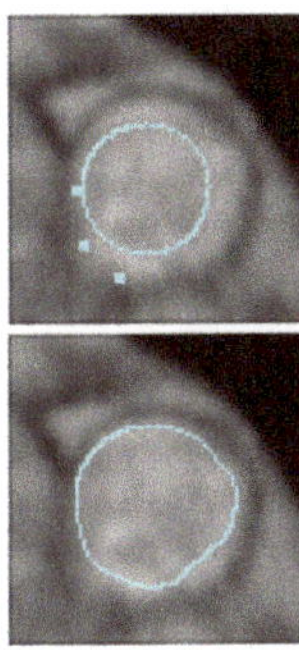
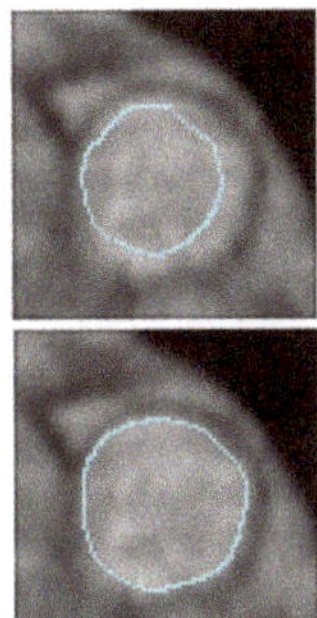

Figure 2. We use Gout et al. [19] algorithm. The initial guess is given inside the region of interest, it is defined by $\Psi = (X - 39)^2 + (Y - 47)^2 - 18^2$, the time step is equal to 0.6, space step is equal to 0.1, distance is computed using the fast marching method (Sethian [14]) and the regularization term is equal to 0.8. **Top left**: Interpolation condition and initial closed contour. **Top right**: iteration 120. **Bottom left**: Iteration 240. **Bottom right**: Iteration 420.

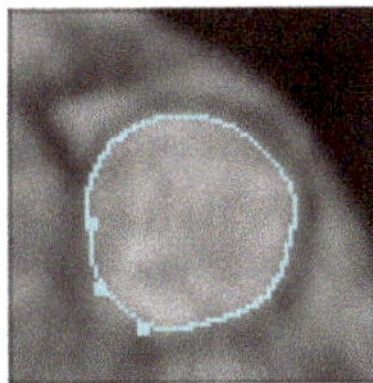

Figure 3. Final result (iteration 420), with the interpolation conditions (3 points) using Gout et al. algorithm [19].

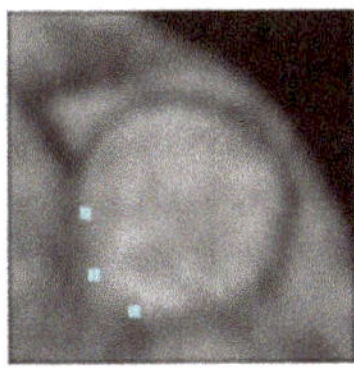
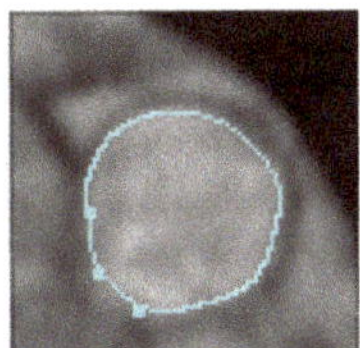

Figure 4. Using our proposed approach, the process does not require an initial guess, the interpolation conditions automatically initiate the process. **Left**: interpolation conditions. **Right**: final result after 80 iterations.

Table 1. Gain of the automatic generated initial guess.

Method	Example 1	Example 2
Gout et al. [19]	420 iterations	260 iterations
Our method	80 iterations	8 iterations

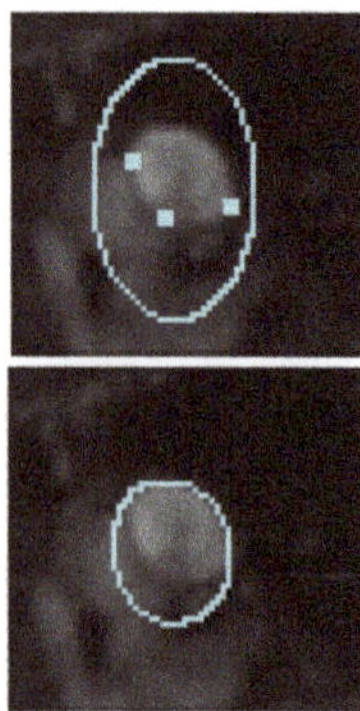
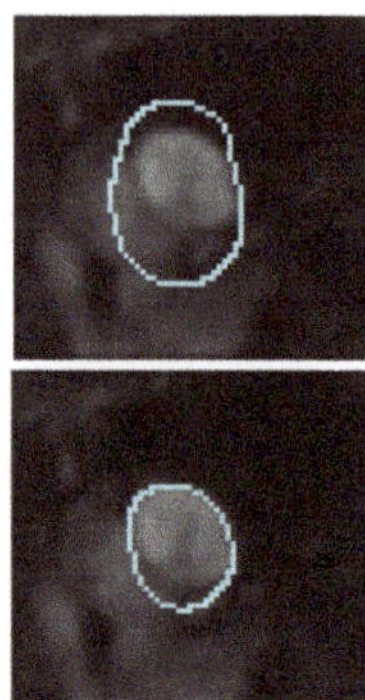
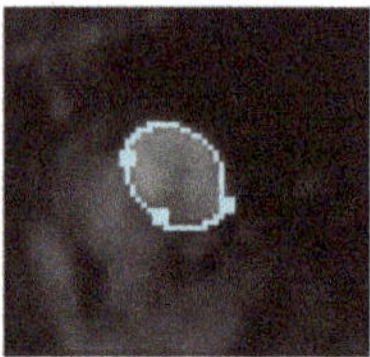

Figure 5. We use Gout et al. [19] algorithm. The initial guess is given outside the region of interest, it is defined by $\Psi = \frac{(X-30)^2}{15^2} + \frac{(Y-30.5)^2}{23^2} - 1$, the time step is equal to 0.3, space step is equal to 0.1, distance is computed using the fast marching method (Sethian [14]) and the regularization term is equal to 0.8. **Top left**: Interpolation condition and initial guess. **Top right**: iteration 80. **Bottom left**: Iteration 100. **Bottom right**: Iteration 160.

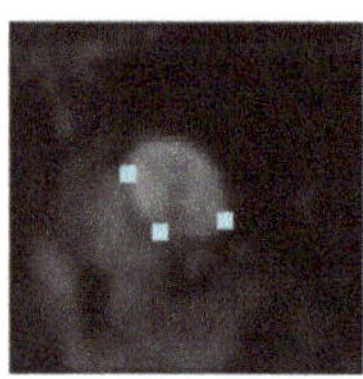

Figure 6. Final result (iteration 260), with the interpolation conditions (3 points) using Gout et al. algorithm [19].

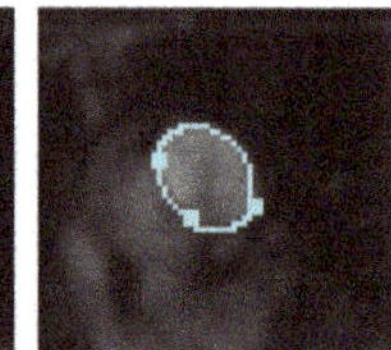

Figure 7. Left: interpolation conditions. **Right**: final result after 8 iterations using our proposed approach.

To get the same final contour, our method clearly requires much less iterations. This result is analogous on all the tests we have done, especially when a closed contour is chosen outside the region of interest (in [19]), in this case, the total number of iterations is reduced by a factor of 5 to 60. If the initial contour in [19] is chosen inside the region of interest, the total number of iterations is reduced by a factor of 4 to 20. The only cases where the number of iterations is equivalent between the two different methods are when we choose as initial guess in [19] a closed curve very close to the final result and interpolating at least 2 given interpolation conditions, this is of course a major constraint.

3.2. *Quantitative Performance*

We use the BraTS Dataset [28], which was collected and shared by the MICCAI Brain Tumor Segmentation Challenge (see [28] for more details) several years ago. We use 270 MR scans, each with four MIR sequences: T1-weighted (native image, sagittal or axial 2D acquisitions with 1–6 mm slice thickness), T1-weighted with contrast-enhanced (Gadolinium) image, T2-weighted image (axial 2D acquisition, with 2–6 mm slice thickness), and FLAIR

(T2-weighted FLAIR image, axial, coronal, or sagittal 2D acquisitions, 2–6 mm slice thickness) (see [28] for more details). The training data have the size $240 \times 240 \times 155$ pixels, and have the manual segmentation labels for many types of brain tumors. We trained the network to generate segmentation masks corresponding to the center slice of the input. We used the well-known Adam optimization method [29]. We trained the deep network using 238 training data, we set the initial learning rate as 10^{-4}, and multiplied by 0.5 after every 20 epochs. Using a single GPU, we trained the models with batch size 4 for 40 epochs.

In order to evaluate the performance of our method, we give comparisons between several algorithms. We here recall several metrics:

– The Jaccard index [30], or Intersection over Union (IoU), is a commonly used metric in segmentation. It is defined as the area of intersection between the predicted segmentation map and the ground truth, divided by the area of union between the predicted segmentation map and the ground truth:

$$\mathrm{IoU} = J(G,P) = \frac{|G \cap P|}{|G \cup P|}, \tag{21}$$

where G is the ground truth and P denotes the predicted segmentation maps. It ranges between 0 and 1. Mean-IoU (mIoU) is defined as the average IoU over all classes. It is widely used in reporting the performance of segmentation algorithms.

– The Dice coefficient (Dice) is a popular metric for image segmentation, especially in medical imaging. This coefficient can be defined as twice the overlap area of predicted and ground-truth maps, divided by the total number of pixels in both images:

$$\mathrm{Dice} = \frac{2|G \cap P|}{|G| + |P|} \tag{22}$$

When applied to binary segmentation maps, and referring to the foreground as a positive class, the Dice coefficient is essentially identical to the F1 score:

$$\mathrm{Dice} = \frac{2TP}{2TP + FP + FN} = \mathrm{F1}, \tag{23}$$

(The F1 score, which is defined as the harmonic mean of precision (Prec) and recall (Rec): F1$= \frac{2\mathrm{Prec\ Rec}}{\mathrm{Prec} + \mathrm{Rec}}$, where Prec $=\frac{TP}{TP+FP}$ and Rec $=\frac{TP}{TP+FN}$.), where TP refers to the true positive fraction, FP refers to the false positive fraction, and FN refers to the false negative fraction.

– The Hausdorff distance (Hd) evaluates the quality of the segmentation boundaries by computing the maximum distance between the prediction and its ground truth:

$$\mathrm{Hd} = \max\left\{ \sup_{p \in \partial P} \inf_{z \in \partial G} \|p - z\|_2, \sup_{z \in \partial G} \inf_{p \in \partial P} \|p - z\|_2 \right\} \tag{24}$$

where ∂P and ∂G are the surface point sets of P (segmentation prediction) and G (ground truth).

Of course, a large Dice coefficient or a small HD means an accurate segmentation result. Figure 8 shows several numerical example (here, we ramdomly sampled 33% of BRATS labeled dataset). In Figure 8, we give the dice score obtained for U-Net and our proposed algorithm. We can see that our approach clearly gives better results.

We also compare with the U-net algorithm ([27], see also [31]), tables show the evaluation results, please note that we have suppressed the best and worst 10% of the results, and that for our proposed approach, we have considered only 2 to 4 points as geometric conditions (the initial guess -curve- is automatically constructed from these points), all that leads to make the global results quite homogeneous. We also compare with the Chan-Vese algorithm [11]: in this case, the initial condition is a closed contour located inside the region of interest.

If we use 100% of the BraTS dataset, we obtain the Table 2.

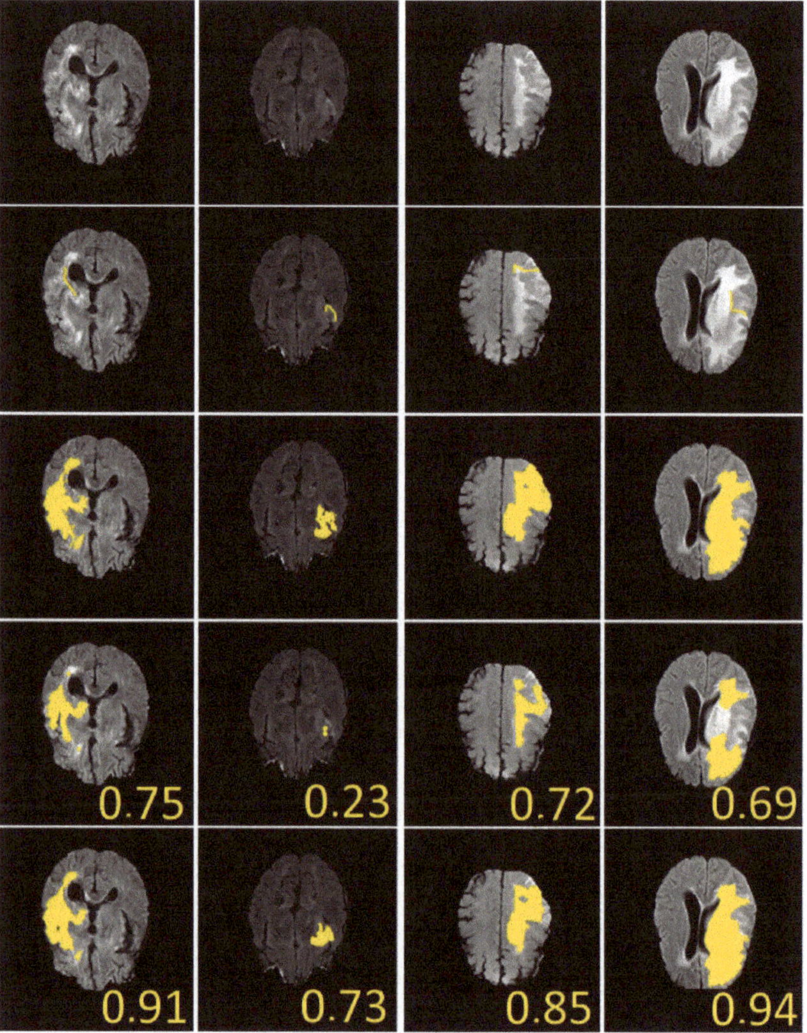

Figure 8. Brain Tumor Segmentation (using BRATS dataset [28]). Several results from our method using labeled data on the BRATS dataset [28]. The tumor is underlined in yellow. First row: center slices of input. Second row: initial conditions for our algorithm. Third row: ground-truth labels. Fourth row: results from the supervised U-Nets learning method introduced by Ronnenberger et al. [27]. Fifth row: results from our proposed method. The scores in the bottom of each results denote Dice score.

Table 2. Accuracies of segmentations models on a third of labeled data from BraTS Dataset [28]. We use a Nvidia GeForce RTX 2080 Ti 11G Turbo Edition (Boost frequence: 1545 MHz, GPU memory: 11 GB).

Method	mIoU	Dice	Hd	GPU Time
U-Net [27]	78.3	87.7	43.5	4.45
Chan-Vese [11]	77.6	88.1	41.5	2.02
Our method	79.6	89.1	39.5	2.12

We see that the GPU time is clearly better with our algorithm (and the Chan-Vese one too). The quality is globally equivalent: a little better with our algorithm when not considering the entire database, and equivalent if not. This is an important point because it validates our algorithm when we compare with MR scans segmented with deep learning algorithms with few labeled data.

3.3. Applications to Medical Imaging

Here, we apply our method to 2D medical images, the goal being to outline the cross-sectional area of a great thoracic vessel, namely the main pulmonary artery, in order to non-invasively assess pulmonary arterial hypertension. Figure 9 refers to a slice set perpendicularly to the MPA axis (arrow) of a 78 years old patient, who was suspected of pulmonary arterial hypertension, suffering from breathlessness. The presented magnitude image was acquired from a velocity encoded sequence (Venc = 1.5 m/s), during a 5 min time duration, with a 30 ms temporal resolution (1T Magnetom Expert imager; Siemens; Germany). The image quality is moderate, and in particular the vessel wall is irregular around the arrow, that is, not anatomically possible, and therefore an interpolation in the outlining by the physician is needed.

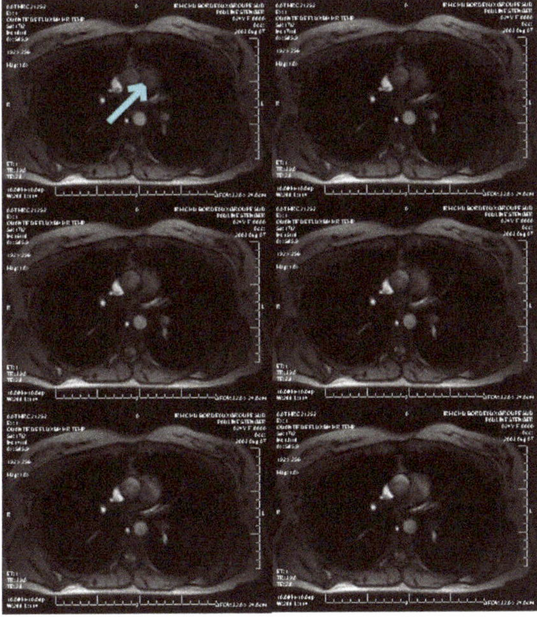

Figure 9. Example of an image sequence (courtesy of CHU Bordeaux, France). The arrow shows the pulmonary artery to segment.

Since the lack of annotated data is one of the most common limitations encountered in many deep learning approaches like the ones we will study in this subsection, deep leaning approaches are not suitable (at least for now!) for our considered applications.

An example of the segmentation of a pulmonary artery with a curve to interpolate is given in Figure 10. A part of this artery is difficult to segment since its bottom left is always located in a blurred zone.

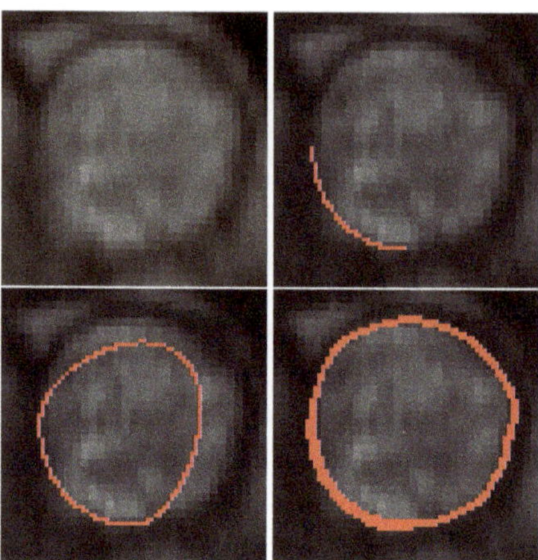

Figure 10. Blood vessel image, user-defined constraint (**top right**), initial contour (**bottom left**) and final (**bottom right**) contours. Let us note that we have successfully segmented the complete image sequence.

Let us note that the MPA CSA variations versus time, throughout a cardiac cycle, obtained from the present automatic segmentation method are in agreement with previous results obtained from a manual outlining (Hôpital of Bordeaux, France). Moreover, the present method enables us to get a reliable assessment of the systolic-diastolic difference in the vessel CSA, which should lead to a further improvement in the accuracy of a non-invasive blood pressure estimation (using well-known methods, see [32,33]). In several medical applications, let us note that a registration process must be added to the segmentation process (see [34–37] for more details).

3.4. Applications to Geophysical Imaging

We now apply our algorithm to the segmentation of a non-continuous layer on a geophysical dataset (Figure 11 (right)) extracted from a 3D dataset (Figure 11 (left)). This is a complicated image to segment since there is a vertical fault (see Figure 12). As a priori condition, we here consider a set of point to interpolate (instead of a curve) in order to help the segmentation process (Figure 12). The algorithm starts with a polygonal line interpolating the set of points (to interpolate) as an initial condition, and then segment the image until the edge(s) of the image. Here, the final contour is a segment and not a closed curve, this is of great interest when one wants to track faults or cracks in segmentation processes.

The obtained result (Figure 13) has been validated by geophysicians, but the next step requires an algorithm for 3D applications in order to segment 3D geophysical datasets. We are also in the process of creating a free labeled geophysical dataset, in order to test deep learning techniques on these images.

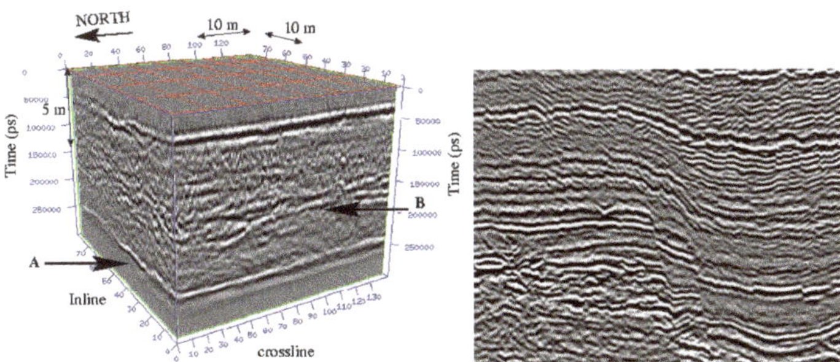

Figure 11. **Left**: 3D seismic dataset (courtesy of TotalEnergies, Pau, France). Two strong and continuous reflectors (Horizon A and Horizon B) appear. **Right**: 2D image (extracted from the 3D seismic dataset) with vertical faults and layers. Considering the complexity of such dataset, it is impossible to segment it without a priori given conditions. The segmentation process consists in finding layers and/or faults.

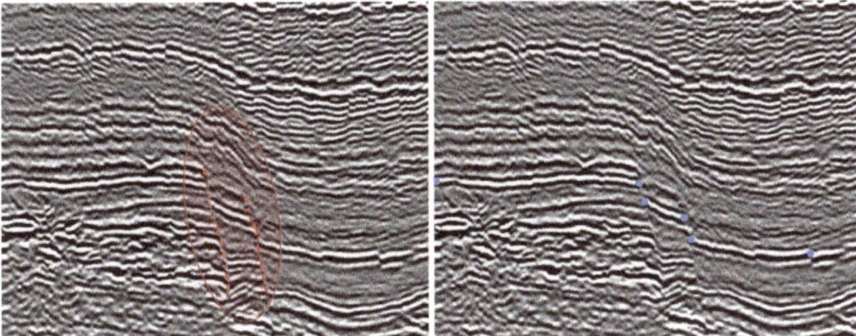

Figure 12. **Left**: The red zone underlines the vertical fault (with dotted segments). **Right**: As a geometric conditions, we consider a set of points given by the user.

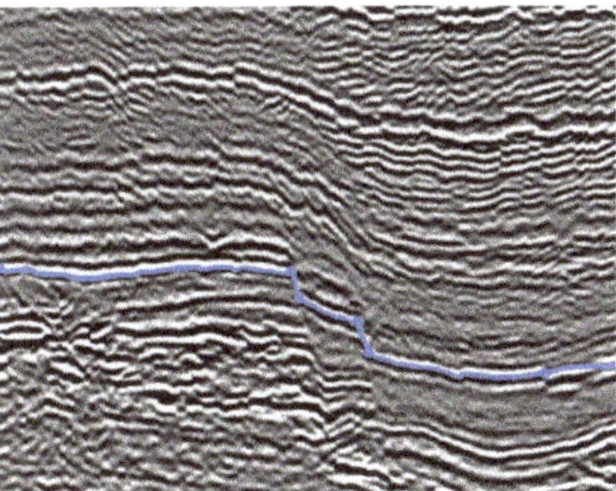

Figure 13. Final result: we can see that the searched layer is correctly segmented. On this specific dataset, the final contour has to reach the edge of the image: the contour goes beyond the last point to interpolate (right side) to join the edge of the image.

4. Conclusions

We have proposed a method to segment medical or geophysical images with a priori conditions. The efficiency of the method has been shown on 2D images, knowing that such images are difficult to segment with usual approaches since the searched contour is not continuous (geophysical dataset) or when the zone to segment is blurred (medical images).

Compared to Gout et al. [19], the added value brought by our proposed method (automatic initial guess) is materialized by a total number of iterations reduced by a factor of 4 to 60.

The GPU time is also a key point to underline about our algorithm, much lower than deep learning based approaches. About the sensitivity of the model, especially for geophysical data, it is not possible to segment correctly without a precise choice of the interpolation conditions: the choice of such condition strongly impacts the final result, this point is less obvious in medical applications where the choice of the initial condition improves the result only on very noisy areas and/or with a blurred edge, otherwise, the result (whatever the initial condition) is very close to classic algorithms (Chan Vese algorithm for example). The implementation of a 3D code is a work in progress, the main problem for a 3D segmentation process consists in choosing the initial (geometric) condition (a surface, a curve and/or points), which is a complicated task on complex 3D images (mainly because there are 3D data visualization difficulties "to see inside" the 3D dataset, especially in geophysics to give the initial condition—see Figure 14)—the mathematical results remain valid in 3D.

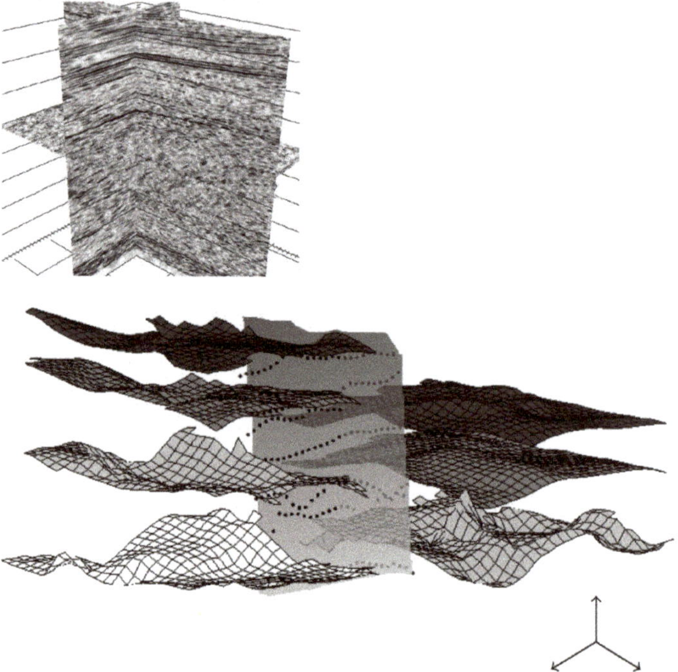

Figure 14. Top: Another 3D view of the 3D geological dataset of Figure 5. **Bottom**: From the 3D dataset, we use a set of points (well data given by geologists), and each surface/layer is then constructed with a D^m-spline operator (see [22] for more details on D^m-spline surface approximation). We here automatically construct 7 surfaces that will be considered as an initial condition. Next step consists in the 3D segmentation process from these initial conditions—this last part is a work in progress.

Author Contributions: Conceptualization, G.K., C.G., T.C.-F. and S.K.; methodology, G.K., C.G. and T.C.-F.; software, G.K. and T.C.-F.; validation, G.K., C.G. and S.K.; formal analysis, G.K.; investigation, G.K., C.G., T.C.-F. and S.K.; writing—original draft preparation, G.K.; writing—review and editing, G.K., C.G., T.C.-F. and S.K.; funding acquisition, C.G. All authors have read and agreed to the published version of the manuscript.

Funding: The research was funded by the European Union with the European regional development fund (ERDF, 18P03390/18E01750/18P02733/21E05300) and by the Région Normandie Council via the M2SINUM and DEFHY3GEO projects, Agence Nationale de la Recherche via MEDISEG project

(ANR-21-CE23-0013), and Maison Normande des Sciences du Numérique (MNSN, under DSG2-2021 project).

Acknowledgments: The authors thank the CRIANN (Centre Régional Informatique et d'Applications Numériques de Normandie, France) for providing computational resources.

Conflicts of Interest: The authors declare no conflict of interest.

References

1. Fukushima, K. Neocognitron: A self-organizing neural network model for a mechanism of pattern recognition unaffected by shift in position. *Biol. Cybern.* **1980**, *36*, 193–202. [CrossRef] [PubMed]
2. Waibel, A.; Hanazawa, T.; Hinton, G.; Shikano, K.; Lang, K.J. Phoneme recognition using time-delay neural networks. *IEEE Trans. Acoust. Speech Signal Process.* **1989**, *37*, 328–339. [CrossRef]
3. LeCun, Y.; Bottou, L.; Bengio, Y.; Haffner, P. Gradient-based learning applied to document recognition. *Proc. IEEE* **1998**, *86*, 2278–2324. [CrossRef]
4. Hochreiter, S.; Schmidhuber, J. Long short-term memory. *Neural Comput.* **1997**, *9*, 1735–1780. [CrossRef]
5. Badrinarayanan, V.; Kendall, A.; Cipolla, R. Segnet: A deep convolutional encoder-decoder architecture for image segmentation. *IEEE Trans. Pattern Anal. Mach. Intell.* **2017**, *39*, 2481–2495. [CrossRef]
6. Goodfellow, I.; Pouget-Abadie, J.; Mirza, M.; Xu, B.; Warde-Farley, D.; Ozair, S.; Courville, A.; Bengio, Y. Generative adversarial nets. In Proceedings of the Advances in Neural Information Processing Systems 27: Annual Conference on Neural Information Processing Systems 2014, Montreal, QC, Canada, 8–13 December 2014; pp. 2672–2680.
7. Minaee, S.; Boykov, Y.Y.; Porikli, F.; Plaza, A.J.; Kehtarnavaz, N.; Terzopoulos, D. Image Segmentation Using Deep Learning: A Survey. *IEEE Trans. Pattern Anal. Mach. Intell.* **2021**, 1. [CrossRef]
8. Fantazzini, A.; Esposito, M.; Finotello, A.; Auricchio, F.; Conti, M.; Pane, B.; Spinella, G. 3D Automatic Segmentation of Aortic Computed Tomography Angiography Combining Multi-View 2D Convolutional Neural Networks. *Cardiovasc. Eng. Tech.* **2020**, *11*, 576–586. [CrossRef]
9. Luo, Q.; Wang, L.; Lv, J.; Xiang, S.; Pan, C. Few-shot learning via feature hallucination with variational inference. In Proceedings of the IEEE Winter Conference on Applications of Computer Vision (WACV), Waikoloa, HI, USA, 3–8 January 2021; pp. 3963–3972.
10. Kass, M.; Witkin, A.; Terzopoulos, D. Snakes: Active contour models. *Int. J. Comput. Vis.* **1988**, *1*, 321–331. [CrossRef]
11. Chan, T.F.; Vese, L.A. Active contours without edges. *IEEE Trans. Image Process.* **2001**, *10*, 266–277. [CrossRef]
12. Mumford, D.; Shah, J. Optimal approximation by piecewise smooth functions and associated variational problems. *Comm. Pure Appl. Math.* **1989**, *42*, 577–685. [CrossRef]
13. Vese, L.; Le Guyader, C. *Variational Methods in Image Processing*; CRC Press: Boca Raton, FL, USA; Taylor and Francis: Oxfordshire, UK, 2015; 410p, ISBN 9781439849736.
14. Sethian, J.A. A fast marching level set method for monotonically advancing fronts. *Proc. Natl. Acad. Sci. USA* **1996**, *93*, 1591–1595. [CrossRef] [PubMed]
15. Forcadel, N.; Le Guyader, C.; Gout, C. Generalized fast marching method: Applications to image segmentation. *Numer. Algorithms* **2008**, *48*, 189–211. [CrossRef]
16. Caselles, V.; Kimmel, R.; Sapiro, G. Geodesic Active Contours. *Int. J. Comput. Vis.* **1997**, *22*, 61–79. [CrossRef]
17. Le Guyader, C.; Lambert, Z.; Petitjean, C. Analysis of the weighted Van der Waals-Cahn-Hilliard model for image segmentation. In Proceedings of the 10th International Conference on Image Processing Theory, Tools and Applications IPTA 2020, Paris, France, 9–12 November 2020.
18. Cremers, D.; Sochen, N.; Schnörr, C. Towards Recognition-Based Variational Segmentation Using Shape Priors and Dynamic Labeling. *Lect. Notes Comput. Sci.* **2003**, *2695*, 388–400.
19. Gout, C.; Le Guyader, C.; Vese, L. Segmentation under geometrical conditions using geodesic active contours and interpolation using level set methods. *Numer. Algorithms* **2005**, *39*, 155–173. [CrossRef]
20. Gout, C.; Le Guyader, C. Geodesic active contour under geometrical conditions: Theory and 3D applications. *Numer. Algorithms* **2008**, *48*, 105–133.
21. Gout, C.;Vieira-Teste, S. Using deformable models to segment complex structures under geometric constraints. In Proceedings of the 4th IEEE Southwest Symposium on Image Analysis and Interpretation, Austin, TX, USA, 2–4 April 2000; pp. 101–105.
22. Gout, C.; Lambert, Z.; Apprato, D. *Data Approximation: Mathematical Modelling and Numerical Simulations*; INSA Rouen Normandie; EDP Sciences: Les Ulis, France, 2019; 168p, ISBN 9978-2-7598-2367-3.
23. Osher, S.; Sethian, J.A. Fronts propagating with curvature dependent speed: Algorithms based on Hamilton-Jacobi formulations. *J. Comput. Phys.* **1988**, *79*, 12–49. [CrossRef]
24. Osher, S.; Fedkiw, R. *Level Set Methods and Dynamic Implicit Surfaces*; Springer: Cham, Switzerland, 2003.
25. Weickert, J.; Kühne, G. Fast methods for implicit active contours models. In *Geometric Level Set Methods in Imaging, Vision, and Graphics*; Springer: New York, NY, USA, 2003.
26. Cohen, L.D. On active contours models and balloons. *Comput. Vision Graph. Image Process. Image Underst.* **1991**, *53*, 211–218. [CrossRef]

27. Ronneberger, O.; Fischer, P.; Brox, T. U-net: Convolutional networks for biomedical image segmentation. In Proceedings of the MICCAI International Conference on Medical Image Computing and Computer-Assisted Intervention, Munich, Germany, 5–9 October 2015; Springer: Cham, Switzerland, 2015; pp. 234–241.
28. Menze, B.H.; Jakab, A.; Bauer, S.; Kalpathy-Cramer, J.; Farahani, K.; Kirby, J.; Burren, Y.; Porz, N.; Slotboom, J.; Wiest, R.; et al. The multimodal brain tumor image segmentation benchmark (BRATS). *IEEE Trans. Med. Imaging* **2014**, *34*, 1993–2024. [CrossRef]
29. Kingma, D.P.; Ba, J. Adam: A method for stochastic optimization, International Conference on Learning Representations. *arXiv* **2015**, arXiv:1412.6980.
30. Jaccard, P. Distribution de la flore alpine dans le bassin des Dranses et dans quelques régions voisines. *Bull. SociéTé Vaudoise Sci. Nat.* **1901**, *37*, 241–272.
31. Kim, B.; Ye, J.C. Cycle-Consistent Adversarial Network with Polyphase U-Nets for Liver Lesion Segmentation. 2018. Available online: https://openreview.net/pdf?id=SyQtAooiz (accessed on 2 March 2021).
32. Laffon, E.; Laurent, F.; Bernard, V.; De Boucaud, L.; Ducassou, D.; Marthan, R. Noninvasive assessment of pulmonary arterial hypertension by MR phase-mapping method. *J. Appl. Physiol.* **2001**, *90*, 2197–2202. [CrossRef] [PubMed]
33. Laffon, E.; Vallet, C.; Bernard, V.; Montaudon, M.; Ducassou, D.; Laurent, F.; Marthan, R. A computed method for non invasive MRI assessment of pulmonary arterial hypertension. *J. Appl. Physiol.* **2004**, *96*, 463–468. [CrossRef]
34. Debroux, N.; Le Guyader, C. A joint segmentation/registration model based on a nonlocal characterization of weighted total variation and nonlocal shape descriptors. *SIAM J. Imaging Sci.* **2018**, *11*, 957–990. [CrossRef]
35. Ozeré, S.; Le Guyader, C.; Gout, C. Joint segmentation/registration model by shape alignment via weighted total variation minimization and nonlinear elasticity. *SIAM J. Imaging Sci.* **2015**, *8*, 1981–2020. [CrossRef]
36. Gout, C.; Le Guyader, C.; Romani, L.; Saint-Guirons, A.-G. Approximation of surfaces with fault(s) and/or rapidly varying data, using a segmentation process, Dm-splines and the finite element method. *Numer. Algorithms* **2008**, *48*, 67–92. [CrossRef]
37. Le Guyader, C.; Apprato, D.; Gout, C. On the Construction of Topology-Preserving Deformation Fields. *IEEE Trans. Image Process.* **2012**, *21*, 1587–1599. [CrossRef] [PubMed]

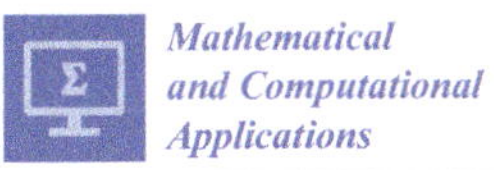
Mathematical and Computational Applications

Article
In Vivo Validation of a Cardiovascular Simulation Model in Pigs

Moriz A. Habigt [1,*], Jonas Gesenhues [2], Maike Stemmler [2], Marc Hein [1], Rolf Rossaint [1] and Mare Mechelinck [1]

1 Anaesthesiology Clinic, Faculty of Medicine, RWTH Aachen University, Pauwelsstr. 30, 52074 Aachen, Germany; mhein@ukaachen.de (M.H.); rrossaint@ukaachen.de (R.R.); mmechelinck@ukaachen.de (M.M.)
2 Institute of Automatic Control, RWTH Aachen University, Campus-Boulevard 30, 52074 Aachen, Germany; j.gesenhues@irt.rwth-aachen.de (J.G.); m.stemmler@irt.rwth-aachen.de (M.S.)
* Correspondence: mhabigt@ukaachen.de; Tel.: +49-241-80-88179

Abstract: Many computer simulation models of the cardiovascular system, of varying complexity and objectives, have been proposed in physiological science. Every model needs to be parameterized and evaluated individually. We conducted a porcine animal model to parameterize and evaluate a computer simulation model, recently proposed by our group. The results of an animal model, on thirteen healthy pigs, were used to generate consistent parameterization data for the full heart computer simulation model. To evaluate the simulation model, differences between the resulting simulation output and original animal data were analysed. The input parameters of the animal model, used to individualize the computer simulation, showed high interindividual variability (range of coefficient of variation: 10.1–84.5%), which was well-reflected by the resulting haemodynamic output parameters of the simulation (range of coefficient of variation: 12.6–45.7%). The overall bias between the animal and simulation model was low (mean: −3.24%, range: from −26.5 to 20.1%). The simulation model used in this study was able to adapt to the high physiological variability in the animal model. Possible reasons for the remaining differences between the animal and simulation model might be a static measurement error, unconsidered inaccuracies within the model, or unconsidered physiological interactions.

Keywords: computer simulation; cardiovascular system; parameterization; validation; animal model

Citation: Habigt, M.A.; Gesenhues, J.; Stemmler, M.; Hein, M.; Rossaint, R.; Mechelinck, M. In Vivo Validation of a Cardiovascular Simulation Model in Pigs. *Math. Comput. Appl.* **2022**, *27*, 28. https://doi.org/10.3390/mca27020028

Received: 25 November 2021
Accepted: 15 March 2022
Published: 18 March 2022

1. Introduction

Computer simulation models of the cardiovascular system are increasingly used in the development and improvement of ventricular assist devices (VADs) and VAD control algorithms [1–3], as well as in many other fields of cardiovascular research. With an increasing accuracy of the computer simulation models, clinical use is also imaginable but demands the possibility to highly individualize the used models. Examples of possible areas of application are the planning of high-risk operations, such as the correction of congenital heart defects or implantation of biomedical devices (e.g., valves or ventricular assist devices), to predict the cardiovascular reserve of the patient or response to a certain drug therapy. Since the 1960s, when computer simulation of the cardiovascular system started [4,5], many different simulation models have been developed. In addition to the different strategies to implement basic physiological conditions, the various models also differ in complexity and focus on specific physiological phenomena. Generally, three different types of models are distinguishable: myocardial activity can be modelled by a time-varying elastance curve [3,6–12], a sarcomere model [13–15], or isovolumetric contractions [16,17]. In the first case, emphasis is placed on the correct simulation of the starling curve; in the second case, the interaction between sarcomere and ventricular cavity mechanics is addressed; in the third case, the afterload dependency of the heart is stressed.

The arterial system has also been simulated in many ways. Lumped parameter models ("Windkessel models") [3,18] can be distinguished from tube [14,19–21] and anatomically-based distributed models [22–25]. Another important difference between the described models is the integration of different physiological regulatory mechanisms. Some models focus on the exact simulation of physiological autoregulatory mechanisms, e.g., the baroreceptor reflex [8]. In contrast, other models neglect this mechanism and emphasize further mechanisms, such as left and right ventricular interaction [11,14], coronary blood flow [15], the venous return curve [12], or cardiopulmonary interaction [26].

It is difficult to determine the most accurate and effective way to simulate specific processes in the cardiovascular system. Every simulation model needs to be evaluated individually, has its own advantages and disadvantages, and probably shows different behaviour under different circumstances. Scientists working with these models need to know the differences, possible flaws, and error susceptibility of the models they are working with.

We wanted to assess a full heart computer simulation model, which was previously described by our group [3,27], especially with regard to its ability to be adapted to individual hemodynamic settings. We, therefore, conducted a porcine animal study and used the data as the basis to parameterize the model. Then, we compared the results of the animal study to the simulative results. In a third step, we varied the input parameters to the simulation, in order to determine by which parameters the results were affected the most and must, therefore, be especially critical to time-varying elastance simulation models of the cardiovascular system.

2. Materials and Methods

2.1. Animal Model

2.1.1. Induction and Maintenance of Anaesthesia

All procedures described below are compliant with the Guide for the Care and Use of Laboratory Animals [28] and reviewed and approved by the local animal care committee and governmental animal care office (No. 84-02.04.2013. A476 and 8.87-50.10.45.08.257; Landesamt für Natur, Umwelt und Verbraucherschutz Nordrhein-Westfalen, Recklinghausen, Germany). Thirteen healthy pigs (*German landrace*, 46.26 kg ± 4.46 kg bodyweight [BW]) received premedication by intramuscular injection of 4 mg/kg azaperone (Elanco Tiergesundheit AG, Basel, Switzerland) and were anaesthetized by intravenous injection of 3 mg/kg BW propofol (Hexal AG, Holzkirchen, Germany) for oral intubation. Anaesthesia was maintained by the insufflation of 0.9–1.2 vol% isoflurane and continuous application of 6–8 µg/kg BW/h fentanyl (Ratiopharm GmbH, Ulm, Germany). Normoventilation was achieved by the application of a tidal volume of 8–10 mL/kg BW and monitored by expiratory CO_2 measurement ($PaCO_2$ 36–42 mmHg) and regular arterial blood gas analysis. Electrolytes and blood glucose were similarly monitored by arterial blood gas analysis and held in a physiological range. The haematocrit was stabilized by the infusion of 6–10 mL/kg BW/h of a balanced crystalloid solution (Sterofundin Iso Braun, B.Braun AG, Melsungen, Germany) solution and application of 500 mL of a balanced colloid solution (Gelafundin Iso Braun) after instrumentation. The body temperature was held constant (38 °C) by the use of an airflow warming blanket.

2.1.2. Surgical Instrumentation

After dissecting the neck vessels on the right side, one central venous catheter was introduced into the right internal jugular vein, and two 12 F sheaths were introduced into the right carotid artery. A median thoracotomy was performed, and the pericardium was opened longitudinally. The aorta and pulmonary arteries were separated, and a perivascular ultrasound transit time flow probe (MA 20 PAX; Transonic Systems Europa, Maastricht, The Netherlands) was positioned centrally around each vessel and connected to a flow meter (T402-PV, Transonic Systems Europa). To measure pulmonary and aortic pressure, a solid-state pressure sensor (CA-61000-PL, CD Leycom, Zoetemeer, The Netherlands) was introduced through a stab wound in the right ventricular outflow tract, with another equal sensor through the sheath in the right carotid artery. The pressure sensors were positioned 3–4 cm distal to the respective valves (i.e., the pulmonary and aortic valves) and connected to a pressure interface (Sentron, CD Leycom). A multi-segment dual-field 7F conductance catheter (SPR-570-7; Millar Instruments, Houston, TX, USA) was placed through the apex of the right ventricle along the outflow track. A second equal catheter was introduced through the second right carotid sheath and aortic valve, with the tip placed in the left ventricular apex. Echocardiographic imaging was performed to verify the correct positioning of the catheter. The volume segments of the catheters were connected to two signal processors (Sigma-5 DF, CD Leycom). To avoid electrical interferences and for simultaneous biventricular measurements, the excitation frequency of one processor was modified from 20 to 15 kHz. The pressure sensors of the catheters were connected to a pressure interface (PCU-2000; Millar Instruments). Through a sheath in the right femoral vein, a 7F balloon catheter was placed in the inferior vena cava, in order to perform a short-term preload reduction during apnoea and constant positive end expiratory pressure (PEEP). A schematic of the instrumentation is shown in Figure 1. After the completion of instrumentation, the animals recovered for 30 min, with continued isoflurane and fentanyl narcosis, in order to achieve stable blood pressure, cardiac output (CO), and normothermia.

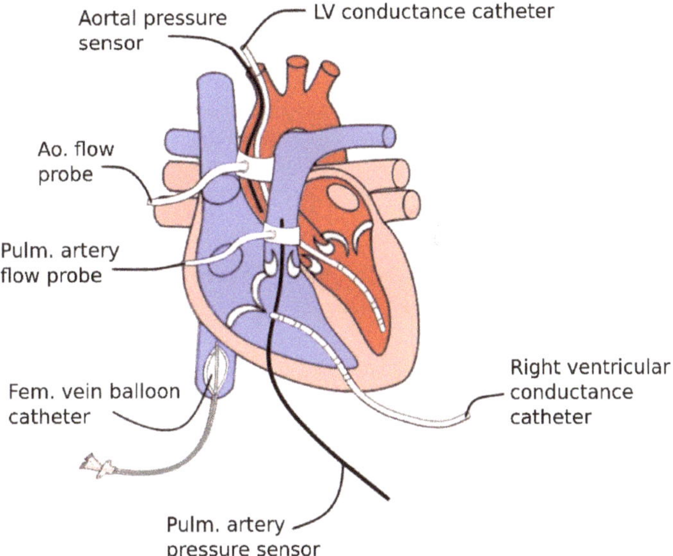

Figure 1. Instrumentation of the heart with left ventricular conductance catheter, aortal pressure sensor, aortal flow probe, pulmonary artery flow probe, femoral vein balloon catheter, pulmonary artery pressure sensor, and right ventricular conductance catheter. Figure is a modification of figure "Heart normal" by Eric Pierce [29]. The original figure can freely be published under CC BY-SA license (https://en.wikipedia.org/wiki/User:Wapcaplet, accessed on 16 March 2022).

2.2. Data Acquisition and Calculations

Signals were acquired continuously, at a sampling rate of 1000 Hz, using a data acquisition device (Powerlab, AD Instruments, Dunedin, New Zealand) and software (LabChart, AD Instruments).

The conductance volume values were calibrated prior to the measurement of each animal. Therefore, the volume signal was corrected by stroke volume (SV), obtained from the aortic flow probe (slope factor α) and parallel conductance, calculated from venous hypertonic saline injections, as described previously [30–32]. The signals were finally analysed off-line with custom-made software (CIRCLAB 2020; Paul Steendijk, Leiden, The Netherlands).

Global haemodynamics were described by heart rate (HR), mean arterial pressure (MAP), CO, SV, and ejection fraction (EF). End-systolic pressure (P_ES), volume (V_ES), end-diastolic pressure (P_ED), and volume (V_ED) were used to describe the ventricular dimensions. These parameters were calculated as averages over 20 s. Systolic function was characterized by the end-systolic pressure volume relationship (ESPVR) and preload recruitable stroke work (PRSW), obtained from pressure-volume loops acquired during short preload reduction by caval occlusion during apnoea. The ESPVR is the linear regression of the end-systolic pressure volume points and characterized by the slope, end-systolic elastance (E_ES), and x-axis intercept (V0_ES). The PRSW is the slope of the linear regression between the ventricular stroke work (SW) and end-diastolic volume (EDV). The passive (stiffness) components of ventricular relaxation were displayed by the exponential regression of the end-diastolic pressure volume points (end-diastolic pressure volume relationship: EDPVR), which was characterized by the indices P0, V0, and λ.

$$\text{P\_ED} = \text{P0\_ES} * (\exp(\lambda(\text{V\_ED} - \text{V0})) - 1 \quad (1)$$

as described by Wang et al. [32]. We iteratively calculated the single indices of the equation. Values from three to five consecutive recordings were averaged.

The compliant characteristics (Cs) of the Windkessel vessels were calculated by dividing the SV by the difference in systolic and diastolic pressure (pulse pressure [PP]), as previously described [33]. Fourier series expressions for pressure and flow signals of 20 s duration were used to calculate systemic vascular resistance and impedance. The impedance modulus at each frequency was calculated as the ratio between pressure and flow moduli (amplitudes). The total resistance (Z0) and characteristic impedance (Zc) were derived from moduli at zero frequency and an average of moduli between 2 and 15 Hz [34].

2.3. Simulation Model

An electrical analogue of the circulation model is given in Figure 2.

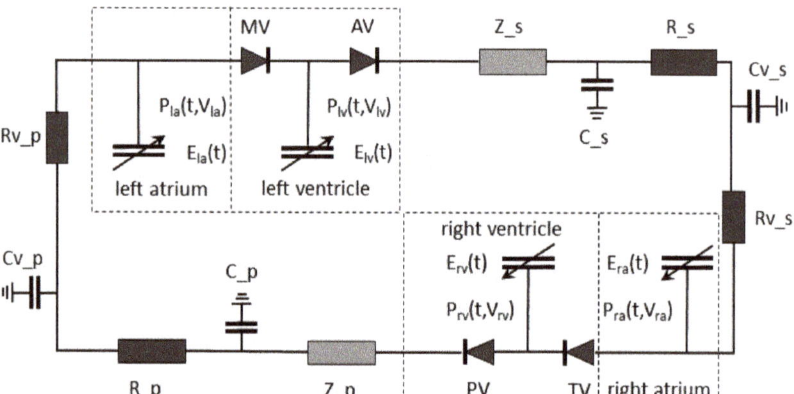

Figure 2. Electrical analogue of the circulation model. The single components are denoted by the following designation. Vascular and specific resistance: R, Z. Vascular compliances: C. Venous system: v. systemic circulatory system: _s. Pulmonary circulatory system: _p. Valves: TV, PV, MV, and AV. Elastance function of the atria and ventricles: E. A closer description of the single components is given in Table S1 in the supplemental materials.

A three-element Windkessel model was used to describe the systemic and pulmonary vascular systems. Each arterial system consisted of two hydraulic resistances with

$$Q_{Resistance}(t) = \frac{\Delta P_{Resistance}(t)}{R_i} \tag{2}$$

and an interconnected compliancy with

$$P_{Compliancy}(t) = \frac{V_{Compliancy}(t)}{C_i} \tag{3}$$

The venous system consisted of one hydraulic resistance and one connected compliance only.

The pulsation of the heart chambers and atria was modelled as a nonlinear time-varying elastance, where the ventricle volume determined the corresponding time-varying pressures $P(t) = f(V(t) - V0)$, with unstressed volumes V0. The time varying-elastance function used in our model was based on the work of Chung et al. [35], whereas concrete values were obtained from previously published animal data [36]. The filling and ejection phases were characterized by the exponential EDPVR

$$P\_ED(V) = P0\_ES^{\lambda(V - V0\_ED)} - 1 \tag{4}$$

and linear ESPVR

$$P\_ES(V) = E\_ES \cdot V + V0\_ES \tag{5}$$

where E_ES is the slope, and V0_ES is the x-axis offset of the specific relationship.

On this basis, the instantaneous pressure could be determined by

$$P(V,t) = \varphi(t) \cdot P\_ES(V) + (1 - \varphi(t)) \cdot P\_ED(V). \tag{6}$$

The activation function, $\varphi(t)$, triggered the systole and ran periodically between 0 and 1. One period of the activation function was defined as the sum of exponential functions

$$\Phi(t) = \sum_{i=1}^{5} = A_i \cdot e^{-\left(\frac{t - B_i}{C_i}\right)^2}. \tag{7}$$

The parameters A_i, B_i, and C_i were obtained using experimental data, as described by Gesenhues et al. [17]. This function was used with a normalized time signal

$$t_N = (t - k_D) \cdot HR/60 - \text{floor}((t - k_D) \cdot HR/60) + k_O \tag{8}$$

to obtain the activation function signal $\varphi(t)$

$$\varphi(t) = \Phi(t = t_N(t)). \tag{9}$$

The normalized time signal, t_N, restarts after each beat was implemented, using a shifted sawtooth wave with a frequency equal to the heart rate. The floor function mapped a real number to the largest previous integer. The shape of the activation function and consecutive left ventricular volume and pressure values for animal 4 are shown in Figure 3.

$$Q_{Valve,ideal}(t) = \begin{cases} \frac{\Delta P_{Valve}(t)}{R_{Valve}} & if\ \Delta P_{Valve}(t) > 0 \\ 0 & if\ \Delta P_{Valve}(t) \leq 0 \end{cases} \tag{10}$$

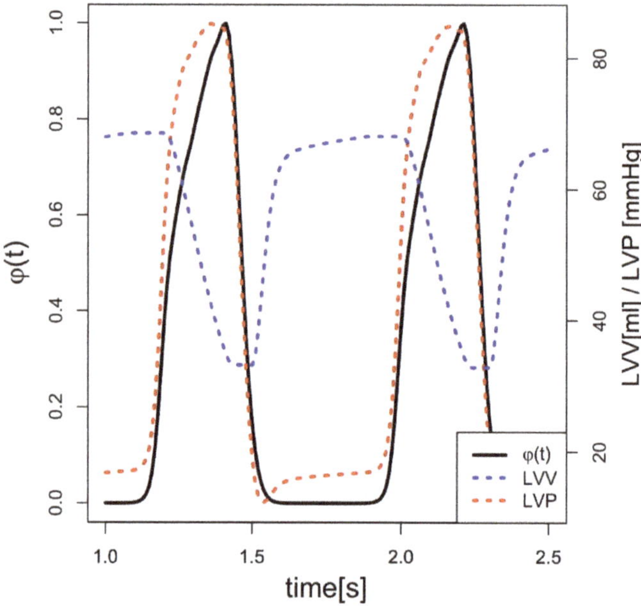

Figure 3. Activation function. Simulated activation function (ϕ(t)—black), consecutive left ventriular volume (LVV—blue), and left ventricular pressure (LVP—red) in animal 4.

The heart valves acted as non-return valves, which meant that they allowed flow at a given hydraulic resistance R_{Valve} in only one direction, depending on the pressure differential ΔP_{Valve} across the valve. A detailed description of the single components is given in Table S1 of the supplemental materials.

This simplified closed circuit model of the whole circulatory system was implemented in the open (object-oriented) modelling language Modelica®. We ran this model in our self-developed simulation tool ModeliChart, which was especially designed for clinical users and allows for independently performing in silico studies in real time [37]. The model was a full circulatory model, as previously described [3,27]. In contrast to our latter study [3], we did not use a VAD in this study.

To parametrize the simulation according to the animal data, we used the general configuration of settings described in [3], which is based on the data described in the literature, and only individually adjusted ten input parameters. HR, E_ES, V0_ES, P0_ED, V0_ED, λ, C, Z0, and Zc were calculated, as described above, and could be directly adopted into the simulation. Each animal was simulated individually. The unstressed volume of the systemic venous compliance (V0_sC) was used to adjust the ventricular preload until the EDV between the simulation and animal matched. A short increase of V0_sC was used as a virtual preload reduction to calculate the slope of the PRSW, as shown in Figure 4. The respective values of P_ES, P_ED, V_ES, V_ED, SV, EF, SW, MAP, and PRSW were used to compare individual values from animals and simulation.

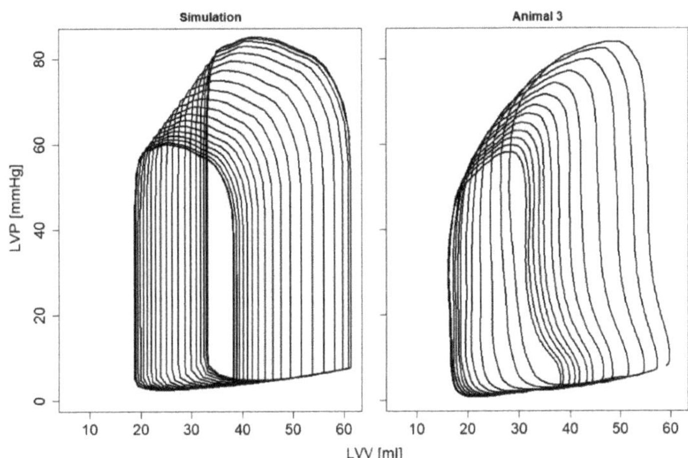

Figure 4. Pressure and volume data of a virtual and in-vivo preload reduction of the simulated and real animal 3. LVP: left ventricular pressure; LVV: left ventricular volume.

2.4. Statistics

The results are displayed as the mean and standard deviation (SD) (PRISM 6.0, GraphPad Software, San Diego, CA, USA). The coefficient of variation (CV) was used to describe interindividual variability and Bland–Altman plots to describe the agreement (bias) and its 95% limits (PRISM 6.0, GraphPad Software).

3. Results

Three out of thirteen pigs could not be analysed: two pigs showed acute bleeding during dissection and the other pig showed an atrial septal defect, which only came obvious during the operation. The input parameters, which were necessary to individualize the simulation model, demonstrated a large interindividual variability (CV = 10.1–84.5%) for the systolic and diastolic function parameters of the left ventricle and parameters characterizing the systemic circulation (Table 1).

The resulting haemodynamic output parameters from the simulation (Figure 4) similarly showed a high interindividual variation (CV = 12.6–45.7%) and low overall bias between the two datasets could be observed (mean: −3.24%; range: −26.5% to 20.1%). The smallest bias between the animal model and simulation was reached for the EF and MAP, with each 5% mean difference between all values. The largest bias, with a mean difference of 27–28% between all values, was observed in the P_ED and PRSW (Table 2).

The best matches between the animal model and simulation (under 5% difference of the values) were reached for SV in animals 9 and 10, in the V_ES in animals 2, 5, and 10, in the EF in animals 9 and 10, in the MAP in animals 3, 5, 7, 9, and 10, and in the PRSW in animal 1. The most pronounced differences between animal study and simulation (more than 25% difference) were observed in the P_ED in animals 2, 8 and 10, the SW in animals 2, 7, 8, and 10, and the PRSW in animals 2, 3, 4, 7, 8, and 10 (Figure 5).

Table 1. Parameters derived from the animal studies, which were used to parametrize the simulation model: HR: heartrate [bpm], E_ES: slope of the ESPVR [mmHg/mL], V0_ES: unstressed volume of the ESPVR [mL], λ: coefficient of the exponential EDPVR, P0_ED: coefficient of the exponential EDPVR [mmHg], V0_ED: unstressed volume of the EDPVR [mL], Cs: systemic arterial compliance [mmHg/mL], Z0: systemic arterial resistance [mmHg*s/mL], Zc: specific aortal resistance [mmHg*s/mL], V0_sC: unstressed volume of the systemic venous compliance [mL], and CV: coefficient of variation [%].

Input Parameter	Animal	HR	E_ES	V0_ES	λ	P0_ED	V0_ED	Cs	Z0	Zc	V0_sC (Simulation Only)
Simulation/Animal	1	101	1.33	−19.02	0.0365	0.2126	−19.36	1.7298	0.7341	0.0894	1100
	2	91	1.14	−28.94	0.0478	0.0394	−50.48	0.9835	1.0017	0.1029	1865
	3	100	1.41	−23.62	0.0263	0.3675	−63.38	1.7140	1.0959	0.1102	1700
	4	75	1.01	−48.25	0.0253	0.7728	−55.30	1.5443	1.3407	0.0944	1680
	5	79	1.13	−51.44	0.0236	0.2142	−64.57	2.2253	1.2959	0.0994	1382
	6	84	1.60	−8.09	0.0254	0.3502	−67.20	1.9238	0.8044	0.0888	580
	7	85	0.80	−48.74	0.0280	0.5369	−38.80	1.7860	0.7217	0.0840	1060
	8	91	1.04	−30.67	0.0444	0.0110	−80.21	1.7580	0.8584	0.1142	1600
	9	98	1.52	−24.77	0.0218	0.5375	−61.07	1.4927	1.1763	0.1074	1730
	10	106	1.34	−14.98	0.0413	0.0053	−113.60	1.2890	0.7172	0.0951	1250
	Mean	49.2	1.23	−29.85	0.0320	0.3047	−61.4	1.6446	0.9746	0.0986	1395
	CV [%]	22.0	20.28	50.4	29.8	84.5	40.5	20.8	24.8	10.1	28.6

Table 2. Results of the Bland–Altman analysis. Shown are the mean of the difference between in vivo data and simulation in percentage (BIAS [%]), 95% limits of agreement, and coefficient of variation for the end-systolic and -diastolic pressure (P_ES, P_ED), stroke volume (SV), end-diastolic and -systolic volume (V_ED, V_ES), ejection fraction (EF), stroke work (SW), mean arterial pressure (MAP), and preload recruitable stroke work (PRSW).

	P_ES	P_ED	SV	V_ED	V_ES	EF	SW	MAP	PRSW
BIAS [%]	13.1	28.0	−21.6	−0.1	9.8	−4.9	−21.3	−5.0	−27.2
95% Limits of Agreement									
From	−3.6	−8.9	−35.4	−1.7	−13.6	−27.4	−49.7	−19.0	−79.5
To	29.9	64.8	−7.7	1.5	33.3	17.6	7.0	8.9	25.1
Coefficient of variation	15.4	45.7	22.7	20.2	24.9	12.6	29.1	17.1	18.0

In animal 2, the percentage difference between the animal model and simulation for P_ED, P_ES, and SW did not lie within an agreement interval of two standard deviations. Similar results were obtained for the PRSW in animal 7 (Figure 6).

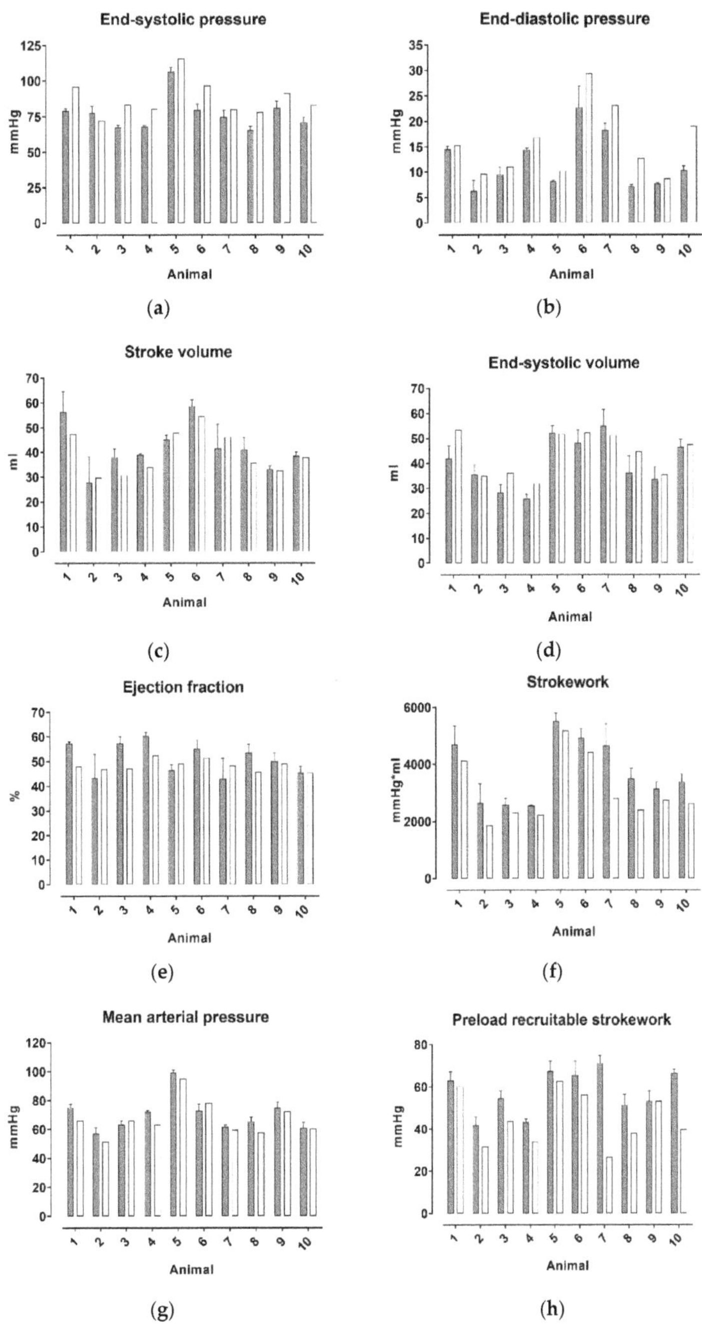

Figure 5. Comparison of animals and simulation. Comparison between the results for the output parameters of animal study (grey) and computer simulation (white) (mean ± SD) of selected haemodynamic parameters: end-systolic pressure (**a**), end-diastolic pressure (**b**), stroke volume (**c**), end-systolic volume (**d**), ejection fraction (**e**), stroke work (**f**), mean arterial pressure (**g**) and preload recruitable stroke work (**h**).

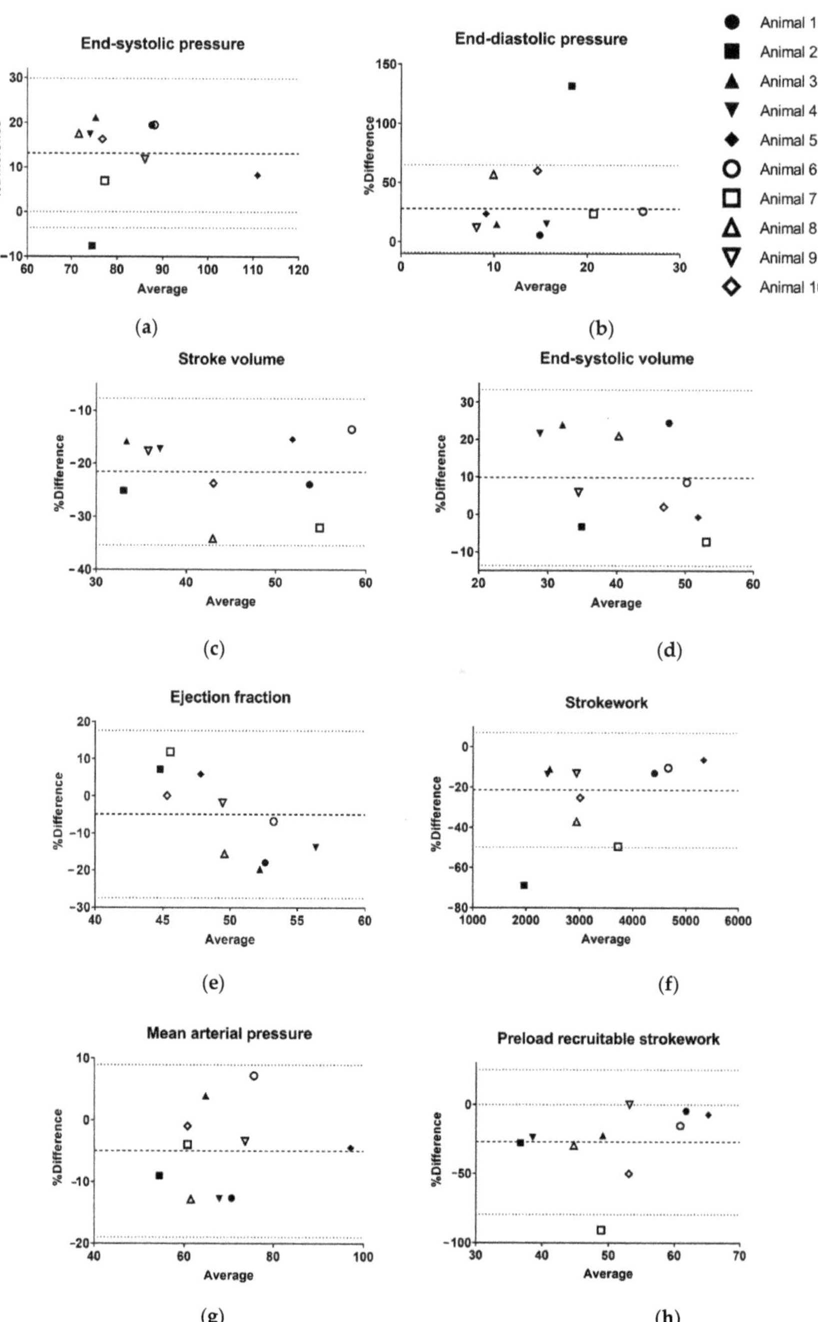

Figure 6. Bland–Altman plots for output parameters of selected haemodynamic parameters: end-systolic pressure (**a**), end-diastolic pressure (**b**), stroke volume (**c**), end-systolic volume (**d**), ejection fraction (**e**), stroke work (**f**), mean arterial pressure (**g**) and preload recruitable stroke work (**h**). The mean difference between the in vivo and simulation data, in percentage, is shown as a detached line. The 95% limits of agreement are indicated by dotted lines.

4. Discussion

Our data show a high interindividual variability of parameters, characterizing cardiac function and systemic circulation, among the single animals, and good agreement between simulation and in vivo data. This high interindividual variability is also present in the simulative results, which suggests that a relatively simple computer simulation, in the form of our approach (time varying elastance, linear ESPVR, exponential EDPVR, and three-element Windkessel model), can adapt to a high physiological variability.

The observed differences between simulation and animal might be a static measurement failure, which might result from unknown inaccuracies within the model (e.g., determination of systemic compliance) or unconsidered physiological interactions. There are two conditions that make our simulation prone to static measurement failure. First, minimal changes in the exponential EDPVR have a substantial impact on the simulation results; second, the parameterization of the three-element Windkessel model has inherent problems in determining the compliance and resulting characteristic impedance.

The problem of determining systemic compliance has been an ongoing issue for several decades. Several methods have been proposed and discussed in earlier publications [33,38–40]. We decided to apply the stroke volume-to-pulse pressure ratio method [38], for reasons of substantive conviction and practicability. How this choice influences the accuracy of our simulation model is difficult to estimate. Comparing an application of the different methods within our simulation model lies beyond the scope of this paper.

Considering the Bland–Altman plot of the simulated EF (Figure 6), the distribution of the observed differences between the animal and simulation models implies a linearity that might be evidence for a systematic error: in the case of a high EF in the animal, the EF in the simulation is systematically underrated and vice versa.

We decided to implement the ESPVR as the linear ratio of pressure and EDV. Considering the literature, it is not clarified whether this strategy is correct. Some authors describe the ESPVR as linear [41–43], while others describe it as parabolic [44–46].

In our model, the contractility of the ventricle is simulated by a time-varying elastance function. It could be discussed whether this strategy is the best approach. Modelling the time-varying elastance as being load-independent might not apply, at least when the cardiac load is changed by a VAD [47]. Due to a lack of other direct comparisons between animal and simulated data, it is not clear whether other approaches to model cardiac activity [13–16] are superior to our approach.

As mentioned above, there are also unconsidered physiological mechanisms not included in this model, which might be underestimated in their importance. In the current simulation model, we do not consider the ventricular interaction through the ventricular septum, as it has been done by other groups [11,14,48]. Simulating the interventricular septum as a rigid ventricular wall might insufficiently mirror phenomena in certain situations, e.g., when the simulation model is used to simulate integrated VAD pumps at a high pump speed, where suction effects might occur. Additionally, we did not model the interaction between the coronary perfusion and contractility of the ventricles [15], cardiopulmonary pressure interaction [26], or venous return curve [12]. Autoregulation mechanisms, such as the baroreceptor reflex [8] or Anrep effect [49], are also neglected. The disregard of these mechanisms might seem careless but offers the advantage of a relatively simple computer model, which seems favourable, regarding teaching purposes and clinical implementation. Moreover, the simplicity of the simulation model enables the augmentation of the model by certain features, according to the respective scientific question. For example, the interventricular wall interaction presented in the model of Smith et al. [11] could also be introduced in our model, as both models are based on the time varying elastance.

Another factor that has a high impact on the simulatory results is the modelling of the vascular system. We chose the three-element Windkessel model [50], as this model has been well-examined and established [9,12,13]. However, there are other approaches [14] that might be favourable, considering certain problems, e.g., the difficult determination of compliance and resulting characteristic impedance, as mentioned above.

Our animal data vary significantly between the different healthy animals. If human data show comparable variability, which can be assumed, especially in ill patients, then simulation algorithms for clinical decision-making must be easily adaptable and simple to parametrize, which will be easily achievable with simple models.

5. Conclusions

The direct clinical use of our model does yet not seem feasible, as the parameterization values must be generated invasively. However, the presented data and approaches might help to develop simple, reliable computer simulation models that might be transferred to the clinic in the future.

Thus, the simulation of cardiovascular processes is a reliable tool, if existing limitations are considered. To further improve the accuracy of cardiovascular computer simulation models and transfer these tools to the clinic, further studies are needed.

Supplementary Materials: The following supporting information can be downloaded at: https://www.mdpi.com/article/10.3390/mca27020028/s1, Table S1: Non-individualized simulation parameters.

Author Contributions: Conceptualization, M.A.H., M.H. and M.M.; software, J.G. and M.A.H.; investigation, M.A.H., J.G., M.S., M.H. and M.M.; formal analysis, M.A.H. and M.H.; data curation, M.A.H. and M.H.; writing—original draft preparation, M.A.H.; writing—review and editing, M.H. and M.M.; visualization, M.A.H.; funding acquisition, R.R. All authors have read and agreed to the published version of the manuscript.

Funding: This research was funded by the German Research Foundation (DFG), as project Smart Life Support 2.0, grant number 65/15–1 RO 2000/17–1.

Institutional Review Board Statement: All procedures described in this study are compliant with the Guide for the Care and Use of Laboratory Animals and reviewed and approved by the local animal care committee and governmental animal care office (No. 84-02.04.2013. A476 and 8.87-50.10.45.08.257; Landesamt für Natur, Umwelt und Verbraucherschutz Nordrhein-Westfalen, Recklinghausen, Germany).

Data Availability Statement: The datasets generated and analyzed during the current study are available from the corresponding author on reasonable request.

Conflicts of Interest: The authors declare no conflict of interest.

References

1. Scardulla, F.; Pasta, S.; D'Acquisto, L.; Sciacca, S.; Agnese, V.; Vergara, C.; Quarteroni, A.; Clemenza, F.; Bellavia, D.; Pilato, M. Shear stress alterations in the celiac trunk of patients with a continuous-flow left ventricular assist device as shown by in-silico and in-vitro flow analyses. *J. Heart Lung Transplant.* **2017**, *36*, 906–913. [CrossRef] [PubMed]
2. Scardulla, F.; Bellavia, D.; D'Acquisto, L.; Raffa, G.M.; Pasta, S. Particle image velocimetry study of the celiac trunk hemodynamic induced by continuous-flow left ventricular assist device. *Med. Eng. Phys.* **2017**, *47*, 47–54. [CrossRef]
3. Habigt, M.; Ketelhut, M.; Gesenhues, J.; Schrodel, F.; Hein, M.; Mechelinck, M.; Schmitz-Rode, T.; Abel, D.; Rossaint, R. Comparison of novel physiological load-adaptive control strategies for ventricular assist devices. *Biomed. Tech. Berl.* **2017**, *62*, 149–160. [CrossRef] [PubMed]
4. Noordergraaf, A.; Verdouw, P.D.; Boom, H.B. The use of an analog computer in a circulation model. *Prog. Cardiovasc. Dis.* **1963**, *5*, 419–439. [CrossRef]
5. Guyton, A.C.; Coleman, T.G.; Granger, H.J. Circulation: Overall regulation. *Annu. Rev. Physiol.* **1972**, *34*, 13–44. [CrossRef] [PubMed]
6. Suga, H.; Sagawa, K.; Shoukas, A.A. Load independence of the instantaneous pressure-volume ratio of the canine left ventricle and effects of epinephrine and heart rate on the ratio. *Circ. Res.* **1973**, *32*, 314–322. [CrossRef]
7. Leaning, M.; Pullen, H.E.; Carson, E.R.; Finkelstein, L. Modelling a complex biological system: The human cardiovascular system—1. Methodology and model description. *Trans. Inst. Meas. Control.* **1983**, *5*, 71–86. [CrossRef]
8. Ursino, M. Interaction between carotid baroregulation and the pulsating heart: A mathematical model. *Am. J. Physiol.* **1998**, *275*, H1733–H1747. [CrossRef]
9. Segers, P.; Stergiopulos, N.; Schreuder, J.J.; Westerhof, B.E.; Westerhof, N. Left ventricular wall stress normalization in chronic pressure-overloaded heart: A mathematical model study. *Am. J. Physiol. Heart Circ. Physiol.* **2000**, *279*, H1120–H1127. [CrossRef]
10. Vollkron, M.; Schima, H.; Huber, L.; Wieselthaler, G. Interaction of the Cardiovascular System with an Implanted Rotary Assist Device: Simulation Study with a Refined Computer Model. *Artif. Organs* **2002**, *26*, 349–359. [CrossRef]

11. Smith, B.W.; Chase, J.G.; Nokes, R.I.; Shaw, G.M.; Wake, G. Minimal haemodynamic system model including ventricular interaction and valve dynamics. *Med. Eng. Phys.* **2004**, *26*, 131–139. [CrossRef] [PubMed]
12. Colacino, F.M.; Moscato, F.; Piedimonte, F.; Arabia, M.; Danieli, G.A. Left ventricle load impedance control by apical VAD can help heart recovery and patient perfusion: A numerical study. *ASAIO J.* **2007**, *53*, 263–277. [CrossRef] [PubMed]
13. Welp, C.; Werner, J.; Böhringer, D.; Hexamer, M. Investigation of cardiac and cardio-therapeutical phenomena using a pulsatile circulatory model. *Biomed. Tech.* **2004**, *49*, 327–331. [CrossRef]
14. Arts, T.; Delhaas, T.; Bovendeerd, P.; Verbeek, X.; Prinzen, F.W. Adaptation to mechanical load determines shape and properties of heart and circulation: The CircAdapt model. *Am. J. Physiol. Heart Circ. Physiol.* **2005**, *288*, H1943–H1954. [CrossRef]
15. Bovendeerd, P.H.; Borsje, P.; Arts, T.; van De Vosse, F.N. Dependence of intramyocardial pressure and coronary flow on ventricular loading and contractility: A model study. *Ann. Biomed. Eng.* **2006**, *34*, 1833–1845. [CrossRef]
16. Ottesen, J.T.; Danielsen, M. Modeling ventricular contraction with heart rate changes. *J. Theor. Biol.* **2003**, *222*, 337–346. [CrossRef]
17. Gesenhues, J.; Hein, M.; Albin, T.; Rossaint, R.; Abel, D. Cardiac Modeling: Identification of Subject Specific Left-Ventricular Isovolumic Pressure Curves. *IFAC-PapersOnLine* **2015**, *48*, 581–586. [CrossRef]
18. Westerhof, N.; Elzinga, G.; Sipkema, P. An artificial arterial system for pumping hearts. *J. Appl. Physiol.* **1971**, *31*, 776–781. [CrossRef]
19. Burattini, R.; Knowlen, G.G.; Campbell, K.B. Two arterial effective reflecting sites may appear as one to the heart. *Circ. Res.* **1991**, *68*, 85–99. [CrossRef]
20. O'Rourke, M.F. Pressure and flow waves in systemic arteries and the anatomical design of the arterial system. *J. Appl. Physiol.* **1967**, *23*, 139–149. [CrossRef]
21. Wetterer, E.; Kenner, T. *Grundlagen der Dynamik des Arterienpulses*; Springer: Berlin/Heidelberg, Germany, 1968.
22. O'Rourke, M.F.; Avolio, A.P. Pulsatile flow and pressure in human systemic arteries. Studies in man and in a multibranched model of the human systemic arterial tree. *Circ. Res.* **1980**, *46*, 363–372. [CrossRef] [PubMed]
23. Stergiopulos, N.; Young, D.F.; Rogge, T.R. Computer simulation of arterial flow with applications to arterial and aortic stenoses. *J. Biomech.* **1992**, *25*, 1477–1488. [CrossRef]
24. Westerhof, N.; Bosman, F.; De Vries, C.J.; Noordergraaf, A. Analog studies of the human systemic arterial tree. *J. Biomech.* **1969**, *2*, 121–143. [CrossRef]
25. Avolio, A.P. Multi-branched model of the human arterial system. *Med. Biol. Eng. Comput.* **1980**, *18*, 709–718. [CrossRef] [PubMed]
26. Lu, K.; Clark, J.W., Jr.; Ghorbel, F.H.; Ware, D.L.; Bidani, A. A human cardiopulmonary system model applied to the analysis of the Valsalva maneuver. *Am. J. Physiol. Heart Circ. Physiol.* **2001**, *281*, H2661–H2679. [CrossRef] [PubMed]
27. Gesenhues, J.; Hein, M.; Habigt, M.; Mechelinck, M.; Albin, T.; Abel, D. Nonlinear object-oriented modeling based optimal control of the heart: Performing precise preload manipulation maneuvers using a ventricular assist device. In Proceedings of the 2016 European Control Conference (ECC), Aalborg, Denmark, 29 June–1 July 2016; pp. 2108–2114.
28. National Research Council Committee for the Update of the Guide for the Care and Use of Laboratory Animals. The National Academies Collection: Reports Funded by National Institutes of Health. In *Guide for the Care and Use of Laboratory Animals*; National Academies Press, National Academy of Sciences: Washington, DC, USA, 2011.
29. Pierce, E. Wikipedia, File:Heart normal.svg. Available online: https://en.wikipedia.org/wiki/File:Heart_normal.svg (accessed on 16 March 2022).
30. Baan, J.; Van Der Velde, E.T.; De Bruin, H.G.; Smeenk, G.J.; Koops, J.; Van Dijk, A.D.; Temmerman, D.; Senden, J.; Buis, B. Continuous measurement of left ventricular volume in animals and humans by conductance catheter. *Circulation* **1984**, *70*, 812–823. [CrossRef]
31. De Vroomen, M.; Cardozo, R.H.L.; Steendijk, P.; van Bel, F.; Baan, J. Improved contractile performance of right ventricle in response to increased RV afterload in newborn lamb. *Am. J. Physiol. Heart Circ. Physiol.* **2000**, *278*, H100–H105. [CrossRef]
32. Steendijk, P.; Baan, J. Comparison of intravenous and pulmonary artery injections of hypertonic saline for the assessment of conductance catheter parallel conductance. *Cardiovasc. Res.* **2000**, *46*, 82–89. [CrossRef]
33. Liu, Z.; Brin, K.P.; Yin, F.C. Estimation of total arterial compliance: An improved method and evaluation of current methods. *Am. J. Physiol.* **1986**, *251 Pt 2*, H588–H600. [CrossRef]
34. Brimioulle, S.; Maggiorini, M.; Stephanazzi, J.; Vermeulen, F.; Lejeune, P.; Naeije, R. Effects of low flow on pulmonary vascular flow-pressure curves and pulmonary vascular impedance. *Cardiovasc. Res.* **1999**, *42*, 183–192. [CrossRef]
35. Chung, D.C.; Niranjan, S.C.; Clark, J.W., Jr.; Bidani, A.; Johnston, W.E.; Zwischenberger, J.B.; Traber, D.L. A dynamic model of ventricular interaction and pericardial influence. *Am. J. Physiol.* **1997**, *272 Pt 2*, H2942–H2962. [PubMed]
36. Hein, M.; Roehl, A.B.; Baumert, J.H.; Bleilevens, C.; Fischer, S.; Steendijk, P.; Rossaint, R. Xenon and isoflurane improved biventricular function during right ventricular ischemia and reperfusion. *Acta Anaesthesiol. Scand.* **2010**, *54*, 470–478. [CrossRef] [PubMed]
37. Gesenhues, J.; Hein, M.; Ketelhut, M.; Habigt, M.; Rüschen, D.; Mechelinck, M.; Albin, T.; Leonhardt, S.; Schmitz-Rode, T.; Rossaint, R.; et al. Benefits of object-oriented models and ModeliChart: Modern tools and methods for the interdisciplinary research on smart biomedical technology. *Biomed. Tech.* **2017**, *62*, 111–121. [CrossRef]
38. Segers, P.; Verdonck, P.; Deryck, Y.; Brimioulle, S.; Naeije, R.; Carlier, S.; Stergiopulos, N. Pulse pressure method and the area method for the estimation of total arterial compliance in dogs: Sensitivity to wave reflection intensity. *Ann. Biomed. Eng.* **1999**, *27*, 480–485. [CrossRef] [PubMed]

39. Stergiopulos, N.; Meister, J.J.; Westerhof, N. Simple and accurate way for estimating total and segmental arterial compliance: The pulse pressure method. *Ann. Biomed. Eng.* **1994**, *22*, 392–397. [CrossRef] [PubMed]
40. Gesenhues, J.; Hein, M.; Ketelhut, M.; Albin, T.; Abel, D. Towards Medical Cyber-Physical Systems: Modelica and FMI based Online Parameter Identification of the Cardiovascular System. In Proceedings of the 12th International Modelica Conference, Prague, Czech Republic, 15–17 May 2017; Linköping University Electronic Press: Linköping, Sweden, 2017; pp. 613–621.
41. Brimioulle, S.; Wauthy, P.; Ewalenko, P.; Rondelet, B.; Vermeulen, F.; Kerbaul, F.; Naeije, R. Single-beat estimation of right ventricular end-systolic pressure-volume relationship. *Am. J. Physiol. Heart Circ. Physiol.* **2003**, *284*, H1625–H1630. [CrossRef] [PubMed]
42. Freeman, G.L.; Little, W.C.; O'Rourke, R.A. The effect of vasoactive agents on the left ventricular end-systolic pressure-volume relation in closed-chest dogs. *Circulation* **1986**, *74*, 1107–1113. [CrossRef] [PubMed]
43. Little, W.C.; Cheng, C.P.; Mumma, M.; Igarashi, Y.; Vinten-Johansen, J.; Johnston, W.E. Comparison of measures of left ventricular contractile performance derived from pressure-volume loops in conscious dogs. *Circulation* **1989**, *80*, 1378–1387. [CrossRef] [PubMed]
44. Burkhoff, D.; Sugiura, S.; Yue, D.T.; Sagawa, K. Contractility-dependent curvilinearity of end-systolic pressure-volume relations. *Am. J. Physiol.* **1987**, *252*, H1218–H1227. [CrossRef]
45. Kass, D.A.; Beyar, R.; Lankford, E.; Heard, M.; Maughan, W.L.; Sagawa, K. Influence of contractile state on curvilinearity of in situ end-systolic pressure-volume relations. *Circulation* **1989**, *79*, 167–178. [CrossRef]
46. Nordhaug, D.; Steensrud, T.; Korvald, C.; Aghajani, E.; Myrmel, T. Preserved myocardial energetics in acute ischemic left ventricular failure—Studies in an experimental pig model. *Eur. J. Cardiothorac. Surg.* **2002**, *22*, 135–142. [CrossRef]
47. Vandenberghe, S.; Segers, P.; Steendijk, P.; Meyns, B.; Dion, R.A.; Antaki, J.F.; Verdonck, P. Modeling ventricular function during cardiac assist: Does time-varying elastance work? *ASAIO J.* **2006**, *52*, 4–8. [CrossRef] [PubMed]
48. Sack, K.L.; Dabiri, Y.; Franz, T.; Solomon, S.D.; Burkhoff, D.; Guccione, J.M. Investigating the Role of Interventricular Interdependence in Development of Right Heart Dysfunction During LVAD Support: A Patient-Specific Methods-Based Approach. *Front. Physiol.* **2018**, *9*, 520. [CrossRef] [PubMed]
49. Cingolani, H.E.; Perez, N.G.; Cingolani, O.H.; Ennis, I.L. The Anrep effect: 100 years later. *Am. J. Physiol. Heart Circ. Physiol.* **2013**, *304*, H175–H182. [CrossRef] [PubMed]
50. Westerhof, N.; Lankhaar, J.W.; Westerhof, B.E. The arterial Windkessel. *Med. Biol. Eng. Comput.* **2009**, *47*, 131–141. [CrossRef] [PubMed]

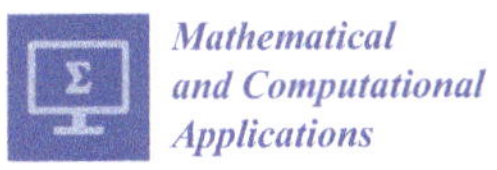
Mathematical and Computational Applications

Article

Resolving Boundary Layers with Harmonic Extension Finite Elements

Harri Hakula

Department of Mathematics and Systems Analysis, Aalto University, Otakaari 1, FI-00076 Aalto, Finland; Harri.Hakula@aalto.fi

Abstract: In recent years, the standard numerical methods for partial differential equations have been extended with variants that address the issue of domain discretisation in complicated domains. Sometimes similar requirements are induced by local parameter-dependent features of the solutions, for instance, boundary or internal layers. The adaptive reference elements are one way with which harmonic extension elements, an extension of the p-version of the finite element method, can be implemented. In combination with simple replacement rule-based mesh generation, the performance of the method is shown to be equivalent to that of the standard p-version in problems where the boundary layers dominate the solution. The performance over a parameter range is demonstrated in an application of computational asymptotic analysis, where known estimates are recovered via computational means only.

Keywords: finite element method; p-version; harmonic extensions

Citation: Hakula, H. Resolving Boundary Layers with Harmonic Extension Finite Elements. *Math. Comput. Appl.* **2022**, *27*, 57. https://doi.org/10.3390/mca27040057

Academic Editor: Gianluigi Rozza

Received: 27 May 2022
Accepted: 4 July 2022
Published: 8 July 2022

1. Introduction

In many practical engineering applications, either the computational domains or indeed the solutions have features that ultimately lead to more expensive finite element simulations than what at first would appear to be necessary. The numerical analysis community has responded to this challenge by introducing different ways for discretising the underlying partial differential equations (PDEs). The methods can either be continuous or discontinuous, even if the problem itself is continuous, or alternatively, the domain discretisations can either be simplicial or polygonal and not necessarily conforming. Notable examples of new developments are virtual elements [1] and the so-called hybrid high-order method [2].

In this paper, the focus is on problems with boundary layers and their resolution with the so-called adaptive reference elements, an hp-finite element method (FEM) (see, for example, [3,4]) where a conforming formulation is constructed on a non-conforming mesh by adapting the shape functions as harmonic extensions [5]. The theoretical foundations for the method can be directly adopted from Weißer's work on boundary element based FEM [6,7]. The work by Ovall is another important reference [8,9].

Examples of solution profiles with boundary layers are shown in Figure 1. From the mesh generation point of view, the issue with standard conforming triangulation is that the effect which is essentially one-dimensional introduces a characteristic length scale to the mesh, which leads to superfluous elements unless the mesh is properly aligned and, simultaneously, the polynomial order is sufficiently high.

The process by which the discretisation is performed so that the parameter-dependent features are properly captured is referred to as the boundary layer resolution. In certain problems, such as shell structures, the solution can be interpreted as a linear combination of features, each with its own characteristic length scale. This includes the so-called smooth component, which spans the entire domain. In order to avoid numerical locking, it is advantageous to use high-order methods for capturing the smooth component accurately [10,11].

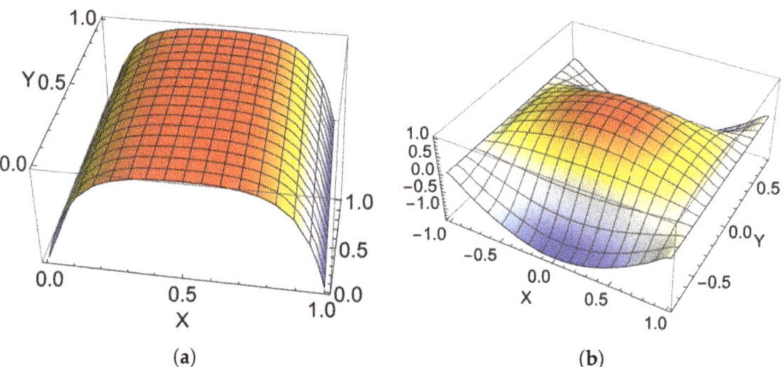

(a) (b)

Figure 1. Solutions of model problems. In elasticity problems, the strongest boundary layers may occur in other vector field components. (**a**) Reaction–diffusion. (**b**) Pitkäranta Cylinder. Detail of transverse deflection (one quarter of the cylinder unfolded).

The novelty of this work is that the efficiency of the adaptive reference elements compared with the standard p-version is demonstrated also in the solution of problems with boundary layers. Furthermore, for problems with linear systems with identical parameter-dependent structures, it is possible to perform computational asymptotic analysis, that is, learn how the overall solution changes as the parameter, for instance, tends to zero. It is known that for thin cylindrical shells depending on the loading and kinematical constraints, it is possible to derive parameter-dependent asymptotic error amplification factors, indicating whether the discretisation has to be adjusted as the parameter changes. In the context of a thin cylindrical benchmark shell, two known error amplification factors are recovered within the numerical experiments below via computation only. This shows that the proposed method maintains its efficiency over the whole practical parameter range.

The rest of this paper is structured as follows: In Section 2, first the boundary layers are introduced formally, and next the basic construction of the harmonic extension shape functions is described before the model problems. The replacement rule-based mesh generation is discussed in more detail in Section 3. Computational asymptotic analysis is developed in Section 4, including the description of the framework for recovering quantities of interest. Numerical examples followed by conclusions are in the next two sections.

2. Preliminaries

In this section, the background concepts, namely boundary layers, construction of the 2D adaptive reference elements, and the model problems, are introduced. Familiarity with the basic concepts of the p- and hp-versions of the finite element method is assumed. The hp-solver is implemented with Mathematica [12]—for its design principles, see [13].

2.1. Boundary Layers

The theory of one-dimensional hp-approximation of boundary layers is due to Schwab [4]. Boundary layer functions are of the form

$$u(x) = \exp(-a\,x/\delta), \quad 0 < x < L, \tag{1}$$

where $\delta \in (0,1]$ is a small parameter, $a > 0$ is a constant, and L is the characteristic length scale of the problem under consideration. Even though the classical p-method, see, for example, [3], is capable of asymptotic superexponential convergence, the judicious choice of a minimal number of elements using *a priori* knowledge of the boundary layers leads to far more efficient solution in the practical range of p. Moreover, in certain classes of problems, it is possible to choose a robust strategy leading to convergence uniformity in δ. However, the distribution of the mesh nodes *depends* on p, and over a range of polynomial degrees $p = 2, \ldots, 8$, say, the mesh is different for every p! In 1D, this is relatively simple, but in 2D, much more difficult since we must allow for the mesh topology to change over the range of polynomial degrees. This is the topic of Section 3 below.

In the absence of exact analysis for all of the problem classes discussed below, the central result is given in the form of a definition independent of the dimension of the problem.

Definition 1 (Layer Element Width). *For every boundary layer in the problem, optimal convergence can be guaranteed if an element of width $O(p\,\delta)$ is aligned in the direction of the decay of the layer.*

Notice, that with c constant, if $c\,p\,\delta \to L$ as p increases, the standard p-method can be interpreted as the limiting method.

Boundary layers can also occur within the domains, i.e., be internal layers, or emanate from a point. For the discussion here, it is useful to define the concept of boundary layer generators (see [14]).

Definition 2 (Layer Generator). *The subset of the domain from which the boundary layer decays exponentially, is called the layer generator. Formally, the layer generator is of measure zero.*

The layer generators are independent of the length scale of the problem under consideration.

2.2. Adaptive Reference Elements

As usual, in finite elements, every element is an instance of some reference element. To fix terminology every *adaptive element* (AE) is said to be an instance of an adaptive reference element (ARE). The adaptive reference element is defined with a set of points or nodes as usual. What is different is that a subset of nodes is used to define the mapping of the element.

Let us consider the example in Figure 2, where an adaptive reference element is shown. The quadrilateral has *five* nodes, divided into four mapping nodes and one edge node. This choice is not unique, however. For instance, the adaptive reference element (ARE) corresponding to element A could be a triangle with mapping nodes {1,2,4} and two edge (hanging) nodes 3 and 5, where the edge $\{1,5,4\}$ would be curved and passing through 5.

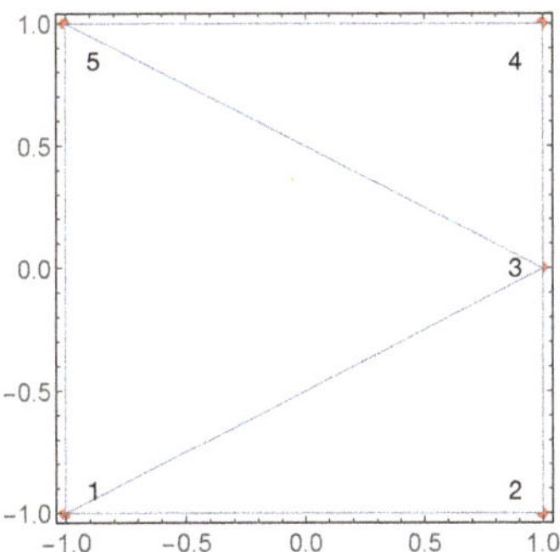

Figure 2. Adaptive reference element A: Quadrilateral with five nodes. Minimal implementation mesh with three triangles. The element mapping is defined using the four corner nodes only.

Definition 3. *(Planar adaptive reference element (ARE))*
Given a set of points K, $|K| \geq 3$, any partition of K into mapping and edge nodes is admissible, if the edge nodes lie on the boundary of some valid (univalent) mapping of the standard reference element defined by the mapping points. This partition defines the adaptive reference element. If the set of edge nodes is empty, the adaptive reference element is equivalent to a standard element.

2.2.1. Shape Functions

The shape functions are computed as harmonic extensions of the restrictions of the standard FEM nodal and edge functions on the element boundary. In other words, for any standard nodal or edge shape function $\phi(x,y)$ we compute its harmonic extension $\varphi(x,y)$:

$$\begin{cases} \Delta\varphi(x,y) = 0, & \text{in } K, \\ \varphi(x,y) = \phi(x,y)|_{\partial K}, & \text{on } \partial K, \end{cases} \tag{2}$$

where K is either a reference quadrilateral or triangle. Some examples of such shape functions are given in Figures 3 and 4. This construction guarantees a continuous formulation combining AEs and standard finite elements. Also notice that the nodal modes automatically form a partition of unity.

For instance, in the case of Figure 2, the nodal shape function associated with the node 3 is computed by setting the restriction $\phi(1,y)|_{\partial K}$ over the edge at $x = 1$ to be the standard linear hat function with $\phi(1,0)|_{\partial K} = 1$.

The associated inner modes $\hat{\varphi}(x,y)$ are the functions satisfying

$$\begin{cases} \Delta\hat{\varphi}(x,y) = q(x,y), & \text{in } K, \\ \hat{\varphi}(x,y) = 0, & \text{on } \partial K, \end{cases} \tag{3}$$

where $q(x,y)$ is some polynomial. For instance, $q(x,y) = 1$ (constant) induces a standard bubble function. The set of elemental inner modes $\hat{\varphi}(x,y)_K$ is constructed with products of Legendre polynomials, that is, all $q(x,y) \in q(x,y)_K$, where $q(x,y)_K = \{P_i(x)P_j(y),\ \mathrm{i} = 0,\ldots,$ $p-2,\ j = 0,\ldots,p-2\}$. With this choice the number of inner modes is the same as with the standard p-version, although the approximation properties are not. There exists a family of polynomials that could be used instead of the computed ones (see [9]).

The computation of the shape functions is done with finite elements (naturally). Hence, the concept of the implementation mesh arises, or more precisely, implementation discretisation. In order to simplify the evaluation of the inner products between the shape functions, every shape function associated with a given element is computed using the same implementation discretisation. One consequence of this is that the same extension may be computed using many different implementation discretisations.

For quadrilaterals, the baseline implementation discretisation is a regular grid with two elements per segment and uniform $p = 20$, and for triangles, a minimal triangulation of the nodes with uniform $p = 16$.

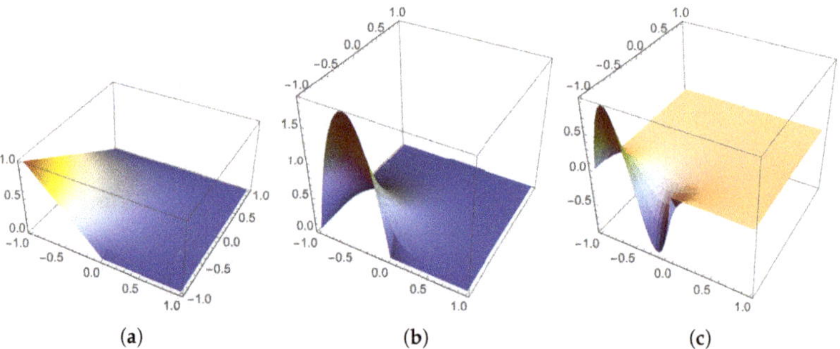

Figure 3. Quadrilateral. (**a**) Nodal mode. (**b**) Edge mode with $p = 2$. (**c**) Edge mode with $p = 3$.

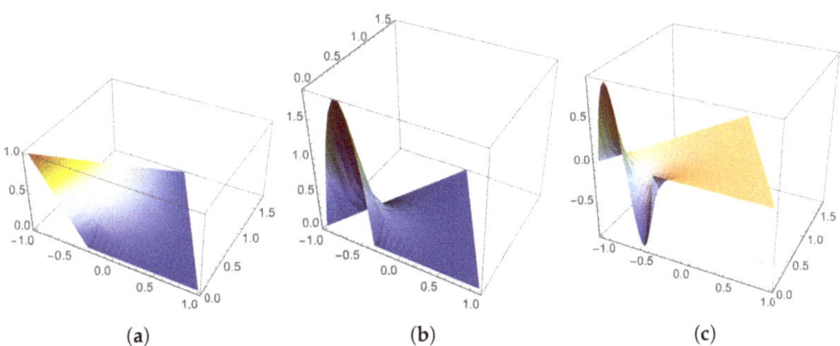

Figure 4. Triangle. (a) Nodal mode. (b) Edge mode with $p = 2$. (c) Edge mode with $p = 3$.

2.2.2. Type of Reference Element

In order to minimise the computational work, it is necessary to introduce a way to maintain bookkeeping for the evaluated shape functions and AREs. Since compatibility with the standard p-version is required (or desired), on split edges, the shape functions must have correct parities.

Legendre polynomials of degree n can be defined using a recursion formula

$$(n+1)P_{n+1}(x) - (2n+1)xP_n(x) + nP_{n-1}(x) = 0, \quad P_0(x) = 1. \tag{4}$$

For our purposes, the central polynomials are the integrated Legendre polynomials for $\xi \in [-1, 1]$

$$\phi_n(\xi) = \sqrt{\frac{2n-1}{2}} \int_{-1}^{\xi} P_{n-1}(t)\, dt, \quad n = 2, 3, \ldots \tag{5}$$

which can be rewritten as linear combinations of Legendre polynomials

$$\phi_n(\xi) = \frac{1}{\sqrt{2(2n-1)}}(P_n(\xi) - P_{n-2}(\xi)), \quad n = 2, 3, \ldots \tag{6}$$

The normalising coefficients are chosen so that

$$\int_{-1}^{1} \frac{d\phi_i(\xi)}{d\xi} \frac{d\phi_j(\xi)}{d\xi}\, d\xi = \delta_{ij}, \quad i, j \geq 2. \tag{7}$$

Therefore, the $\phi_n(\xi)$ inherit the property of the Legendre polynomials,

$$\phi_n(\xi) = (-1)^n \phi_n(-\xi). \tag{8}$$

This means that every edge has to be oriented in such a way that the shape function has a consistent sign or parity on both elements sharing it.

For every element a *type* is assigned in the following way: first a mapping node with the smallest identifier is chosen and the simplex is rotated so that the selected node is in a fixed position (normalisation); next for each edge, the parameter range of its support on the reference element is derived; finally each edge segment is assigned a parity by comparing the identifiers of the end points. Thus, every edge, split or not, has its contribution to the type of the ARE in form of a tuple $(s, [a, b])$, where $s = \pm 1$, and $-1 \leq a < b \leq 1$. For instance, the ARE of Figure 2 has the type S_Q— assuming that the nodes are identified as in the picture—

$$S_Q = ((1, [-1, 1]), ((1, [-1, 0]), (1, [0, 1])), (1, [-1, 1]), (-1, [-1, 1])). \tag{9}$$

The convention is that the positive direction is from the node with the smallest identifier. For standard p-version quadrilaterals, there are four types, and for triangles, two types, in both cases assuming that the inner modes need not have an orientation. The inner modes are always assumed to be oriented in the same way, that is, they do not affect the type.

2.2.3. Mesh Generation and Refinement

Since the proposed method is an extension of the p-version, mesh generation is identical to that of p-version. The power of the method lies in its ability to adapt to (seemingly) non-conforming mesh refinements. In Figure 5, two examples of strategies for refining the mesh to a corner singularity are illustrated.

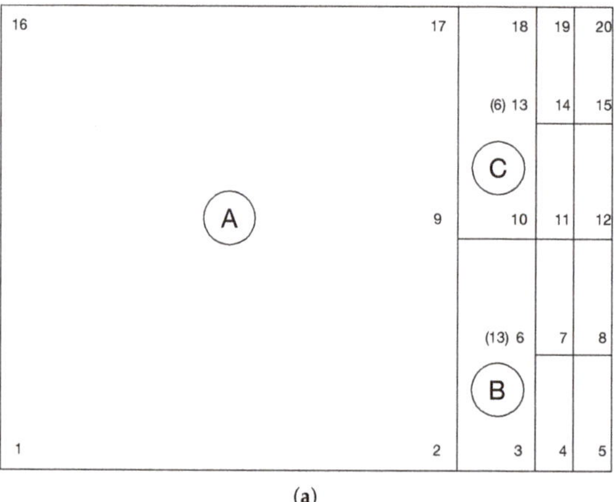

(a)

Tag	e_1	e_2	e_3	e_4	Type
A	(1,2)	(2,9,17)	(17,16)	(16,1)	1
B	(2,3)	(3,6,10)	(10,9)	(9,2)	1
C	(9,10)	(10,13,18)	(18,17)	(17,9)	1

(b)

Tag	e_1	e_2	e_3	e_4	Type
A	(1,2)	(2,9,17)	(17,16)	(16,1)	1
B	(2,3)	(3,13,10)	(10,9)	(9,2)	2
C	(9,10)	(10,6,18)	(18,17)	(17,9)	3

(c)

Figure 5. Boundary layer mesh example. Four levels of elements toward the right hand boundary. Normalised AEs are given as node lists, every split edge has the same parametrisation on the reference element. Types are labelled in the order of occurrence. (**a**) Boundary layer mesh. (**b**) Uniform: List of AEs and AREs (types). The node identifiers 6 and 13 have been swapped and hence all three labelled elements have different types. (**c**) Non-uniform: (6) $\leftrightarrow$ (13): List of AE and ARE (types).

The nodes are given identifiers in the order of creation. Thus, in both instances there are four AEs, but only two different types; hence, only two AREs have to be computed. Moreover, if one wanted to solve some problem using both meshes one at the time, the representative reference elements of each type would have to be computed only once. In Figure 5a, the AEs are given using nodes on edges.

2.3. Model Problems

Two classes of model problems are considered: a standard reaction–diffusion problem, and a thin cylindrical shell. Both of these problems are classical and have been discussed by many authors. It is of particular interest that the resulting linear systems have an identical parameter-dependent structure that can be exploited in the solution for multiple parameter values with a given right-hand-side.

2.3.1. Reaction-Diffusion Problem

The standard model problem is the reaction–diffusion problem. In the case of constant diffusion, it has the form

$$\begin{aligned} -\epsilon\Delta u(x,y) + u(x,y) &= f(x,y), \quad \text{in } \Omega, \\ u(x,y) &= 0, \quad \text{on } \partial\Omega. \end{aligned} \tag{10}$$

For the full 2D problem on a unit square, one can derive an exact solution by restricting the diffusion to some specific direction, for instance, along the x-axis only. In the numerical experiments, such a case is considered (see Figure 1a above). The expected characteristic length scale of the solution is $\sqrt{\epsilon}$.

2.3.2. Cylindrical Shells

Shells are thin three-dimensional structures that are very expensive to simulate, unless one performs dimension reduction in the model, that is, the original problem is written as lower-dimensional one, where the reduced dimension is represented by a parameter [15]. Here the resulting two-dimensional shell model has a vector field with five components $\mathbf{u} = (u, v, w, \theta, \psi)$, where the first three are the displacements and the latter two are the rotations in the axial and angular directions, respectively. The convention is that the computational domain D is given by the surface parametrisation and the axial/angular coordinates are denoted by x and y.

The total energy is given by a quadratic functional

$$\mathcal{F}(\mathbf{u}) = \frac{1}{2}Y\mathbf{A}(\mathbf{u},\mathbf{u}) - \mathbf{Q}(\mathbf{u}), \tag{11}$$

where $\mathbf{A}$ represents strain energy and $\mathbf{Q}$ is the load potential. The constant factor $Y = E/(12(1-\nu^2))$, where E is the Young modulus of the material, and ν is the Poisson ratio.

Deformation energy $\mathcal{A}(\mathbf{u},\mathbf{u})$ is divided into bending, membrane, and shear energies, denoted by subscripts B, M, and S, respectively:

$$\mathcal{A}(\mathbf{u},\mathbf{u}) = t^2\mathcal{A}_B(\mathbf{u},\mathbf{u}) + \mathcal{A}_M(\mathbf{u},\mathbf{u}) + \mathcal{A}_S(\mathbf{u},\mathbf{u}), \tag{12}$$

where t is the dimensionless thickness, $t = d/L$, where d is the actual thickness and L is some characteristic length scale, for instance, the diameter of the domain.

Let us consider a cylindrical shell with a midsurface ω generated by a function $f_1(x) = R$, $x \in [-x_0, x_0]$, $x_0 > 0$. In this case, the product of the Lamé parameters (metric), $A_1(x)A_2(x) = R$, and the reciprocal curvature radii are $1/R_1(x) = 0$ and $1/R_2(x) = 1/R$, since

$$A_1(x) = \sqrt{1 + [f_1'(x)]^2}, \quad A_2(x) = f_1(x), \tag{13}$$

and

$$R_1(x) = -\frac{A_1(x)^3}{f_1''(x)}, \quad R_2(x) = A_1(x)A_2(x). \tag{14}$$

Bending, membrane, and shear energies are given as

$$\begin{aligned} t^2\mathcal{A}_B(\mathbf{u},\mathbf{u}) &= t^2 \int_\omega \big[\nu(\kappa_{11}(\mathbf{u}) + \kappa_{22}(\mathbf{u}))^2 \\ &\quad +(1-\nu)\sum_{i,j=1}^{2} \kappa_{ij}(\mathbf{u})^2\big]\, A_1(x,y)A_2(x,y)\,dx\,dy, \end{aligned} \tag{15}$$

$$\mathcal{A}_M(\mathbf{u},\mathbf{u}) = 12\int_\omega \left[\nu(\beta_{11}(\mathbf{u})+\beta_{22}(\mathbf{u}))^2 + (1-\nu)\sum_{i,j=1}^{2}\beta_{ij}(\mathbf{u})^2\right] A_1(x,y)A_2(x,y)\,dx\,dy, \tag{16}$$

$$\mathcal{A}_S(\mathbf{u},\mathbf{u}) = 6(1-\nu)\int_\omega \left[(\rho_1(\mathbf{u})^2+\rho_2(\mathbf{u}))^2\right] \times A_1(x,y)A_2(x,y)\,dx\,dy. \tag{17}$$

We have omitted the scaling Y.

In the special case of constant radius, using the identities above, the bending, membrane, and shear strains [16], κ_{ij}, β_{ij}, and ρ_i, respectively, can be written as

$$\begin{aligned}
\kappa_{11} &= \frac{\partial\theta}{\partial x}, \quad \kappa_{22} = \frac{1}{R}\frac{\partial\psi}{\partial y}, \quad \kappa_{12} = \frac{1}{2}\left(\frac{\partial\psi}{\partial x} + \frac{1}{R}\frac{\partial\theta}{\partial y} - \frac{1}{R}\frac{\partial v}{\partial x}\right), \\
\beta_{11} &= \frac{\partial u}{\partial x}, \quad \beta_{22} = \frac{1}{R}\frac{\partial v}{\partial x} + \frac{w}{R}, \quad \beta_{12} = \frac{1}{2}\left(\frac{1}{R}\frac{\partial u}{\partial y} + \frac{\partial v}{\partial x}\right), \\
\rho_1 &= \frac{\partial w}{\partial x} - \theta, \quad \rho_2 = \frac{1}{R}\frac{\partial w}{\partial y} - \frac{v}{R} - \psi.
\end{aligned} \tag{18}$$

By minimising the the total energy Equation (11) the variational problem becomes the following: Find $\mathbf{u} \in \mathcal{U} \subset [H^1(\omega)]^5$ such that

$$\mathbf{A}(\mathbf{u},\mathbf{v}) = \mathbf{Q}(\mathbf{v}) \quad \forall \mathbf{v} \in \mathcal{U}. \tag{19}$$

and the corresponding finite element problem: Find $\mathbf{u}_h \in \mathcal{U}_h$ such that

$$\mathbf{A}(\mathbf{u}_h,\mathbf{v}) = \mathbf{Q}(\mathbf{v}) \quad \forall \mathbf{v} \in \mathcal{U}_h. \tag{20}$$

The load potential has the form

$$\mathbf{Q}(\mathbf{v}) = \int_\omega \mathbf{f}(x,y)\,\mathbf{v}\,dx\,dy. \tag{21}$$

If the load acts in the transverse direction of the shell surface, it has the form $\mathbf{f}(x,y) = [0,0,f_w(x,y),0,0]^T$. It can be shown that if for the load $\mathbf{f} \in [L^2(\omega)]^5$ holds, the problem Equation (19) has a unique weak solution $\mathbf{u} \in [H^1(\omega)]^5$. The corresponding result holds in the finite dimensional case, when the finite element method is employed.

3. Boundary Layer Resolution

What makes adaptive reference elements particularly appealing in boundary layer dominated problems is the possibility to use replacement rules as the basic refinement strategy. This basic idea has been discussed earlier (see [5]), but the fact that any predicate can be used to select the elements or cells for refinement has not been addressed before. In Figure 6a, a sequence of meshes generated by applying the same rule three times is shown. The rule simply first verifies that the element has an edge on the boundary before the three new elements are generated, one ARE and two ordinary. In this simple case, the process is greedy in the sense that the elements not participating in the refinement step are not affected. In a general case, the replacement rules could be specified on the edge level with additional bookkeeping required to maintain the data structures for the elements themselves.

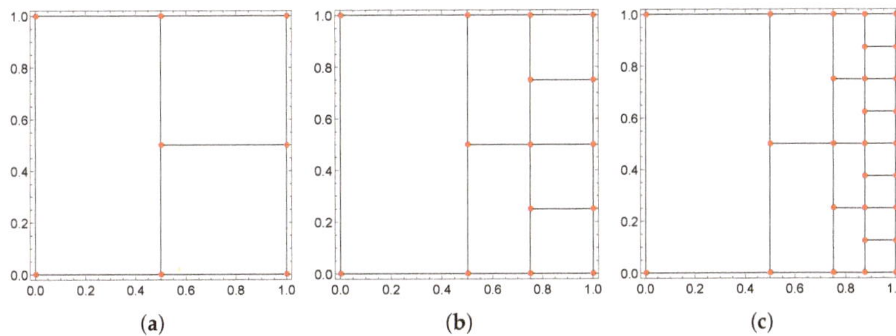

Figure 6. Sequence of refined meshes. Same replacement rule applied three times with a unit square as the initial configuration. The underlying finite element discretisation is continuous, despite the apparent hanging nodes. (**a**) Rule applied once. (**b**) Rule applied twice. (**c**) After three applications of the rule.

In Figure 6, the rule is kept constant for illustration purposes. The refinement rule first divides the selected elements along the x-axis, and then the elements touching the boundaries aligned with the y-axis. The axial edge length ratio is 1:1. Of course, nothing prevents one from modifying the rules from one application or layer to another. The crucial requirement is that the set of rules at one step has to be consistent, that is, every topological component has to be refined in the same way, regardless of the element containing it. For example, in Figure 6a, there exists an edge connecting nodes $(1/2, 1/2)$ and $(1, 1/2)$. In Figure 6b, that edge is split into two segments and node $(3/4, 1/2)$ is added. During the process, both elements touching the boundary at $x = 1$ have created the same node. The final step of the refinement algorithm verifies that all destructive changes are consistent and removes possible duplicates.

The shell problems have a rich boundary layer structure, including internal layers. In the axial direction, the characteristic length scales are $\sqrt{t}$ and t, where the latter is the short scale often omitted from the models. In the angular direction, there exists a relatively long layer of $\sqrt[4]{t}$, which plays an important role in the free vibration of shells of revolution. This is not an exhaustive list; for a detailed discussion on boundary layer structures of cylindrical shells and their resolution in standard hp-FEM setting, see [17]. In Figure 7, both a schematic of the axial layer dependencies and a sample sequence of meshes adapted to a single layer generator represented by the entire boundary at $x = 1$. The axial edge length ratio is approximately 6:1. Without any loss of generality in the numerical examples, only problems with simple boundary segment layer generators are considered. It has to be emphasised that, strictly speaking, the smooth component spanning the entire domain can be interpreted as a boundary layer, and it is precisely this feature that necessitates the use of higher order methods, i.e., the p-version, unless special so-called shell elements are used.

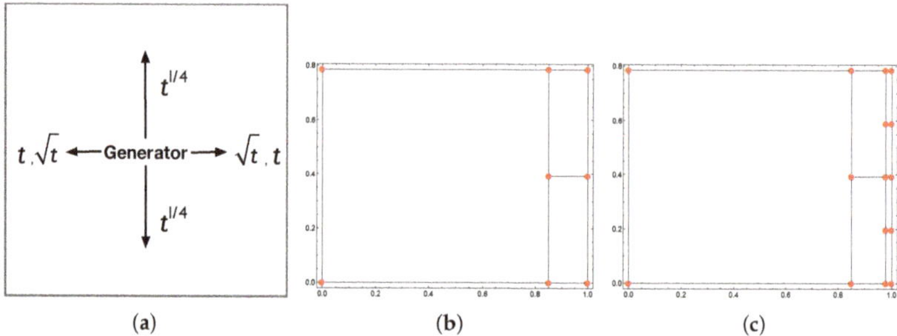

(**a**) (**b**) (**c**)

Figure 7. Cylindrical shell: Layer structure. Examples of minimal meshes for a case with a generator at $x = 1$ with $t = 1/100$. (**a**) Layer structure for the axial directions, there also exists a possible smooth component spanning the entire domain. (**b**) Mesh adapted for $\sqrt{t}$. (**c**) Mesh adapted for $\sqrt{t}$ and t.

4. Computational Asymptotic Analysis

Asymptotic analysis has been a standard tool for understanding the parameter-dependence of, for instance, mechanical systems. The goal is to understand what happens when the critical parameter tends to some limit, in the context of this paper, for instance, when the dimensionless thickness of a shell tends to zero, i.e., $t \to 0$. Rather than utilising some a priori understanding of the behaviour of the systems, computational asymptotic analysis relies on accurate solution of individual realisations of the problem and the subsequent recovery of the quantities of interest from the computed solutions.

4.1. Solving Parameter-Dependent Sequences of Linear Systems

It is necessary to solve systems in terms of **A** and **B** (for shells bending, and membrane and shear summed together, respectively)

$$\mathbf{S}(\sigma)v = (\mathbf{B} + \sigma\mathbf{A})\, v = b, \tag{22}$$

for every $\sigma > 0$. In order to avoid the factorisation of $\mathbf{S}(\sigma)$ at every σ, it would be natural to consider iterative methods. Unfortunately, parameter-independent preconditioning of singularly perturbed systems is an open problem. However, if one considers a *sequence of problems*, it is possible to transform the problem as follows: Let $\mathbf{B} = \mathbf{L}\mathbf{L}^T$ (Cholesky decomposition); then,

$$\mathbf{L}(\mathbf{I} + \sigma\mathbf{L}^{-1}\mathbf{A}\mathbf{L}^{-T})\mathbf{L}^T v = b, \tag{23}$$

where the subspace defined by $\mathbf{L}^{-1}\mathbf{A}\mathbf{L}^{-T}$ is invariant over all parameters σ. The inner systems $(\mathbf{I} + \sigma\mathbf{L}^{-1}\mathbf{A}\mathbf{L}^{-T})\, \hat{v} = \hat{b}$, have their spectra bounded by 1 from below, making it sufficient to collect the largest eigenvectors into a subspace W, say. Once the subspace W is constructed, the deflated conjugate gradient method can be applied [18,19]. Here, collecting Lanczos vectors is also sufficient since it is assumed that loading remains constant and independent of σ.

Remark 1. *Since the system can be singularly perturbed, i.e.,* **B** *is not necessarily invertible, it is possible to use the stabilised version*

$$((\mathbf{B} + \epsilon_\sigma\mathbf{A}) + (\sigma - \epsilon_\sigma)\mathbf{A})\, v = b, \tag{24}$$

where $\epsilon_\sigma \in [0, \sigma]$. The choice of optimal shift ϵ_σ is problem-dependent.

4.2. Recovering Quantities of Interest

Once a sequence of problems is solved, in particular, in the case of elasticity, it is interesting to recover the parameter-dependence of the energy components, e.g., bending and membrane energies for shells. In short, rather than trying to a priori analytically predict the behaviour, one simply derives the quantities of interest by numerical estimation a posteriori.

For instance, in the context of shells, given a sequence of solutions, the procedure is as follows: (i) first the displacements are normalised so that $w \sim 1$, (ii) then strains are evaluated with normalised displacements, (iii) rates of change for strains are derived through regression, and (iv) energy dependence is discovered by evaluating the energy expressions using the estimated rates for strains.

In the case of the cylindrical shell benchmark below, both known theoretical asymptotic dependencies on the dimensionless thickness are recovered using this procedure.

5. Numerical Experiments

The numerical experiments were designed to cover two types of PDEs with boundary layers. The goal here is to establish the convergence properties of the proposed method in comparison with the standard p-version. The idea is to extend the p-version, but of course with the approximative shape functions, there are non-standard sources of error that are very difficult to analyse, except through computational studies, such as these. The reference values of the quantities of interest are listed in Table 1 (see, for example, [20]). All convergence graphs presented in this section are log–log plots.

Table 1. Reference values used in numerical experiments. Reaction–diffusion with $\epsilon = 1/100$ and Pitkäranta cylinder with $t = 1/100$.

Case	Norm	Value Used	Exact
Reaction-Diffusion	L^2	0.83673056017854	$\frac{\sqrt{\frac{1}{10}(13+40e^{10}+7e^{20})}}{1+e^{10}}$
Cylinder (Clamped)	Squared energy	2.6882879572571783	-
Cylinder (Free)	Squared energy	7043.3120530934690	-

5.1. Reaction–Diffusion

The first example is the standard model problem for simple boundary layers, the reaction–diffusion problem (10). The computational domain is the unit square and homogeneous Dirichlet boundary conditions are set on all boundaries. The typical shape of the potential function is shown above in Figure 1a. Here, the diffusion is restricted to x variable only with unit loading. The exact solution when setting $\epsilon = 1/100$ is

$$u(x,y) = \frac{-e^{10-10x} - e^{10x} + 1 + e^{10}}{1 + e^{10}}. \tag{25}$$

The mesh and the observed convergence in the L^2-norm is given in Figure 8. The results are very good indeed. In particular, the convergence pattern has staircasing, which in fact is to be expected in the context of p-version since the exact solution is symmetric (that is, even locally), and as $p = 2, \ldots, 8$ one should not expected significant improvement as p changes from even to odd.

No attempt to find an optimal distribution of degrees of freedom has been made. This question is beyond the scope of this study. The mesh of Figure 8a has 36 nodes, 56 edges, and 21 quadrilateral (15 regular, six AREs with five nodes). Therefore at $p = 8$ the number of degrees of freedom is $36 + 56(p-1) + 21(p-1)(p-1) = 1457$.

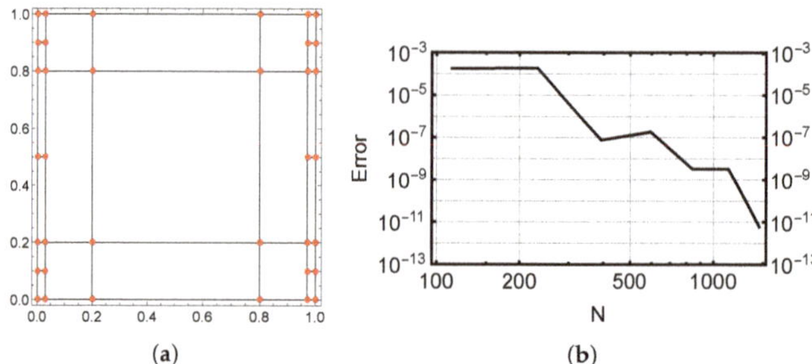

Figure 8. Reaction–diffusion with $\epsilon = 1/100$. Convergence graph has a characteristic staircasing behaviour, where the even nature of the exact solution over the domain results in no convergence as the polynomial order changes from even to odd, $p = 2, \ldots, 8$. (**a**) Symmetric mesh. (**b**) L^2-convergence (absolute error).

5.2. Pitkäranta Cylinder

Pitkäranta cylinder (see [20]) is a well-known and widely used benchmark problem. The computational domain is reduced to one sixteenth, $\Omega = [0,1] \times [0, \pi/4]$, through clever application of symmetry–antisymmetry boundary conditions for a given Fourier mode type loading $f(x,y) = \cos(2y)$, which is constant in the axial direction. In the clamped case, the boundary conditions are $y = 0 : v = \psi = 0$, $x = 1$ clamped, $y = \pi/4 : u = w = \theta = 0$, and $x = 0 : u = \theta = 0$, and in the free case there are no conditions on $x = 1$. Sample displacement fields are shown in Figure 9.

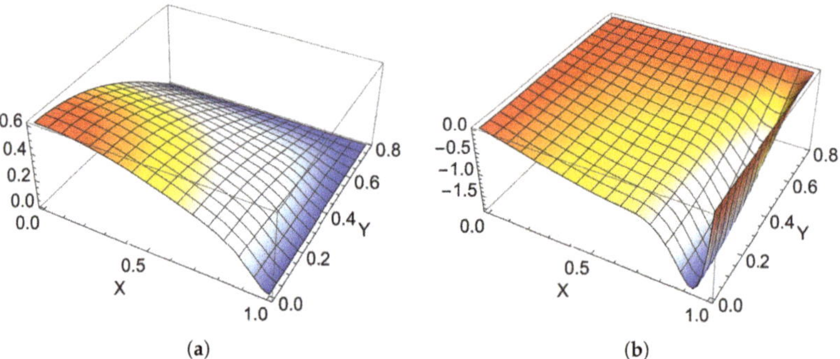

Figure 9. Pitkäranta cylinder: clamped case. Detail of the domain with symmetry/antisymmetry boundary conditions applied. The boundary layer in the rotation component with the characteristic length scale of $\sqrt{t} = 1/10$ is clearly visible. (**a**) Transverse deflection. (**b**) Rotation θ.

5.2.1. On Numerical Locking

It has been already discussed above that the p-version can be used to alleviate problems associated with numerical locking within the standard FEM framework. Of course, there exist many special shell elements that try to resolve the issue by modifying the underlying variational formulation [15].

In Figure 10, two sets of convergence data is presented. In the first set, the mesh is fixed, but the polynomial order p varies, $p = 1,2,3,4$. In the second set the polynomial order is fixed at $p = 2$ over a set of $g \times g$-grids, $g = 20, 40, 60$. The numerical locking is evident. For both cases, clamped and free, there is a clear difference in performance as the polynomial order is increased from $p = 2$ to $p = 3$. Figure 10f indicates that the standard h-version converges very slowly indeed, even at $p = 2$.

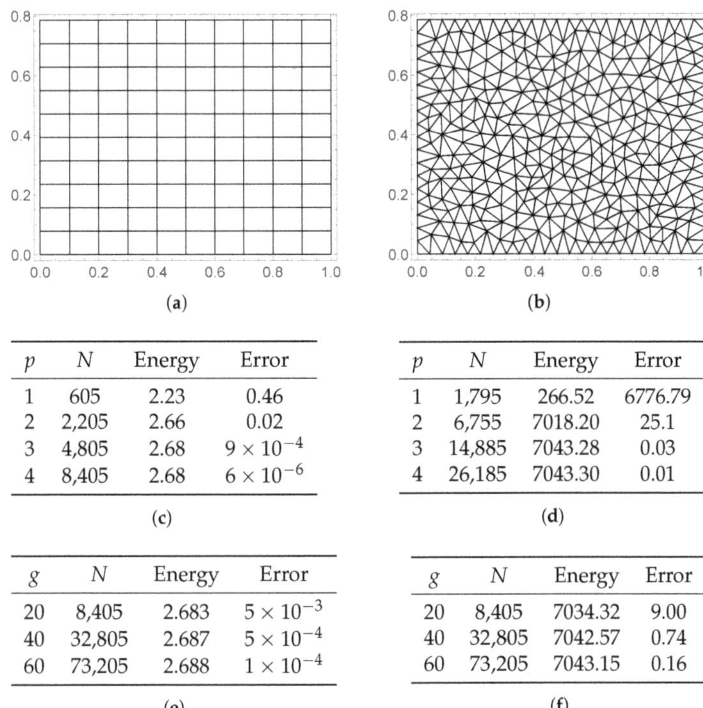

p	N	Energy	Error
1	605	2.23	0.46
2	2,205	2.66	0.02
3	4,805	2.68	9×10^{-4}
4	8,405	2.68	6×10^{-6}

(c)

p	N	Energy	Error
1	1,795	266.52	6776.79
2	6,755	7018.20	25.1
3	14,885	7043.28	0.03
4	26,185	7043.30	0.01

(d)

g	N	Energy	Error
20	8,405	2.683	5×10^{-3}
40	32,805	2.687	5×10^{-4}
60	73,205	2.688	1×10^{-4}

(e)

g	N	Energy	Error
20	8,405	7034.32	9.00
40	32,805	7042.57	0.74
60	73,205	7043.15	0.16

(f)

Figure 10. Pitkäranta cylinder, $t = 1/100$. (**a**) Clamped case on 10×10-grid: Mesh. (**b**) Free boundary: Mesh. (**c**) Clamped case on 10×10-grid: Errors. Squared energy norm convergence (absolute error) in $p = 1, \ldots, 4$ using different discretisations. (**d**) Free case: Errors. Squared energy norm convergence (absolute error) in $p = 1, \ldots, 4$ using different discretisations. (**e**) Clamped case on $g \times g$-grid: Errors. Squared energy norm convergence (absolute error) at $p = 2$ over a series of uniform discretisations. (**f**) Free case on $g \times g$-grid: Errors. Squared energy norm convergence (absolute error) at $p = 2$ over a series of uniform discretisations. N is the number of degrees of freedom.

5.2.2. Convergence Results

As in the reaction–diffusion problem, the convergence results are very good. In Figure 11, the proposed method practically matches that of the comparable p-version solution. As expected the number of degrees of freedom in the best observed case is significantly smaller than those given in Figure 10. One word of caution though, since the special form of loading actually allows for a 1D solution, for the p-version, there exists a mesh with exactly two elements which does not have a corresponding ARE configuration, and the proposed method degenerates to standard p-version. If one takes the number of significant digits into consideration or uses relative error, the convergence rate in the free case is in fact slightly better since the bulk of the energy is carried by the smooth component, which is resolved using a high polynomial order.

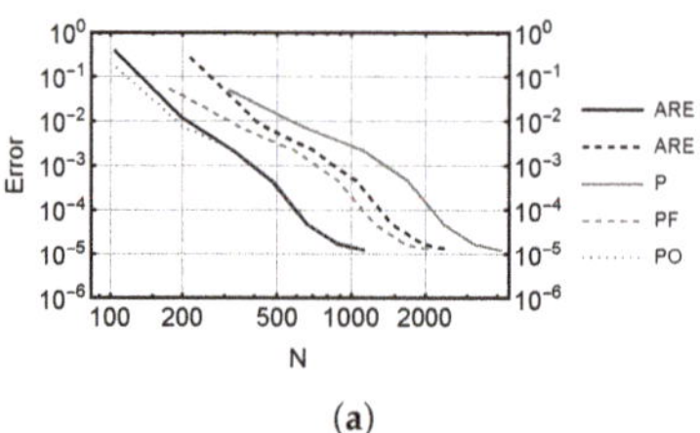

(a)

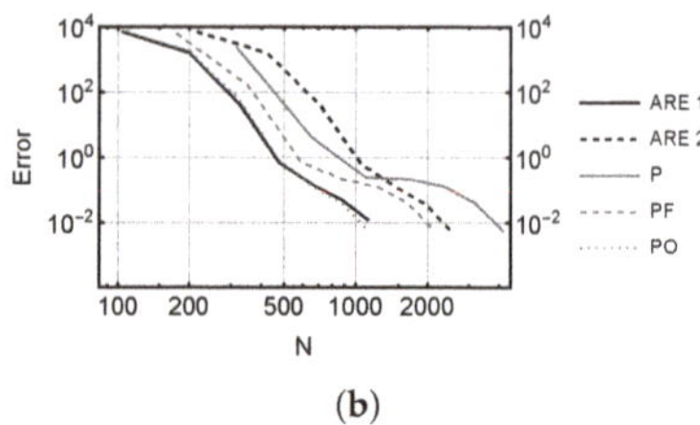

(b)

Figure 11. ARE convergence of Pitkäranta cylinder, $t = 1/100$. Squared energy norm convergence (absolute error) in $p = 2, \ldots, 8$ using different discretisations: ARE 1: one layer, ARE 2: two layers, P: full tensor product grid with two layers, PF: full tensor product grid with one layer, PO: minimal full tensor product grid conforming to ARE 1. The observed convergence of the proposed method in both test cases agrees with the standard p-version. N is the number of degrees of freedom. (**a**) Clamped case. (**b**) Free boundary.

5.2.3. Energy Dependence

The final and perhaps the most interesting experiment is the attempt to recover the energy dependencies via computational asymptotic analysis. Two sets of results are shown in Figures 12 and 13. Some of the strains diverge as $t \to 0$. This in itself is not a cause of worry if the overall energy remains constrained. However, it is possible that there remains a parameter-dependent error amplification factor $C(t)$, and indeed, for the free case, the expected $C(t) \sim 1/t$ is recovered. This is the primary source of numerical locking, defined as an unavoidable loss of convergence rate.

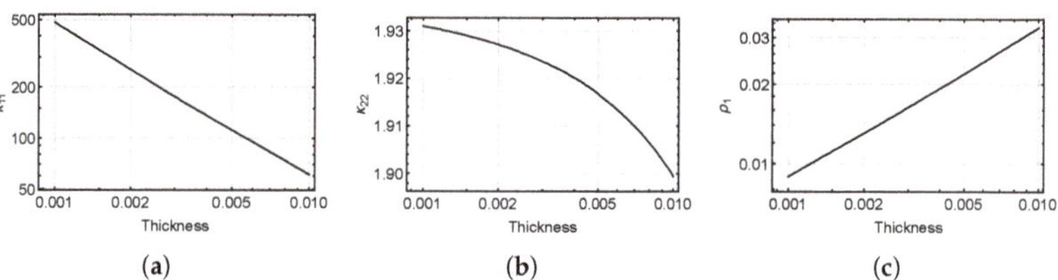

(a) (b) (c)

Figure 12. Clamped case. Practically perfect agreement with the a priori predictions. Notice, that κ_{22} is essentially constant. Additionally, $\rho_1 \to 0$ indicates that the Kirchhoff–Love condition is satisfied without imposing it within the model explicitly. (**a**) $\kappa_{11} \sim 1/t$. (**b**) $\kappa_{22} \sim 1$. (**c**) $\rho_1 \sim t$.

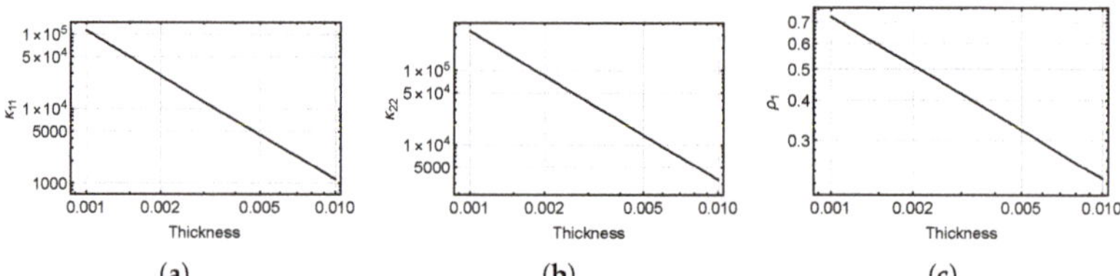

(a) (b) (c)

Figure 13. Free boundary. Notice, that ρ_1 is diverging, however, with a very small constant. This indicates that the boundary layer resolution is not accurate enough to conform to Kirchhoff–Love condition. The aggregate parameter-dependence is correct, the bending terms dominate the energy completely. (**a**) $\kappa_{11}(\mathbf{u}) \sim 1/t^2$. (**b**) $\kappa_{22}(\mathbf{u}) \sim 1/t^2$. (**c**) $\rho_1(\mathbf{u}) \sim 1/\sqrt{t}$.

The strains are computed using the procedure outlined in Section 4.2 above. Notice that in the clamped case $\rho_1 \to 0$, but not in the free case. This convergence implies the so-called Kirchhoff–Love condition. In the free case, the divergence is mild and completely dominated by the strongly increasing bending terms. Nevertheless, this is a point where the traditional analysis would simply assume such convergence also in the free case. Of course, this is ultimately a question of validity of the chosen model and cannot be resolved here. More pragmatically, one can interpret the result as an indication that the discretisation does not capture the short t-layer in the free case.

Once the strains are recovered, the error amplification factor $C(t)$ can be deduced by inserting the rates into the energy formulation and finding the most significant power of t in the resulting expression. More precisely, simply evaluating Equations (15) and (16), and adding the resulting expressions, one gets an approximate dependence for the square of the energy norm. For the clamped case, the recovered strains (ignoring the shear terms) are

$$\begin{aligned} &\beta_{11}(\mathbf{u}) \sim 1,\ \beta_{22}(\mathbf{u}) \sim 1,\ \beta_{12}(\mathbf{u}) \sim 1, \\ &\kappa_{11}(\mathbf{u}) \sim 1/t,\ \kappa_{22}(\mathbf{u}) \sim 1,\ \kappa_{12}(\mathbf{u}) \sim 1/\sqrt{t}, \\ &C(t) \sim 1, \end{aligned} \tag{26}$$

and for the free case, equivalently

$$\begin{aligned} &\beta_{11}(\mathbf{u}) \sim 1/t,\ \beta_{22}(\mathbf{u}) \sim 1/t,\ \beta_{12}(\mathbf{u}) \sim 1\sqrt{t}, \\ &\kappa_{11}(\mathbf{u}) \sim 1/t^2,\ \kappa_{22}(\mathbf{u}) \sim 1/t^2,\ \kappa_{12}(\mathbf{u}) \sim 1/t^{3/2}, \\ &C(t) \sim 1/t, \end{aligned} \tag{27}$$

where the $C(t)$ are square roots conforming to the definition of the energy norm (squared energy is used in experiments). The $C(t) \sim 1/t$ is exactly the predicted worst-case error amplification factor. For more details, see [11]. The parameter-dependent error amplification simply means that as the parameter changes, the discretisation may have to be adjusted in order to maintain the same level of accuracy.

6. Conclusions

Harmonic extension finite elements provide an extension to the standard p-version. Indeed, if the discretisation is such that no adapted shape functions are needed, the method is the p-version. Adaptive reference elements are one way to implement the concept, and currently, the names can be used interchangeably. Until now, the method has been shown to perform well on problems with strong singularities. Here, the efficacy also on problems with boundary layers in established. For cylindrical shells, the numerical experiments indicate that the method converges in p as expected, and furthermore, the performance is maintained, even over a range of parameters. Via computational asymptotic analysis, the expected error amplification factors can be recovered also with the proposed method.

There are many open questions still when one considers implementation of the method. All numerical experiments considered here had constant coefficients. This simplifies numerical integration greatly and allows for the reuse of integrated reference elements. Similarly, the implementation meshes for the computed or adapted shape functions is the same for all shape functions in order to simplify the overall implementation and evaluation of the inner products. There are many opportunities for optimisation in this regard. Finally, even though adaptation of p-version specific error estimators should be straightforward, this work has not even started yet.

Discretisation of problems on complicated domains is a challenge that will not disappear any time soon. There are many options for non-standard approaches and many of them show great promise. Harmonic extension finite elements are a conservative, non-intrusive approach aimed for retaining as much as possible of the standard implementations and investments in the existing solvers. They also fit in within the standard a posteriori error analysis frameworks; however, rigorous proofs are still missing.

Funding: This research received no external funding.

Acknowledgments: We acknowledge the computational resources provided by the Aalto Science-IT project.

Conflicts of Interest: The author declares no conflict of interest.

References

1. Beirão da Veiga, L.; Brezzi, F.; Cangiani, A.; Manzini, G.; Marini, L.D.; Russo, A. Basic principles of virtual element methods. *Math. Model. Methods Appl. Sci.* **2013**, *23*, 199–214. [CrossRef]
2. Di Pietro, D.A.; Droniou, J. *The Hybrid High-Order Method for Polytopal Meshes*; Springer: Berlin/Heidelberg, Germany, 2020.
3. Szabo, B.; Babuska, I. *Finite Element Analysis*; Wiley: Hoboken, NJ, USA, 1991.
4. Schwab, C. *p- and hp-Finite Element Methods*; Oxford University Press: Oxford, UK, 1998.
5. Hakula, H. Adaptive reference elements via harmonic extensions and associated inner modes. *Comput. Math. Appl.* **2020**, *80*, 2272–2288. [CrossRef]
6. Hofreither, C.; Langer, U.; Weißer, S. Convection-adapted BEM-based FEM. *ZAMM—J. Appl. Math. Mech./Z. Angew. Math. Mech.* **2016**, *96*, 1467–1481. [CrossRef]
7. Weißer, S. Arbitrary order Trefftz-like basis functions on polygonal meshes and realization in BEM-based FEM. *Comput. Math. Appl.* **2014**, *67*, 1390–1406. [CrossRef]
8. Ovall, J.S.; Reynolds, S.E. A High-Order Method for Evaluating Derivatives of Harmonic Functions in Planar Domains. *SIAM J. Sci. Comput.* **2018**, *40*, A1915–A1935. [CrossRef]
9. Anand, A.; Ovall, J.S.; Reynolds, S.E.; Weißer, S. Trefftz Finite Elements on Curvilinear Polygons. *SIAM J. Sci. Comput.* **2020**, *42*, A1289–A1316. [CrossRef]
10. Pitkäranta, J. The problem of membrane locking in finite element analysis of cylindrical shells. *Numer. Math.* **1992**, *61*, 523–542. [CrossRef]
11. Hakula, H.; Leino, Y.; Pitkäranta, J. Scale resolution, locking, and high-order finite element modelling of shells. *Comput. Methods Appl. Mech. Eng.* **1996**, *133*, 157–182. [CrossRef]
12. Wolfram Research, Inc. *Mathematica*, version 13.0.1; Wolfram Research, Inc.: Champaign, IL, USA, 2021.
13. Hakula, H.; Tuominen, T. Mathematica implementation of the high order finite element method applied to eigenproblems. *Computing* **2013**, *95*, 277–301. [CrossRef]
14. Pitkäranta, J.; Matache, A.M.; Schwab, C. Fourier mode analysis of layers in shallow shell deformations. *Comput. Methods Appl. Mech. Eng.* **2001**, *190*, 2943–2975. [CrossRef]
15. Chapelle, D.; Bathe, K.J. *The Finite Element Analysis of Shells*; Springer: Berlin/Heidelberg, Germany, 2003.
16. Malinen, M. On the classical shell model underlying bilinear degenerated shell finite elements: General shell geometry. *Int. J. Numer. Methods Eng.* **2002**, *55*, 629–652. [CrossRef]
17. Hakula, H. hp-boundary layer mesh sequences with applications to shell problems. *Comput. Math. Appl.* **2014**, *67*, 899–917. [CrossRef]
18. Saad, Y.; Yeung, M.; Erhel, J.; Guyomarch, F. A deflated version of the conjugate gradient algorithm. *SIAM J. Sci. Comput.* **2000**, *21*, 1909–1926. [CrossRef]
19. Meerbergen, K. The solution of parametrized symmetric linear systems. *SIAM J. Matrix Anal. Appl.* **2003**, *24*, 1038–1059. [CrossRef]
20. Pitkäranta, J.; Leino, Y.; Ovaskainen, O.; Piila, J. Shell Deformation states and the Finite Element Method: A Benchmark Study of Cylindrical Shells. *Comput. Methods Appl. Mech. Eng.* **1995**, *128*, 81–121. [CrossRef]

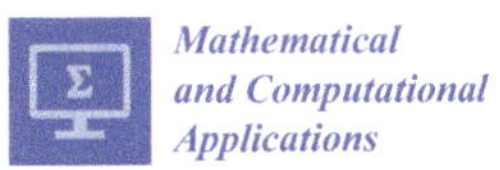
Mathematical and Computational Applications

Article

Using the Theory of Functional Connections to Solve Boundary Value Geodesic Problems

Daniele Mortari

Aerospace Engineering, Texas A&M University, 3141 TAMU, College Station, TX 77843, USA; mortari@tamu.edu

Abstract: This study provides a least-squares-based numerical approach to estimate the boundary value geodesic trajectory and associated parametric velocity on curved surfaces. The approach is based on the Theory of Functional Connections, an analytical framework to perform functional interpolation. Numerical examples are provided for a set of two-dimensional quadrics, including ellipsoid, elliptic hyperboloid, elliptic paraboloid, hyperbolic paraboloid, torus, one-sheeted hyperboloid, Moëbius strips, as well as on a generic surface. The estimated geodesic solutions for the tested surfaces are obtained with residuals at the machine-error level. In principle, the proposed approach can be applied to solve boundary value problems in more complex scenarios, such as on Riemannian manifolds.

Keywords: functional interpolation; ordinary differential equations; numerical methods

Citation: Mortari, D. Using the Theory of Functional Connections to Solve Boundary Value Geodesic Problems. *Math. Comput. Appl.* **2022**, 27, 64. https://doi.org/10.3390/mca27040064

Academic Editor: Nicholas Fantuzzi

Received: 12 July 2022
Accepted: 25 July 2022
Published: 27 July 2022

1. Introduction

Geodesic trajectories represent the paths on a curved surface, whose acceleration has no component on the local tangent plane to the surface. In many common scenarios, a geodesic represents the straightest path between two points. In particular, in a Riemannian manifold, geodesics are characterized by the property of having no geodesic curvature [1,2].

Most of the general studies on geodesics' trajectories are provided as initial value problems, while most of the boundary value problems are related to the biaxial and triaxial ellipsoid. In [3], the boundary value problem on an ellipsoid with boundary (Dirichlet) conditions is replaced by an initial value problem with Dirichlet and Neumann conditions. In particular, the Neumann condition is obtained iteratively by numerically integrating a system of four first-order differential equations. Triaxial ellipsoids are considered in [4,5], while [6] has provided an analytical approach to solve this boundary value problem with symmetry. Reference [7] contains, because of its importance to terrestrial geodesy, a rich literature survey on the geodesic equations for low eccentric ellipsoid in both Cartesian and polar coordinates. Since the problem of fitting ellipsoid is important in geodesy (all geodetic calculations are performed on a reference ellipsoid), Refs. [5,8] analyzes the computational differences in the fitting ellipsoid using biaxial ellipsoid instead of triaxial ellipsoid and by performing least-squares ellipsoid fitting. Noteworthy, Reference [7] has improved (in terms of computational time) the existing solutions using differential equations in Cartesian coordinates and Taylor series expansions by simplifying previous formulations.

In this study, the geodesic boundary value problem is numerically investigated for *any* curved surface using the general geodesic equations from differential geometry. These geodesic equations are then solved by nonlinear least-squares using the method that the Theory of Functional Connections [9] (TFC) has introduced to solve differential equations [10,11]. The existence of a solution for the geodesic boundary value problem (boundary value problems, in general, may have a unique solution, no solution, or infinite solutions) is guaranteed by the Hopf–Rinow theorem [12,13]:

> *If a length-metric space* (M, d) *is complete and compact then any two points,* $(\boldsymbol{p}_1, \boldsymbol{p}_2 \in M)$*, can be connected by a minimizing geodesic.*

In this theorem, M indicates a manifold and d the metric distance. Roughly speaking, *complete* indicates a space where there are no "points missing" from it (inside or at the boundary), while *compact* indicates a space that is *closed* (space bounds included) and *bounded* (distance between any two points limited).

Motivated by this theorem, this study proposes a general numerical and accurate approach to solve boundary value geodesic problems in curved surfaces. First, this study briefly provides the geodesic equations for the general Riemannian spaces, followed by a short background on TFC. Then, it shows how to apply TFC to solve boundary value geodesic problems by nonlinear least-squares. In particular, an ad hoc algorithm to avoid local minima is presented.

The proposed approach is then numerically tested on various quadric surfaces, such as triaxial ellipsoid, elliptic hyperboloid, elliptic paraboloid, hyperbolic paraboloid, torus, one-sheeted hyperboloid, Moëbius strip, and on a generic surface. In all these tests, machine error estimation of the geodesic trajectory is obtained along with the indirect numerical proof that the associated parametric velocity is constant.

2. Background on Geodesics Equations on Riemannian Manifold

A Riemannian manifold is a smooth curved space in which the infinitesimal distance, $\mathrm{d}s$, satisfies

$$\mathrm{d}s^2 = g_{ij}\,\mathrm{d}x^i\,\mathrm{d}x^j \tag{1}$$

where g_{ij} is the covariant metric tensor of the space and where Einstein's notation has been used in Equation (1).

The metric tensor is the fundamental tool used in differential geometry to study curved spaces. Specifically, for Euclidean spaces, the metric tensor is diagonal ($g_{ij} = 0$ if $i \neq j$) and, in particular, is the identity, $g_{ij} = \delta_{ij}$, for the Cartesian metric tensor. The metric tensor, g_{ij}, is the matrix composed by all inner products between all partials of the vector defining the Riemannian surface, $\boldsymbol{p}(x_1, x_2, \ldots, x_n)$

$$g_{ij} = \left(\frac{\partial \boldsymbol{p}}{\partial x_i}\right)^{\mathrm{T}} \frac{\partial \boldsymbol{p}}{\partial x_j}$$

In Riemannian geometry, geodesics are not the same as "shortest curves" between two points, though the two concepts are closely related. The difference is that geodesics are only locally the shortest distance between points. Going the "long way round" on a great circle between two points on a sphere is a geodesic but not the shortest path between the points. In addition, geodesic paths need not be unique.

On a curved surface, the length of a parametric trajectory, defined by the coordinates $x^i = x^i(t)$ and connecting the position vectors, $\boldsymbol{p}(t_0)$ and $\boldsymbol{p}(t_f)$, respectively, is,

$$L = \int_{t_0}^{t_f} \mathrm{d}s = \int_{t_0}^{t_f} \sqrt{g_{ij}\frac{\mathrm{d}x^i}{\mathrm{d}t}\frac{\mathrm{d}x^j}{\mathrm{d}t}}\,\mathrm{d}t \tag{2}$$

This parametric trajectory has no normal acceleration if it satisfies the following n differential equations [1,2,14],

$$\frac{\mathrm{d}^2x^i}{\mathrm{d}t^2} + \Gamma^i_{jk}\frac{\mathrm{d}x^j}{\mathrm{d}t}\frac{\mathrm{d}x^k}{\mathrm{d}t} = 0. \tag{3}$$

These n second-order ordinary differential equations are the *geodesic equations*. They define the geodesic trajectories on a manifold with metric tensor, g_{ij}. In Equation (3), the Γ^i_{jk} terms are the Christoffel symbols of second kind, defined as,

$$\Gamma^i_{jk} = \frac{1}{2}g^{mi}\left(\frac{\partial g_{km}}{\partial x^j} + \frac{\partial g_{mj}}{\partial x^k} - \frac{\partial g_{jk}}{\partial x^m}\right) \tag{4}$$

where g^{mi} is the controvariant inverse of the covariant metric tensor (or conjugate or dual metric),

$$g^{mi} = (g_{mi})^{-1}$$

Note that the Christoffel symbols satisfy the relationship, $\Gamma^i_{jk} = \Gamma^i_{kj}$ and are related to the Christoffel symbols of the first kind, $[m, jk]$, by the relationship, $[m, jk] = g_{mi}\,\Gamma^i_{jk}$.

Geodetic (Parametric) Velocity

A curve on a surface is called geodesic if, at every point, the acceleration is either zero or parallel to the normal. A geodesic trajectory, $\mathbf{p}(t)$, on a curved surface has constant parametric speed. The proof is immediate,

$$\frac{\mathrm{d}\dot{p}^2}{\mathrm{d}t} = \frac{\mathrm{d}(\dot{\boldsymbol{p}}^{\mathrm{T}}\dot{\boldsymbol{p}})}{\mathrm{d}t} = 2\ddot{\boldsymbol{p}}^{\mathrm{T}}\dot{\boldsymbol{p}} = 0.$$

The word "velocity," here, is not the instantaneous ratio between distance and time, but the instantaneous ratio between distance and the parameter t selected to describe the trajectory (which can also be the time).

Let us consider a two-dimensional surface in a three-dimensional space. Let the surface be described by the parametic equations,

$$x = f_x(u, v), \qquad y = f_y(u, v) \quad \text{and} \quad z = f_z(u, v),$$

then, by computing,

$$\dot{p}_x = \frac{\partial f_x}{\partial u}\,\dot{u} + \frac{\partial f_x}{\partial v}\,\dot{v}, \qquad \dot{p}_y = \frac{\partial f_y}{\partial u}\,\dot{u} + \frac{\partial f_y}{\partial v}\,\dot{v}, \quad \text{and} \quad \dot{p}_z = \frac{\partial f_z}{\partial u}\,\dot{u} + \frac{\partial f_z}{\partial v}\,\dot{v},$$

the velocity in a trajectory defined by $\dot{p}(u(t), v(t))$, is provided by $\dot{p}^2 = \dot{p}_x^2 + \dot{p}_y^2 + \dot{p}_z^2$.

To give a trivial example, the parametric description of a unit-radius sphere is provided by,

$$\boldsymbol{p}^{\mathrm{T}} = \{\sin u\, \cos v, \quad \sin u\, \sin v, \quad \cos u\}$$

where $u \in [0, \pi]$ and $v \in [0, 2\pi)$. The covariant metric tensor and the geodesic equations for the sphere are,

$$g_{ij} = \begin{bmatrix} 1 & 0 \\ 0 & \sin^2 u \end{bmatrix} \quad \text{and} \quad \begin{cases} \mathcal{L}_u = \ddot{u} - \dot{v}^2 \sin u\, \cos u = 0 \\ \mathcal{L}_v = \ddot{v} + 2\dot{u}\,\dot{v}\, \cot u = 0 \end{cases},$$

respectively, and the geodetic parametric velocity on the sphere is,

$$\dot{p}^2 = 2\dot{u}^2 \cos^2 u + \dot{v}^2 \sin^2 u.$$

Note that $u = 0$ and $u = \pi$ make singular the g_{ij} matrix and, consequently, make singular the geodesic equations. This singularity changes location if another set of polar coordinates is selected to describe a sphere.

3. Background on the Theory of Functional Connections

The Theory of Functional Connections (TFC) is a mathematical framework to perform *functional interpolation* [9,15]. This is performed by analytically deriving some functionals, called *constrained expressions*, representing *all* functions satisfying a set of linear constraints in n-dimensional space. This way, constrained optimization problems subject to linear constraints, such as differential equations, are transformed into unconstrained problems. The optimization problem consists then of deriving the expression of an unconstrained *free function*, $g(x)$ (Note that $g(x)$ can be discontinuous, partially defined, and even the Dirac delta function, as long as it is defined on where the constraints are defined.).

In general, for a univariate function, $y(x)$, subject to n linear constraints (e.g., points, derivatives, integrals, and any linear combination of them), two equivalent definitions of constrained expressions can be introduced [15–17],

$$y(x, g(x)) = g(x) + \sum_{k=1}^{n} \eta_k(x, g(x))\, s_k(x) \tag{5}$$

$$y(x, g(x)) = g(x) + \sum_{k=1}^{n} \rho_k(x, g(x))\, \phi_k(x, s(x)) \tag{6}$$

In the formal definition of Equation (5), the $\eta_k(x, g(x))$ are *functional coefficients* whose expressions are derived by imposing the n constraints on (5), and the $s_k(x)$ are a set of n user-defined linearly independent *support functions*. In Equation (6), the $\phi_k(x, s(x))$ are *switching functions* which imply changing between the constraints, and the $\rho_k(x, g(x))$ are *projection functionals*, which project the free function $g(x)$ to the kth constraint. See Ref. [9] for a complete and detailed explanation of these terms.

For example, consider a function $y(x)$ subject to the constraints

$$\left.\frac{\mathrm{d}y}{\mathrm{d}x}\right|_{x_1} = \dot{y}_1 \quad \text{and} \quad \left.\frac{\mathrm{d}y}{\mathrm{d}x}\right|_{x_2} = \dot{y}_2. \tag{7}$$

Using the form given in Equation (5) with $s_1(x) = x$ and $s_2(x) = x^2$, the constraints in Equation (7) can be expressed as

$$y(x, g(x)) = g(x) + \eta_1\, s_1(x) + \eta_2\, s_2(x) = g(x) + \eta_1\, x + \eta_2\, x^2 \tag{8}$$

where the two constants η_1 and η_2 are computed from Equations (7) and (8) as follows:

$$\begin{Bmatrix} \dot{y}_1 - \dot{g}_1 \\ \dot{y}_2 - \dot{g}_2 \end{Bmatrix} = \begin{bmatrix} \dot{s}_1(x_1) & \dot{s}_2(x_1) \\ \dot{s}_1(x_2) & \dot{s}_2(x_2) \end{bmatrix} \begin{Bmatrix} \eta_1 \\ \eta_2 \end{Bmatrix} \quad \to \quad \begin{Bmatrix} \eta_1 \\ \eta_2 \end{Bmatrix} = \begin{bmatrix} 1 & 2x_1 \\ 1 & 2x_2 \end{bmatrix}^{-1} \begin{Bmatrix} \dot{y}_1 - \dot{g}_1 \\ \dot{y}_2 - \dot{g}_2 \end{Bmatrix}$$

from which,

$$\begin{Bmatrix} \eta_1 \\ \eta_2 \end{Bmatrix} = \frac{1}{x_2 - x_1} \begin{Bmatrix} x_2(\dot{y}_1 - \dot{g}_1) - x_1(\dot{y}_2 - \dot{g}_2) \\ -(\dot{y}_1 - \dot{g}_1) + (\dot{y}_2 - \dot{g}_2) \end{Bmatrix}.$$

Substituting the η_1 and η_2 expressions into Equation (8), the following *constrained expression*,

$$y(x, g(x)) = g(x) + \underbrace{\frac{x\,(2x_2 - x)}{2(x_2 - x_1)}}_{\phi_1(x, s(x))} \underbrace{(\dot{y}_1 - \dot{g}_1)}_{\rho_1(x, g(x))} + \underbrace{\frac{x\,(x - 2x_1)}{2(x_2 - x_1)}}_{\phi_2(x, s(x))} \underbrace{(\dot{y}_2 - \dot{g}_2)}_{\rho_2(x, g(x))}, \tag{9}$$

is obtained for the constraints given in Equation (7). Equation (9) satisfies the constraints (7), *no matter what the free function* $g(x)$ *is*. Moreover, Equation (9) indicates the expressions of the switching functions, $\phi_k(x, s(x))$, and the projection functionals, $\rho_k(x, g(x))$. Here, the meaning of the switching functions becomes more clear: when the first constraint, $\dot{y}(x_1) = \dot{y}_1$, holds then $\phi_1(x_1) = 1$ and $\phi_2(x_1) = 0$; when the second constraint, $\dot{y}(x_2) = \dot{y}_2$, holds then $\phi_1(x_2) = 0$ and $\phi_2(x_2) = 1$. The projection functionals are scalars in this univariate case, but they become functionals in the multivariate case (see [9] for full explanation).

The multivariate TFC [9,16] extends the original univariate theory [15] to n dimensions and to any linear (boundary and/or internal) constraints. This extension can be summarized by the expression,

$$y(\boldsymbol{x}, g(\boldsymbol{x})) = h(c(\boldsymbol{x})) + g(\boldsymbol{x}) - h(g(\boldsymbol{x})),$$

where $\boldsymbol{x} = (x_1, x_1, \ldots, x_n)^{\mathsf{T}}$ is the vector of n coordinates, $c(\boldsymbol{x})$ is a function specifying the linear constraints, $h(c(\boldsymbol{x}))$ is *any* interpolating function satisfying the linear constraints, and

$g(x)$ is the free function. Several examples on how to derive constrained expressions can be found in [9,17,18].

The Theory of Functional Connections has been developed for points, derivatives, integrals, infinite, component constraints, and any linear combination of them for univariate functions [15] as well as multivariate functions [16] in rectangular domains, while in generic domains some initial results have been obtained via domain mapping [19]. The main feature of these functionals (constrained expressions) is that they allow for restricting the whole function space of a constrained optimization problems to just the space of its feasible solutions, fully satisfying the constraints. This way, a large number of constrained optimization problems can be transformed into unconstrained ones, which can be solved by more simple, efficient, robust, reliable, fast, and accurate methods.

The first application of TFC was in solving linear [10] and nonlinear [11] ODEs. This has been conducted by expanding the free function $g(x)$ in terms of a set of basis functions (e.g., orthogonal polynomials, Fourier, neural networks, etc.). Linear or nonlinear least-squares method is then used to find the coefficients of the expansion. This TFC approach for solving ODEs has many advantages over traditional methods: (1) it consists of a unified framework to solve IVP, BVP, or multi-valued problems, (2) it provides an analytically approximated solution that can be used for subsequent manipulation (derivatives, integrals), (3) the solution is usually obtained in millisecond and at machine error accuracy, (4) the procedure is numerically robust (small condition number), and (5) it can solve differential equations subject to a variety of different constraint types. Additionally, TFC has been also applied to solve other mathematical optimization problems [20] such as in: homotopy continuation for control problems [21], epidemiological models [22], radiative transfer problems [23], rarefied-gas dynamics [24], Timoshenko-Ehrenfest beam [25], hybrid systems [26], machine learning [27–30], quadratic and nonlinear programming problems subject to linear equality constraints [31], orbit transfer and propagation [18,32,33], optimal control problems via indirect methods, relative motion [34], landing on small and large planetary bodies [35], and intercept problems [30].

4. Solving the Geodesic Equations Using the Theory of Functional Connections

Any trajectory on a two-dimensional surface in three-dimensional space can always be described by two coordinates, $[u(t), v(t)]$, depending on a parameter t. Specifically, let us consider a trajectory from an initial point $[u_0, v_0]$ to a final point $[u_f, v_f]$, while the parameter range is $t \in [-1, +1]$.

All possible trajectories, $[u(t), v(t)]$, connecting $[u_0, v_0]$ to $[u_f, v_f]$, can be represented by the following two functionals (constrained expressions) [9]

$$\begin{cases} u(t, g_u(t)) = g_u(t) + \dfrac{1-t}{2}(u_0 - g_{u0}) + \dfrac{1+t}{2}(u_f - g_{uf}) \\ v(t, g_v(t)) = g_v(t) + \dfrac{1-t}{2}(v_0 - g_{v0}) + \dfrac{1+t}{2}(v_f - g_{vf}) \end{cases} \tag{10}$$

where $g_u(t)$ and $g_v(t)$ are two free functions and where it has been set $g_{u0} = g_u(-1)$, $g_{v0} = g_v(-1)$, $g_{uf} = g_u(+1)$, and $g_{vf} = g_v(+1)$. *No matter what the functions, $g_u(t)$ and $g_u(t)$ are,* the Equation (10) always generate trajectories moving from $[u_0, v_0]$ to $[u_f, v_f]$, as t increases from $t = -1$ to $t = +1$.

The parametric derivatives of Equation (10) are,

$$\begin{cases} \dot{u}(t, g_u(t)) = \dot{g}_u(t) - \dfrac{1}{2}(u_0 - g_{u0}) + \dfrac{1}{2}(u_f - g_{uf}) \\ \ddot{u}(t, g_u(t)) = \ddot{g}_u(t) \\ \dot{v}(t, g_v(t)) = \dot{g}_v(t) - \dfrac{1}{2}(v_0 - g_{v0}) + \dfrac{1}{2}(v_f - g_{vf}) \\ \ddot{v}(t, g_v(t)) = \ddot{g}_v(t) \end{cases} \tag{11}$$

Let us express the free functions, $g_u(t)$ and $g_v(t)$, as a linear combination of linearly independent basis functions, $\boldsymbol{h}(t)$, (e.g., orthogonal polynomials). Then,

$$\begin{cases} g_u(t) = \boldsymbol{\xi}_u^{\mathrm{T}} \boldsymbol{h}(t) \\ g_v(t) = \boldsymbol{\xi}_v^{\mathrm{T}} \boldsymbol{h}(t), \end{cases} \quad \begin{cases} \dot{g}_u(t) = \boldsymbol{\xi}_u^{\mathrm{T}} \dot{\boldsymbol{h}}(t) \\ \dot{g}_v(t) = \boldsymbol{\xi}_v^{\mathrm{T}} \dot{\boldsymbol{h}}(t), \end{cases} \quad \text{and} \quad \begin{cases} \ddot{g}_u(t) = \boldsymbol{\xi}_u^{\mathrm{T}} \ddot{\boldsymbol{h}}(t) \\ \ddot{g}_v(t) = \boldsymbol{\xi}_v^{\mathrm{T}} \ddot{\boldsymbol{h}}(t). \end{cases} \quad (12)$$

The geodesic equations provided in Equation (3) can be solved using the expressions given in Equations (10) and (11), with the free functions expanded as in Equation (12), and by discretizing the parameter t from $t = -1$ to $t = +1$. (When expressing the free functions in terms orthogonal polynomial, the best discretization of t for the least-squares problem is obtained by the Chebyshev–Gauss–Lobatto points distribution.) The resulting equations are two nonlinear algebraic equations in the unknown coefficient vectors, $\boldsymbol{\xi}_u$ and $\boldsymbol{\xi}_v$, that can be solved by nonlinear least-squares,

$$\begin{Bmatrix} \boldsymbol{\xi}_u \\ \boldsymbol{\xi}_v \end{Bmatrix}_{k+1} = \begin{Bmatrix} \boldsymbol{\xi}_u \\ \boldsymbol{\xi}_v \end{Bmatrix}_k - (\mathcal{J}_k^{\mathrm{T}} \mathcal{J}_k)^{-1} \mathcal{J}_k^{\mathrm{T}} \begin{Bmatrix} \mathcal{L}_u \\ \mathcal{L}_v \end{Bmatrix}_k \quad \text{where} \quad \mathcal{J}_k = \begin{bmatrix} \dfrac{\partial \mathcal{L}_u}{\partial \boldsymbol{\xi}_u}, & \dfrac{\partial \mathcal{L}_u}{\partial \boldsymbol{\xi}_v} \\ \dfrac{\partial \mathcal{L}_v}{\partial \boldsymbol{\xi}_u}, & \dfrac{\partial \mathcal{L}_v}{\partial \boldsymbol{\xi}_v} \end{bmatrix}_k \quad (13)$$

is the Jacobian of the system.

Note that, the constrained expressions given in Equation (10) are derived using constant and linear terms of support functions. This implies that the set of basis functions adopted in the least-squares process for the free functions, $g_u(t)$ and $g_v(t)$, cannot include both the constant and the linear terms. For instance, if Chebyshev orthogonal polynomials are selected as support functions, then $T_0 = 1$ and $T_1 = t$ must be excluded, otherwise the matrix to invert in the least-squares process becomes singular. This is because a least-squares process of a linear combination of functions is singular if not all the functions are linearly independent.

Initial Guess and the Local Minima Problem

Since the constrained expressions provide trajectories always satisfying the constraints (for any expression of the free functions), then the most natural (and simplest) initial guess is to start the iterative least-squares solving Equation (13) by setting $\boldsymbol{\xi}_u = \boldsymbol{\xi}_v = \boldsymbol{0}$. This is equivalent to initially select $g_u(t) = g_v(t) = 0$ and, consequently—see Equation (10)—, selecting an initial linear variation of the parametric variables, u and v, from the initials (u_0, v_0) to the final values (u_f, v_f).

Nonlinear least-squares applied to find the geodesic trajectory is, unfortunately, a process affected by local minima. Typically, when an iterative procedure enters into the convergence phase, then each following step is smaller than the previous,

$$L_2(\Delta\boldsymbol{\xi}_{k+1}) < L_2(\Delta\boldsymbol{\xi}_k) \qquad \text{where} \qquad \Delta\boldsymbol{\xi}_k = \begin{Bmatrix} \Delta\boldsymbol{\xi}_u \\ \Delta\boldsymbol{\xi}_v \end{Bmatrix}_k \quad (14)$$

This convergence criteria has two interesting aspects: (1) it can be used to push the convergence to the maximum accuracy and (2) no tolerance is needed to stop the iterations. In fact, when the convergence reaches the maximum accuracy, the procedure cannot anymore improve the estimation of $\Delta\boldsymbol{\xi}_k$ and the inequality, $L_2(\Delta\boldsymbol{\xi}_{k+1}) > L_2(\Delta\boldsymbol{\xi}_k)$, happens because of numerical errors (convergence saturation).

Figure 1 shows the flowchart of the algorithm adopted to avoid local minima. Starting with $\boldsymbol{\xi}_0 = \boldsymbol{0}$, the least-squares iterations continue until Equation (14) is verified for N consecutive times (initial convergence loop). During this sequence, if the least-squares diverges, then the algorithm restarts with a new random initial guess. If Equation (14) is verified for N consecutive times, then the algorithm enters into the convergence loop and exits when $L_2(\Delta\boldsymbol{\xi}_{k+1}) > L_2(\Delta\boldsymbol{\xi}_k)$ is experienced. Then, the L_2 norm of the residuals is computed. The L_2 norm of the residuals allows one to discriminate local minima and global minima (associated with the geodesic trajectory) using a small tolerance, such as $\varepsilon = 10^{-15}$. If $L_2(\boldsymbol{r}_k) > \varepsilon$ then the algorithm restarts with a new random initial guess.

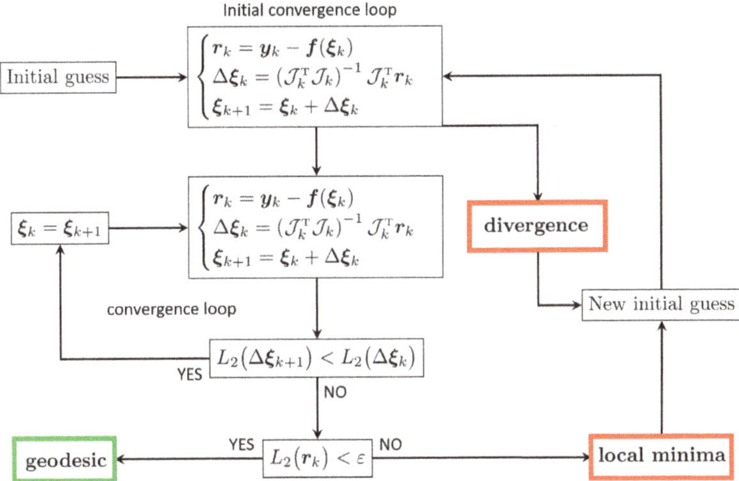

Figure 1. Local minima avoidance algorithm flowchart.

5. Numerical Validations

In this section, some boundary value geodesic problems are numerically solved to validate the proposed methodology on 2-dimensional surfaces, for visualization purposes. The proposed approach has been tested on eight different kind of surfaces: triaxial ellipsoid, elliptic hyperboloid, elliptic paraboloid, hyperbolic paraboloid, one-sheeted hyperboloid, torus, Moëbius strip, and a generic surface. Six of them are shown in Figure 2.

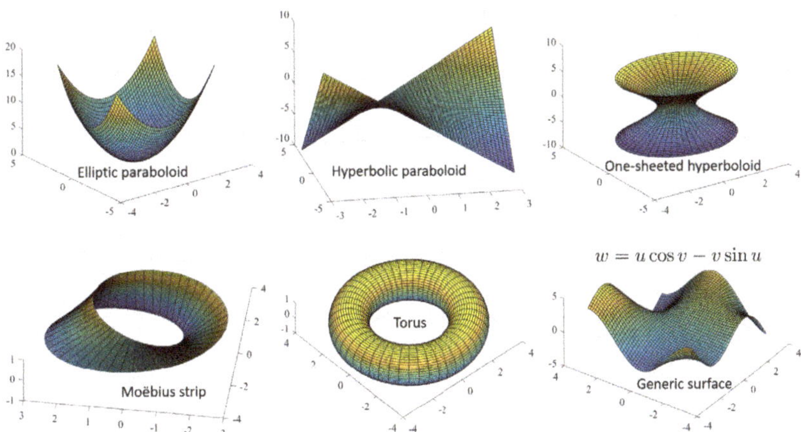

Figure 2. Examples of six tested surfaces.

5.1. Triaxial Ellipsoid

A (non-oriented) triaxial ellipsoid is described by,

$$\boldsymbol{p}^{\mathrm{T}} = \{a \sin u \cos v, \quad b \sin u \sin v, \quad c \cos u\}$$

where $u \in [0, \pi]$ and $v \in [0, 2\pi)$, and (a, b, c) are the three ellipsoid semi-major axes. The covariant metric tensor for the ellipsoid is,

$$g_{ij} = \begin{bmatrix} \cos^2 u(a^2\cos^2 v + b^2 \sin^2 v) + c^2 \sin^2 u & -\dfrac{\sin(2u)\sin(2v)}{4}(a^2 - b^2) \\ -\dfrac{\sin(2u)\sin(2v)}{4}(a^2 - b^2) & \sin^2 u\big[\sin^2 v(a^2 - b^2) + b^2\big] \end{bmatrix}$$

Since the ellipsoid is an affine image of the unit sphere, it is affected by the same singularity problem, occurring for $u = 0$ and $u = \pi$.

By setting $a = 2$, $b = 3$, and $c = 1$, the geodesic equations for this ellipsoid are [36],

$$\begin{cases} \ddot{u} + \dfrac{(5\cos^2 v - 32)\cos u \sin u}{\sin^2 u(5\cos^2 v - 32) + 36}\dot{u}^2 - \dfrac{36 \cos u \sin u}{\sin^2 u(5\cos^2 v - 32) + 36}\dot{v}^2 = 0 \\ \ddot{v} - \dfrac{5\cos v \sin v}{\sin^2 u(5\cos^2 v - 32) + 36}\dot{u}^2 + \dfrac{2\cos u}{\sin u}\dot{u}\dot{v} - \dfrac{5\sin^2 u \cos v \sin v}{\sin^2 u(5\cos^2 v - 32) + 36}\dot{v}^2 = 0 \end{cases}$$

To solve these differential equations using nonlinear least-squares, the following partials must be evaluated to compute the Jacobian of the system,

$$\begin{cases} \dfrac{\partial \mathcal{L}_u}{\partial \xi_u} = \dfrac{\partial \mathcal{L}_u}{\partial u} \cdot \dfrac{\mathrm{d}u}{\mathrm{d}\xi_u} + \dfrac{\partial \mathcal{L}_u}{\partial \dot{u}} \cdot \dfrac{\mathrm{d}\dot{u}}{\mathrm{d}\xi_u} + \dfrac{\partial \mathcal{L}_u}{\partial \ddot{u}} \cdot \dfrac{\mathrm{d}\ddot{u}}{\mathrm{d}\xi_u} \\ \dfrac{\partial \mathcal{L}_u}{\partial \xi_v} = \dfrac{\partial \mathcal{L}_u}{\partial v} \cdot \dfrac{\mathrm{d}v}{\mathrm{d}\xi_v} + \dfrac{\partial \mathcal{L}_u}{\partial \dot{v}} \cdot \dfrac{\mathrm{d}\dot{v}}{\mathrm{d}\xi_v} \\ \dfrac{\partial \mathcal{L}_v}{\partial \xi_u} = \dfrac{\partial \mathcal{L}_v}{\partial u} \cdot \dfrac{\mathrm{d}u}{\mathrm{d}\xi_u} + \dfrac{\partial \mathcal{L}_v}{\partial \dot{u}} \cdot \dfrac{\mathrm{d}\dot{u}}{\mathrm{d}\xi_u} \\ \dfrac{\partial \mathcal{L}_v}{\partial \xi_v} = \dfrac{\partial \mathcal{L}_v}{\partial v} \cdot \dfrac{\mathrm{d}v}{\mathrm{d}\xi_v} + \dfrac{\partial \mathcal{L}_v}{\partial \dot{v}} \cdot \dfrac{\mathrm{d}\dot{v}}{\mathrm{d}\xi_v} + \dfrac{\partial \mathcal{L}_v}{\partial \ddot{v}} \cdot \dfrac{\mathrm{d}\ddot{v}}{\mathrm{d}\xi_v} \end{cases} \tag{15}$$

Figure 3 shows the numerical results using 40 basis functions (Legendre orthogonal polynomials) to describe the free functions and 200 discretization points. This test validates the approach in terms of fast convergence (just six iterations using a noisy initial guess) and in terms of finding constant the parametric velocity.

The geodetic velocity on a generic triaxial ellipsoid has components,

$$\dot{\boldsymbol{p}} = \begin{Bmatrix} a(\dot{u}\cos u \cos v - \dot{v}\sin u \sin v) \\ b(\dot{u}\cos u \sin v + \dot{v}\sin u \cos v) \\ -c\dot{u}\sin u \end{Bmatrix}$$

5.2. Elliptic Paraboloid

For the elliptic paraboloid, the geodesic equations and the TFC solution procedure are provided. The parametric equations of the elliptic paraboloid, $z = x^2 + y^2$, can be described by the vector,

$$\boldsymbol{p}^{\mathrm{T}} = \{u, \quad v, \quad u^2 + v^2\}$$

where $u \in [u_{\min}, u_{\max}]$ and $v \in [v_{\min}, v_{\max}]$. The partials of the generic point, $\boldsymbol{p}$, are

$$\frac{\partial \boldsymbol{p}}{\partial u} = \boldsymbol{p}_u = \begin{Bmatrix} 1 \\ 0 \\ 2u \end{Bmatrix} \quad \text{and} \quad \frac{\partial \boldsymbol{p}}{\partial v} = \boldsymbol{p}_v = \begin{Bmatrix} 0 \\ 1 \\ 2v \end{Bmatrix}$$

from which the metric tensor,

$$g_{ij} = \begin{bmatrix} \boldsymbol{p}_u^{\mathrm{T}}\boldsymbol{p}_u & \boldsymbol{p}_u^{\mathrm{T}}\boldsymbol{p}_v \\ \boldsymbol{p}_v^{\mathrm{T}}\boldsymbol{p}_u & \boldsymbol{p}_v^{\mathrm{T}}\boldsymbol{p}_v \end{bmatrix} = \begin{bmatrix} 1 + 4u^2 & 4uv \\ 4uv & 1 + 4v^2 \end{bmatrix}$$

and its inverse,

$$g^{ij} = \left(g_{ij}\right)^{-1} = \frac{1}{4(u^2+v^2)+1}\begin{bmatrix} 1+4v^2 & -4uv \\ -4uv & 1+4u^2 \end{bmatrix}$$

are derived. The non-null Christoffel symbols are derived using Equation (4),

$$\Gamma^1_{11} = \frac{4\,u}{4(u^2+v^2)+1}, \qquad \Gamma^1_{22} = \frac{4\,u}{4(u^2+v^2)+1},$$
$$\Gamma^2_{11} = \frac{4\,v}{4(u^2+v^2)+1}, \qquad \text{and} \qquad \Gamma^2_{22} = \frac{4\,v}{4(u^2+v^2)+1}.$$

The geodesic equations, given in Equation (3), become

$$\begin{cases} \ddot{u} + \Gamma^1_{11}\,\dot{u}^2 + \Gamma^1_{22}\,\dot{v}^2 = 0 \\ \ddot{v} + \Gamma^2_{11}\,\dot{u}^2 + \Gamma^2_{22}\,\dot{v}^2 = 0 \end{cases}$$

After substituting the Christoffel symbols and rearranging the equations, the geodesic equations for the elliptic paraboloid become,

$$\begin{cases} \mathcal{L}_u = (4u^2+4v^2+1)\ddot{u} + 4u\,\dot{u}^2 + 4u\,\dot{v}^2 = 0 \\ \mathcal{L}_v = (4u^2+4v^2+1)\ddot{v} + 4v\,\dot{u}^2 + 4v\,\dot{v}^2 = 0 \end{cases} \tag{16}$$

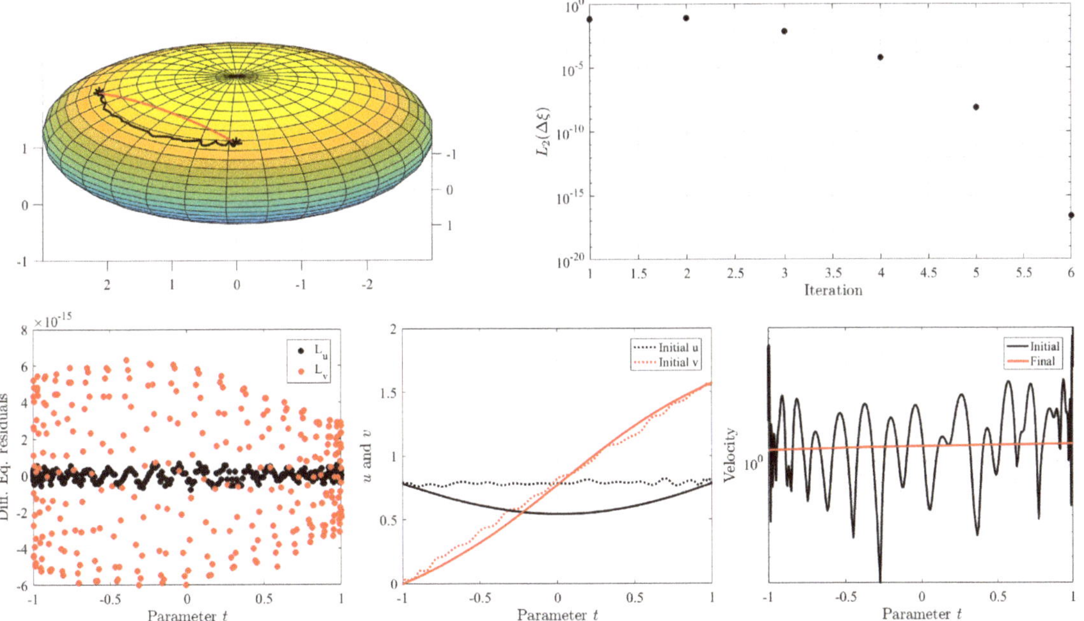

Figure 3. Numerical validation test on triaxial ellipsoid.

To solve these differential equations using nonlinear least-squares, the same structure of partials provided in Equation (15) must be evaluated to compute the Jacobian of the system, where

$$\begin{aligned}
&\frac{\partial \mathcal{L}_u}{\partial u} = 8u\ddot{u} + 4(\dot{u}^2 + \dot{v}^2), && \frac{\partial \mathcal{L}_u}{\partial \dot{u}} = 8u\dot{u}, && \frac{\partial \mathcal{L}_u}{\partial \ddot{u}} = \frac{\partial \mathcal{L}_v}{\partial \ddot{v}} = 4(u^2 + v^2) + 1,\\
&\frac{\partial \mathcal{L}_u}{\partial \dot{v}} = 8u\dot{v}, && \frac{\partial \mathcal{L}_v}{\partial u} = 8u\ddot{v}, && \frac{\partial \mathcal{L}_v}{\partial \dot{u}} = 8v\dot{u},\\
&\frac{\partial \mathcal{L}_v}{\partial v} = 8v\ddot{v} + 4(\dot{u}^2 + \dot{v}^2), && \frac{\partial \mathcal{L}_v}{\partial \dot{v}} = 8v\dot{v}, && \frac{\partial \mathcal{L}_u}{\partial v} = 8v\ddot{u},
\end{aligned}$$

Note that, the elliptic paraboloid can also be expressed by, $\boldsymbol{p} = \{u \sin v,\ u \cos v,\ u^2\}^{\mathsf{T}}$, where u and v are angles. In this case, the geodesic equations provided in Equation (16) can be rewritten as,

$$\begin{cases} \mathcal{L}_u = (4u^2 + 1)\,\ddot{u} + 4u\,\dot{u}^2 - u\,\dot{v}^2 = 0 \\ \mathcal{L}_v = u\,\ddot{v} + 2\dot{u}\,\dot{v} = 0 \end{cases}$$

and the following partials must be computed to solve the boundary value problem,

$$\begin{aligned}
\frac{\partial \mathcal{L}_u}{\partial \xi_u} &= \frac{\partial \mathcal{L}_u}{\partial u} \cdot \frac{\mathrm{d}u}{\mathrm{d}\xi_u} + \frac{\partial \mathcal{L}_u}{\partial \dot{u}} \cdot \frac{\mathrm{d}\dot{u}}{\mathrm{d}\xi_u} + \frac{\partial \mathcal{L}_u}{\partial \ddot{u}} \cdot \frac{\mathrm{d}\ddot{u}}{\mathrm{d}\xi_u}\\
\frac{\partial \mathcal{L}_u}{\partial \xi_v} &= \frac{\partial \mathcal{L}_u}{\partial \dot{v}} \cdot \frac{\mathrm{d}\dot{v}}{\mathrm{d}\xi_v}\\
\frac{\partial \mathcal{L}_v}{\partial \xi_u} &= \frac{\partial \mathcal{L}_v}{\partial u} \cdot \frac{\mathrm{d}u}{\mathrm{d}\xi_u} + \frac{\partial \mathcal{L}_v}{\partial \dot{u}} \cdot \frac{\mathrm{d}\dot{u}}{\mathrm{d}\xi_u}\\
\frac{\partial \mathcal{L}_v}{\partial \xi_v} &= \frac{\partial \mathcal{L}_v}{\partial \dot{v}} \cdot \frac{\mathrm{d}\dot{v}}{\mathrm{d}\xi_v} + \frac{\partial \mathcal{L}_v}{\partial \ddot{v}} \cdot \frac{\mathrm{d}\ddot{v}}{\mathrm{d}\xi_v}
\end{aligned}$$

where,

$$\begin{aligned}
&\frac{\partial \mathcal{L}_u}{\partial u} = 8u\ddot{u} + 4\dot{u}^2 - \dot{v}^2, && \frac{\partial \mathcal{L}_u}{\partial \dot{u}} = 8u\dot{u}, && \frac{\partial \mathcal{L}_u}{\partial \ddot{u}} = 4u^2 + 1, && \frac{\partial \mathcal{L}_u}{\partial \dot{v}} = -2u\dot{v},\\
&\frac{\partial \mathcal{L}_v}{\partial u} = \ddot{v}, && \frac{\partial \mathcal{L}_v}{\partial \dot{u}} = 2\dot{v}, && \frac{\partial \mathcal{L}_v}{\partial \dot{v}} = 2\dot{u}, && \frac{\partial \mathcal{L}_v}{\partial \ddot{v}} = u,
\end{aligned}$$

and,

$$\begin{aligned}
\frac{\partial u}{\partial \xi_u} &= \frac{\partial v}{\partial \xi_v} = \boldsymbol{h} - \frac{1-t}{2}\boldsymbol{h}_0 - \frac{t+1}{2}\boldsymbol{h}_f,\\
\frac{\partial \dot{u}}{\partial \xi_u} &= \frac{\partial \dot{v}}{\partial \xi_v} = \dot{\boldsymbol{h}} + \frac{1}{2}\boldsymbol{h}_0 - \frac{1}{2}\boldsymbol{h}_f,\\
\frac{\partial \ddot{u}}{\partial \xi_u} &= \frac{\partial \ddot{v}}{\partial \xi_v} = \ddot{\boldsymbol{h}},
\end{aligned}$$

This second parametrization of the elliptic paraboloid is provided because a particular attention must be given to the boundary value of angles to avoid discontinuous angle evolution. For example, if the absolute difference between the two angles is $|v_0 - v_f| > \pi$, then the value of the smallest of these two angles is increased by 2π. This avoids the discontinuity at $v = 0$ while unchanging the problem. For instance, if $v_0 = 3\pi/2$ and $v_f = \pi/4$, then the value of v_f is set as $v_f = \pi/4 + 2\pi = 9\pi/4$.

Geodesic (Parametric) Velocity on an Elliptic Paraboloid

The derivative of the position vector is,

$$\dot{\boldsymbol{p}}^{\mathsf{T}} = \{\dot{p}_x,\ \dot{p}_y,\ \dot{p}_z\} = \{\dot{u},\ \dot{v},\ 2u\dot{u} + 2v\dot{v}\}$$

Therefore, the velocity is provided by,

$$\dot{p}(t) = \sqrt{\dot{u}^2 + u^2 + 4(u\dot{u} + v\dot{v})^2}$$

Since the velocity on a geodesic is constant, then the (instantaneous) value $\dot{p}(t)$ for a geodesic is not a function of t and, therefore, the expression of $\dot{p}(t)$ must coincide with the average value of the velocity which, in turn, can be computed using Equation (2),

$$\bar{\dot{p}} = \frac{L}{t_f - t_0} = \frac{1}{t_f - t_0} \int_{t_0}^{t_f} \sqrt{(v^2+1)\dot{u}^2 + 2uv\dot{u}\dot{v} + (u^2+1)\dot{v}^2} \, \mathrm{d}t$$

therefore, for a geodesic trajectory on the elliptic paraboloid, we have a closed form expression for the integral,

$$\int_{t_0}^{t_f} \sqrt{(v^2+1)\dot{u}^2 + 2uv\dot{u}\dot{v} + (u^2+1)\dot{v}^2} \, \mathrm{d}t = (t_f - t_0)\sqrt{\dot{u}^2 + u^2 + 4(u\dot{u} + v\dot{v})}$$

5.3. Elliptic Hyperboloid

The parametric description of the elliptic hyperboloid is,

$$\boldsymbol{p}^{\mathrm{T}} = \{a \sinh u \cos v, \quad b \sinh u \sin v, \quad c \cosh u\}$$

and his covariant metric tensor is,

$$g_{ij} = \begin{bmatrix} \cosh^2 u(a^2\cos^2 v + b^2 \sin^2 v) + c^2 \sinh^2 u & -\sinh(2u)\sin(2v)\dfrac{a^2 - b^2}{4} \\ -\sinh(2u)\sin(2v)\dfrac{a^2-b^2}{4} & \sinh^2 u((a^2 - b^2)\sin^2 v + b^2) \end{bmatrix}$$

By setting $a = 1$, $b = 2$, and $c = 3$, the geodesic equations are,

$$\begin{aligned} \ddot{u} &+ \frac{\cosh u \sinh u(13 + 27\cos^2 v)}{27\cosh^2 u \cos^2 v + 13\cosh^2 u - 9 - 27\cos^2 v}\dot{u}^2 + \\ &+ \frac{4\cosh u \sinh u}{27\cosh^2 u \cos^2 v + 13\cosh^2 u - 9 - 27\cos^2 v}\dot{v}^2 = 0 \\ \ddot{v} &- \frac{27\cos v \sin v}{27\cosh^2 u \cos^2 v + 13\cosh^2 u - 9 - 27\cos^2 v}\dot{u}^2 + \\ &+ \frac{2\cosh u}{\sinh u}\dot{u}\,\dot{v} - \frac{27\cos v \sin v\left(\cosh^2 u - 1\right)}{27\cosh^2 u \cos^2 v + 13\cosh^2 u - 9 - 27\cos^2 v}\dot{v}^2 = 0 \end{aligned}$$

The geodesic velocity components on the elliptic hyperboloid are,

$$\begin{cases} \dot{p}_x = a\,(\dot{u}\cosh u \cos v - \dot{v}\sinh u \sin v) \\ \dot{p}_y = b\,(\dot{u}\cosh u \sin v + \dot{v}\sinh u \cos v) \\ \dot{p}_z = c\dot{u}\sinh u \end{cases}$$

Figure 4 provides detailed information of the test results using 40 basis functions (for the free functions) and 300 points discretization. In this case, a very noisy initial guess is selected (black trajectory on the left/top figure), instead of the simple, $\boldsymbol{\xi}_0 = \mathbf{0}$. The right/top figure shows, for subsequent iterations, the L_2 norm of the error. The procedure entered into the convergence phase after about 35 iterations. The convergence loop pushed the accuracy to the limit, when $L_2(\Delta\boldsymbol{\xi}_{k+1}) > L_2(\Delta\boldsymbol{\xi}_k)$ occurred. The final solution residuals are shown in the left/bottom figure, while the central/bottom figure shows the initial and final solutions for $u(t)$ and $v(t)$. Finally, the right/bottom figure gives the parametric velocity of the initial guess (black) and of the final (red) solution (geodesic trajectory). This validates the parametric velocity with a constant value. For this example, no local minima were ever experienced (convex problem).

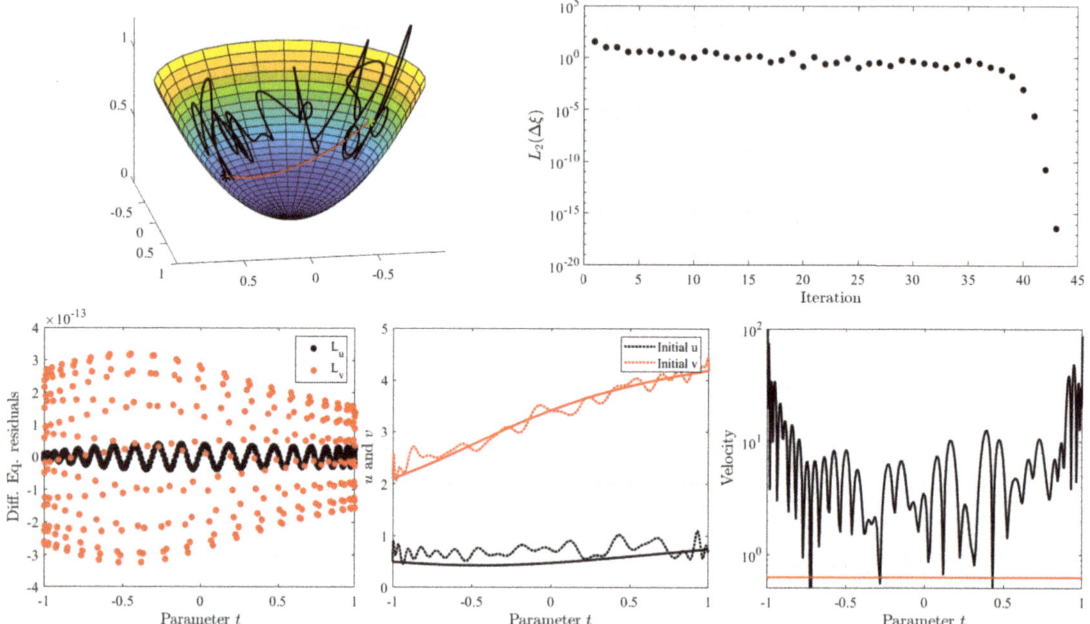

Figure 4. Elliptic paraboloid test results.

5.4. Hyperbolic Paraboloid

The parametric description of the hyperbolic paraboloid is,

$$p^{\top} = \{u, \quad v, \quad u\,v\}$$

the covariant metric tensor for the hyperbolic paraboloid is

$$g_{ij} = \begin{bmatrix} v^2+1 & uv \\ uv & u^2+1 \end{bmatrix}$$

and the Christoffel symbols are, $\Gamma^2_{11} = \dfrac{v}{u^2+v^2+1}$, $\Gamma^1_{12} = \dfrac{v}{u^2+v^2+1}$, $\Gamma^2_{21} = \dfrac{u}{u^2+v^2+1}$, and $\Gamma^1_{22} = \dfrac{u}{u^2+v^2+1}$, while the geodesic equations are,

$$\begin{cases} \mathcal{L}_u = (1+u^2+v^2)\ddot{u} + 2v\,\dot{u}\,\dot{v} = 0 \\ \mathcal{L}_v = (1+u^2+v^2)\ddot{v} + 2\,u\,\dot{u}\,\dot{v} = 0 \end{cases}$$

The expression of the geodesic velocity on hyperbolic paraboloid is,

$$\dot{p}^2 = \dot{u}^2 + \dot{v}^2 + (v\dot{u} + u\dot{v})^2.$$

Figure 5 contains the results obtained on hyperbolic paraboloid (same meaning than those provided for the elliptic paraboloid), while Table 1 shows the L_2 norm of the error step, $L_2(\Delta\xi_k)$, and the associated L_2 norm of the differential equations residuals, $L_2(\boldsymbol{r}_k)$, for the first 8 iterations.

Table 1. First eight iterations for the hyperbolic paraboloid.

Iteration k	$L_2(\Delta\xi_k)$	$L_2(\boldsymbol{r}_k)$
0	-	78,019.3597
1	0.014083	682.5215
2	0.13725	51.5259
3	0.10102	4.9834
4	0.029109	0.10211
5	0.00097301	7.6775×10^{-5}
6	9.8617×10^{-7}	7.4134×10^{-11}
7	1.0705×10^{-12}	7.3451×10^{-17}
8	2.3949×10^{-19}	8.1203×10^{-17}

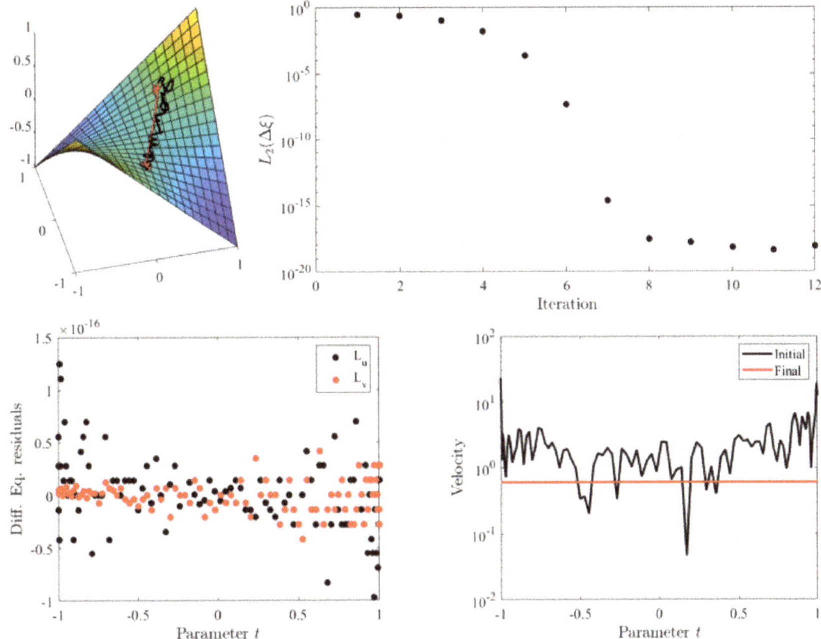

Figure 5. Hyperbolic paraboloid test results.

5.5. Generic Surface

Let us consider the surface identified by,

$$\boldsymbol{p}^{\mathrm{T}} = \{u, \quad v, \quad u\cos v - v\sin u\}$$

in the range $u, v \in [-5, +5]$. By setting, $\alpha = \sin u + u\sin v$ and $\beta = \cos v - v\cos u$, the metric tensor and its inverse for this surface are,

$$g_{ij} = \begin{bmatrix} \beta^2 + 1 & -\alpha\beta \\ -\alpha\beta & \alpha^2 + 1 \end{bmatrix} \quad \text{and} \quad g^{ij} = \frac{1}{\alpha^2 + \beta^2 + 1} \begin{bmatrix} \alpha^2 + 1 & \alpha\beta \\ \alpha\beta & \beta^2 + 1 \end{bmatrix}$$

and the nonzero Christoffel symbols are,

$$\Gamma^1_{11} = \frac{v \sin u}{\alpha^2 + \beta^2 + 1}\beta, \quad \Gamma^1_{12} = \Gamma^1_{21} = -\frac{\cos u + \sin v}{\alpha^2 + \beta^2 + 1}\beta, \quad \Gamma^1_{22} = -\frac{u \cos v}{\alpha^2 + \beta^2 + 1}\beta,$$

$$\Gamma^2_{11} = -\frac{v \sin u}{\alpha^2 + \beta^2 + 1}\alpha, \quad \Gamma^2_{12} = \Gamma^2_{21} = \frac{\cos u + \sin v}{\alpha^2 + \beta^2 + 1}\alpha, \quad \Gamma^2_{22} = \frac{u \cos v}{\alpha^2 + \beta^2 + 1}\alpha$$

The geodesic equations can be written in the compact form,

$$\begin{cases} \mathcal{L}_u = p\,\ddot{u} + \beta\,q = 0 \\ \mathcal{L}_v = p\,\ddot{v} - \alpha\,q = 0 \end{cases}$$

where,

$$p = \alpha^2 + \beta^2 + 1 \quad \text{and} \quad q = (v \sin u)\dot{u}^2 - 2(\cos u + \sin v)\dot{u}\dot{v} - (u \cos v)\dot{v}^2$$

The partials forming the Jacobian matrix in the least-squares process are,

$$\frac{\partial \mathcal{L}_u}{\partial \xi_u} = \ddot{u}\frac{\partial p}{\partial u} \cdot \frac{\partial u}{\partial \xi_u} + p\frac{\partial \ddot{u}}{\partial \xi_u} + \beta\left(\frac{\partial q}{\partial u} \cdot \frac{\partial u}{\partial \xi_u} + \frac{\partial q}{\partial \dot{u}} \cdot \frac{\partial \dot{u}}{\partial \xi_x}\right) + q\frac{\partial \beta}{\partial u} \cdot \frac{\partial u}{\partial \xi_u}$$
$$\frac{\partial \mathcal{L}_u}{\partial \xi_v} = \ddot{u}\frac{\partial p}{\partial v} \cdot \frac{\partial v}{\partial \xi_v} + \beta\left(\frac{\partial q}{\partial v} \cdot \frac{\partial v}{\partial \xi_v} + \frac{\partial q}{\partial \dot{v}} \cdot \frac{\partial \dot{v}}{\partial \xi_v}\right) + q\frac{\partial \beta}{\partial v} \cdot \frac{\partial v}{\partial \xi_v}$$
$$\frac{\partial \mathcal{L}_v}{\partial \xi_u} = \ddot{v}\frac{\partial p}{\partial u} \cdot \frac{\partial u}{\partial \xi_u} - \alpha\left(\frac{\partial q}{\partial u} \cdot \frac{\partial u}{\partial \xi_u} + \frac{\partial q}{\partial \dot{u}} \cdot \frac{\partial \dot{u}}{\partial \xi_u}\right) - q\frac{\partial \alpha}{\partial u} \cdot \frac{\partial u}{\partial \xi_u}$$
$$\frac{\partial \mathcal{L}_v}{\partial \xi_v} = \ddot{v}\frac{\partial p}{\partial v} \cdot \frac{\partial v}{\partial \xi_v} + p\frac{\partial \ddot{v}}{\partial \xi_v} - \alpha\left(\frac{\partial q}{\partial v} \cdot \frac{\partial v}{\partial \xi_v} + \frac{\partial q}{\partial \dot{v}} \cdot \frac{\partial \dot{v}}{\partial \xi_v}\right) - q\frac{\partial \alpha}{\partial v} \cdot \frac{\partial v}{\partial \xi_v}$$

where,

$$\frac{\partial \alpha}{\partial u} = \cos u + \sin v, \quad \frac{\partial \alpha}{\partial v} = u \cos v, \quad \frac{\partial \beta}{\partial u} = v \sin u, \quad \text{and} \quad \frac{\partial \beta}{\partial v} = -\sin v - \cos u$$

and,

$$\frac{\partial p}{\partial u} = 2\alpha\frac{\partial \alpha}{\partial u} + 2\beta\frac{\partial \beta}{\partial u} \quad \text{and} \quad \frac{\partial p}{\partial v} = 2\beta\frac{\partial \beta}{\partial v} + 2\alpha\frac{\partial \alpha}{\partial v}$$

and,

$$\frac{\partial q}{\partial u} = (v \cos u)\dot{u}^2 + 2(\sin u)\dot{u}\dot{v} - (\cos v)\dot{v}^2$$
$$\frac{\partial q}{\partial v} = (\sin u)\dot{u}^2 - 2(\cos v)\dot{u}\dot{v} + (u \sin v)\dot{v}^2$$
$$\frac{\partial q}{\partial \dot{u}} = 2(v \sin u)\dot{u} - 2(\cos u + \sin v)\dot{v}$$
$$\frac{\partial q}{\partial \dot{v}} = -2(\cos u + \sin v)\dot{u} - 2(u \cos v)\dot{v}$$

Figure 6 gives the results obtained for this specific generic surface. The convergence is obtained in just 8 iterations, using 30 basis functions for the free functions and a 300 point discretization. The constancy of the parametric velocity is an alternative way validating the solution as a geodesic trajectory.

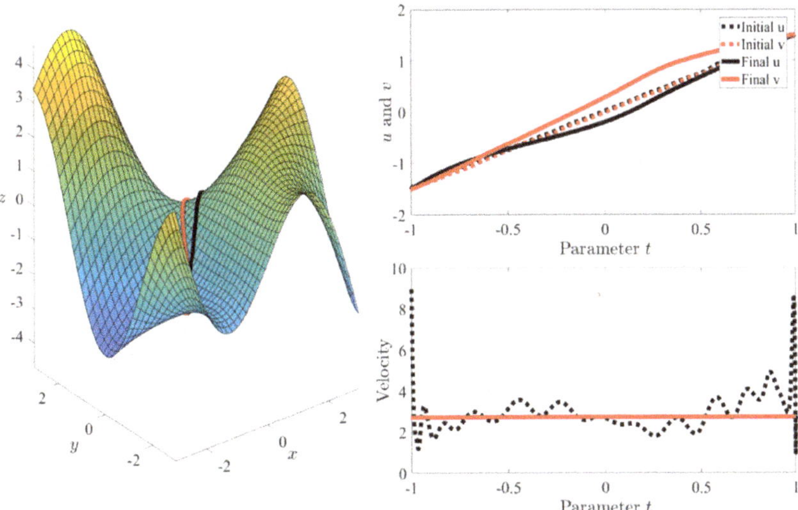

Figure 6. Surface $\boldsymbol{p}^{\mathsf{T}} = \left\{u, \quad v, \quad u\cos v - v\sin u\right\}$ test results.

5.6. One-Sheeted Hyperboloid

This surface is described by the parametric vector,

$$\boldsymbol{p}^{\mathsf{T}}(u,v) = \left\{\sqrt{1+u^2}\,\cos v, \quad \sqrt{1+u^2}\,\sin v, \quad 2\,u\right\}$$

The covariant and the contravariant associated metric tensors are,

$$g_{ij} = \begin{bmatrix} \dfrac{5\,u^2+4}{u^2+1} & 0 \\ 0 & u^2+1 \end{bmatrix} \quad \text{and} \quad g^{ij} = \begin{bmatrix} \dfrac{u^2+1}{5\,u^2+4} & 0 \\ 0 & \dfrac{1}{u^2+1} \end{bmatrix}$$

and the nonzero Christoffel symbols are,

$$\Gamma^1_{11} = \frac{u}{(5u^2+4)(u^2+1)}, \quad \Gamma^1_{22} = -\frac{u\,(u^2+1)}{5\,u^2+4}, \quad \text{and} \quad \Gamma^2_{12} = \Gamma^2_{21} = \frac{u}{u^2+1}$$

This implies the geodesic equations for one-sheeted hyperboloid,

$$\begin{cases} \mathcal{L}_u = (5u^2+4)\,\ddot{u} + \dfrac{u}{u^2+1}\dot{u}^2 - u(u^2+1)\dot{v}^2 = 0 \\ \mathcal{L}_v = (u^2+1)\ddot{v} + 2u\,\dot{u}\,\dot{v} = 0 \end{cases}$$

To apply the nonlinear least-squares, the following partials must be computed,

$$\begin{aligned}
\frac{\partial \mathcal{L}_u}{\partial \boldsymbol{\xi}_u} &= \frac{\partial \mathcal{L}_u}{\partial \ddot{u}} \cdot \frac{\partial \ddot{u}}{\partial \boldsymbol{\xi}_u} + \frac{\partial \mathcal{L}_u}{\partial \dot{u}} \cdot \frac{\partial \dot{u}}{\partial \boldsymbol{\xi}_u} + \frac{\partial \mathcal{L}_u}{\partial u} \cdot \frac{\partial u}{\partial \boldsymbol{\xi}_u} \\
\frac{\partial \mathcal{L}_u}{\partial \boldsymbol{\xi}_v} &= \frac{\partial \mathcal{L}_u}{\partial \dot{v}} \cdot \frac{\partial \dot{v}}{\partial \boldsymbol{\xi}_v} \\
\frac{\partial \mathcal{L}_v}{\partial \boldsymbol{\xi}_u} &= \frac{\partial \mathcal{L}_v}{\partial \dot{u}} \cdot \frac{\partial \dot{u}}{\partial \boldsymbol{\xi}_u} + \frac{\partial \mathcal{L}_v}{\partial u} \cdot \frac{\partial u}{\partial \boldsymbol{\xi}_u} \\
\frac{\partial \mathcal{L}_v}{\partial \boldsymbol{\xi}_v} &= \frac{\partial \mathcal{L}_v}{\partial \ddot{v}} \cdot \frac{\partial \ddot{v}}{\partial \boldsymbol{\xi}_v} + \frac{\partial \mathcal{L}_v}{\partial \dot{v}} \cdot \frac{\partial \dot{v}}{\partial \boldsymbol{\xi}_v}
\end{aligned}$$

The expressions of these partials are,

$$\frac{\partial \mathcal{L}_u}{\partial \ddot{u}} = 5u^2 + 4 \qquad \frac{\partial \mathcal{L}_u}{\partial \dot{u}} = \frac{2u}{u^2+1}\dot{u}$$
$$\frac{\partial \mathcal{L}_u}{\partial u} = 10u^2\ddot{u} + \frac{1-u^2}{(u^2+1)^2}\dot{u}^2 - \left(3u^2+1\right)\dot{v}^2 \qquad \frac{\partial \mathcal{L}_u}{\partial \dot{v}} = -2u\left(u^2+1\right)\dot{v}$$
$$\frac{\partial \mathcal{L}_v}{\partial \ddot{v}} = u^2 + 1 \qquad \frac{\partial \mathcal{L}_v}{\partial \dot{v}} = 2u\,\dot{u}$$
$$\frac{\partial \mathcal{L}_v}{\partial \dot{u}} = 2u\,\dot{v} \qquad \frac{\partial \mathcal{L}_v}{\partial u} = 2u\ddot{v} + 2\dot{u}\,\dot{v}$$

The parametric velocity is given by

$$\dot{p}^2 = \frac{u^2\dot{u}^2}{1+u^2} + 4\dot{u}^2 + (1+u^2)\dot{v}.^2$$

The results of the test conducted on the one-sheeted hyperboloid are reported in Figure 7 where the sub-figures provide full information of the least-squares process. The free functions were expanded by 40 basis functions (Legendre orthogonal polynomials) and the discretization was conducted by 300 points.

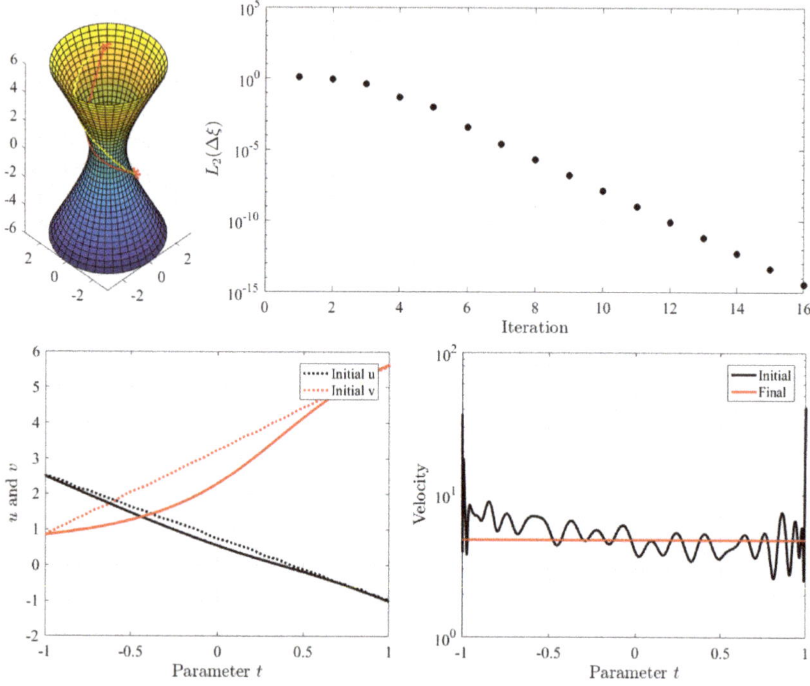

Figure 7. One-sheeted hyperboloid test results.

5.7. *Torus*

The implicit and parametric equation of the torus are,

$$\boldsymbol{p}^{\mathrm{T}} = \left\{(R + r\cos u)\cos v, \quad (R + r\cos u)\sin v, \quad r\,\sin u\right\}$$

where R and r are the two torus radii. The covariant metric tensor for the torus is,

$$g_{ij} = \begin{bmatrix} r^2 & 0 \\ 0 & (R + r\cos u)^2 \end{bmatrix}$$

while the geodesic equations are,

$$\begin{cases} \mathcal{L}_u = r\,\ddot{u} + \dot{v}^2\,(R + r\cos u)\sin u = 0 \\ \mathcal{L}_v = (R + r\cos u)\ddot{v} - 2\dot{u}\dot{v}\,r\sin u = 0 \end{cases}$$

The partials needed to populate the Jacobian are,

$$\begin{aligned}
\frac{\partial \mathcal{L}_u}{\partial \boldsymbol{\xi}_u} &= r\frac{\partial \ddot{u}}{\partial \boldsymbol{\xi}_u} + \dot{v}^2\Big[(R + r\cos u)\cos u - r\sin^2 u\Big]\frac{\partial u}{\partial \boldsymbol{\xi}_u} \\
\frac{\partial \mathcal{L}_u}{\partial \boldsymbol{\xi}_v} &= 2\dot{v}\frac{\partial \dot{v}}{\partial \boldsymbol{\xi}_v}(R + r\cos u)\sin u \\
\frac{\partial \mathcal{L}_v}{\partial \boldsymbol{\xi}_u} &= -\Big[r\ddot{v}\sin u + 2\dot{u}\dot{v}\,r\cos u\Big]\frac{\partial u}{\partial \boldsymbol{\xi}_u} - 2\frac{\partial \dot{u}}{\partial \boldsymbol{\xi}_u}\dot{v}\,r\sin u \\
\frac{\partial \mathcal{L}_v}{\partial \boldsymbol{\xi}_v} &= (R + r\cos u)\frac{\partial \ddot{v}}{\partial \boldsymbol{\xi}_v} - 2\dot{u}\frac{\partial \dot{v}}{\partial \boldsymbol{\xi}_v}\,r\sin u
\end{aligned}$$

and the nonlinear least-squares can be performed using Equation (13). The geodetic parametric velocity on the torus is,

$$\dot{p}^2 = r^2\,\dot{u}^2 + (R + r\cos u)^2\dot{v}^2$$

The sub-figures given in Figure 8 show the details of the least-squares process to estimate the geodesic trajectory on the torus surface. The number of basis functions were 40 and the number of discretization points were 300. The constancy of the velocity as well as the machine-level L_2 norm of the differential equations residuals validate the estimated geodesic trajectory (in red).

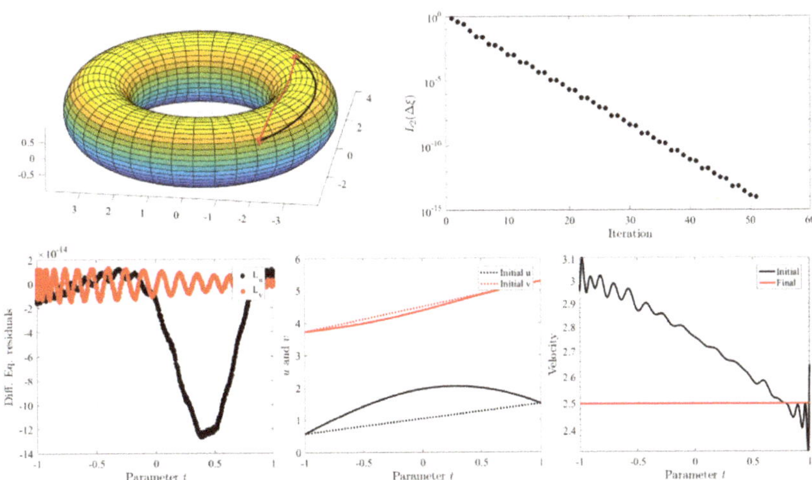

Figure 8. Torus: test results.

5.8. Moëbius Strip

The last test of the proposed least-squares approach is performed on the Moëbius strip, which can be described by,

$$\boldsymbol{p}^{\mathrm{T}} = \left\{\left[2 - u\sin\left(\frac{v}{2}\right)\right]\cos(v),\quad \left[2 - u\sin\left(\frac{v}{2}\right)\right]\sin(v),\quad u\cos\left(\frac{v}{2}\right)\right\}$$

where $u \in [-1,+1]$ and $v \in [0,2\pi)$. The metric tensor and its inverse for this surface can be given in a compact form by setting, $d(u,v) = 4u\sin\left(\frac{v}{2}\right)\left[u\sin\left(\frac{v}{2}\right) - 4\right] + u^2 + 16$,

$$g_{ij} = \begin{bmatrix} 1 & 0 \\ 0 & \dfrac{d(u,v)}{4} \end{bmatrix} \quad \text{and} \quad g^{ij} = \begin{bmatrix} 1 & 0 \\ 0 & \dfrac{4}{d(u,v)} \end{bmatrix}$$

while the nonzero Christoffel symbols are,

$$\Gamma^1_{22} = \sin\left(\frac{v}{2}\right)\left[2 - u\,\sin\left(\frac{v}{2}\right)\right] - \frac{u}{4}$$

$$\Gamma^2_{21} = \Gamma^2_{12} = \frac{u - 4\,\sin\left(\frac{v}{2}\right)\left[2 - u\,\sin\left(\frac{v}{2}\right)\right]}{d(u,v)}$$

$$\Gamma^2_{22} = -\frac{2u\,\cos\left(\frac{v}{2}\right)\left[2 - u\,\sin\left(\frac{v}{2}\right)\right]}{d(u,v)}$$

Hence, the geodesic equations of the Moëbius strip are,

$$\ddot{u} + \Gamma^1_{22}\,\dot{v}^2 = 0 \quad \text{and} \quad \ddot{v} + 2\Gamma^2_{21}\,\dot{u}\,\dot{v} + \Gamma^2_{22}\,\dot{v}^2 = 0$$

which can be written as,

$$\begin{cases} \ddot{u} + c_1(u,v)\,\dot{v}^2 = 0 \\ d(u,v)\,\ddot{v} + c_2(u,v)\,\dot{u}\,\dot{v} + c_3(u,v)\,\dot{v}^2 = 0 \end{cases}$$

where

$$\begin{cases} c_1(u,v) = 2\sin\left(\frac{v}{2}\right) - u\,\sin^2\left(\frac{v}{2}\right) - \frac{u}{4} \\ c_2(u,v) = 2u - 16\,\sin\left(\frac{v}{2}\right) + 8u\,\sin^2\left(\frac{v}{2}\right) \\ c_3(u,v) = -4u\,\cos\left(\frac{v}{2}\right) + 4u^2\,\sin v \end{cases}$$

Figure 9 provide all information about this last test validation case. Table 2 details the iterative results: the L_2 norm of the accuracy gain and the L_2 norm of the differential equation at each iteration. The iterative process ended because $L_2(\Delta\xi_{23}) > L_2(\Delta\xi_{22})$, due to numerical convergence saturation.

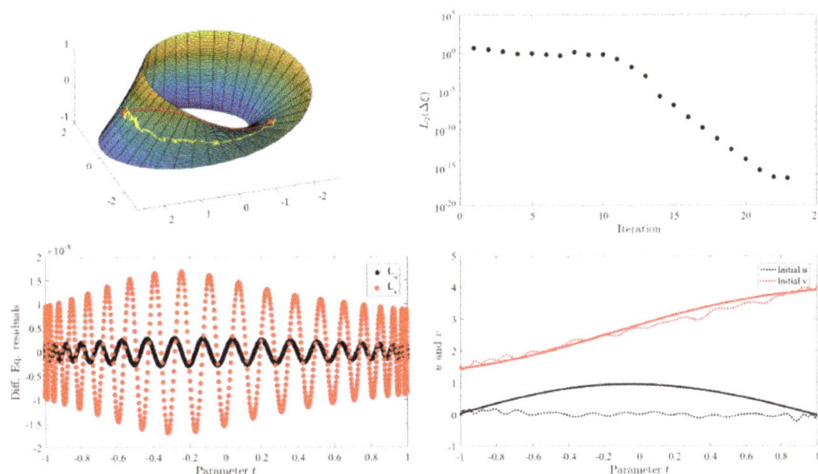

Figure 9. Moëbius strip test results.

Table 2. Moëbius strip: convergence L_2 norms terms used in the least-squares approach.

Iteration k	$L_2(\Delta\xi_k)$	$L_2(r_k)$
1	-	4.4143
2	2,590,717.6067	2.7918
3	125,575.9139	1.4362
4	324.6177	0.7541
5	27,398.8611	0.87001
6	2583.4652	0.60306
7	2142.7967	0.39897
8	2330.2236	1.2032
9	1485.2018	0.55737
10	1197.0006	0.61434
11	311.3524	0.14478
12	487.9178	0.012141
13	35.4515	0.0008965
14	1.1049	2.1828×10^{-6}
15	0.0088866	1.433×10^{-7}
16	0.00014298	3.7817×10^{-9}
17	7.4714×10^{-6}	1.5985×10^{-10}
18	1.5249×10^{-7}	6.5714×10^{-12}
19	3.3531×10^{-10}	2.6535×10^{-13}
20	5.5449×10^{-13}	1.0555×10^{-14}
21	4.0001×10^{-15}	4.1414×10^{-16}
22	2.8094×10^{-15}	4.9073×10^{-17}
23	5.1466×10^{-15}	3.4713×10^{-17}

5.9. Discussions

In this article, a new general method to numerical solve boundary value geodesic problems in curved surfaces is presented. The approach takes advantage of the ability to derive special functionals, called constrained expressions, which always satisfy assigned boundary conditions. The proposed approach has been validated by performing numerical tests on several two-dimensional surfaces. In theory, this approach can be extended to manifolds in higher Riemannian spaces. This will require more computational capability than that used for this article and, most likely, optimized and compiled code, instead of the MATLAB interpreter adopted.

The problem of finding the geodesic by least-squares introduces the risk of getting stuck on some local minima. A simple algorithm has been developed to mitigate this problem. Thanks to this algorithm, all boundary geodesic problems considered were quickly solved with machine-error level accuracy, which is an acceptable definition of an exact solution by engineers.

This article restricts the research on finding geodesic trajectories on surfaces that are continuous and differentiable. The problem of finding geodesic-type trajectories on discretized surfaces, which is solved by combinatorial/computational algorithms and methods, is not taken here into consideration. The performed tests have the only purpose to validate the least-squares approach. A complete analysis quantifying the responses of this methodology to various surface shapes, boundary conditions (e.g., singular boundary points for the Ellipsoid), as well as comparisons with competing approaches, will be the subject of future studies.

As for future research directions, it should be natural to derive the real velocity (instead of the parametric velocity) in geodesic problems of a mass particle, and to use the constancy of the velocity—in addition to or instead of—the geodesic equations. Another future research activity is to investigate what the optimal range of basis functions for the free functions is and to investigate the optimal number of discretization points. These two analyses will help the proposed least-squares solution approach to provide optimal performances. Topics more interesting in physics, such as using the Schwarzschild metric (uncharged, non-rotating black holes) or more general space metrics are welcome to be investigated by researchers with good knowledge on general relativity.

Funding: This research received no external funding.

Conflicts of Interest: The author declares no conflict of interest.

Acronyms

The following abbreviations are used in this manuscript:

TFC	Theory of Functional Connections
BVP	Boundary Value Problems
IVP	Initial Value Problems
FI	Functional Interpolation

References

1. Misner, C.W.; Thorne, K.S.; Wheeler, J.A. *Gravitation*; Macmillan: London, UK, 1973.
2. Landau, L.D. *The Classical Theory of Fields*; Elsevier: Amsterdam, The Netherlands, 2013; Voume 2.
3. Panou, G.; Delikaraoglou, D.; Korakitis, R. Solving the geodesics on the ellipsoid as a boundary value problem. *J. Geod. Sci.* **2013**, *3*, 40–47. [CrossRef]
4. Panou, G. The geodesic boundary value problem and its solution on a triaxial ellipsoid. *J. Geod. Sci.* **2013**, *3*, 240–249. [CrossRef]
5. Bektas, S. Geodetic Computations on triaxial ellipsoid. *Int. J. Min. Sci.* **2015**, *1*, 25–34.
6. Cotter, C.J.; Holm, D.D. Geodesic boundary value problems with symmetry. *arXiv* **2009**, arXiv:0911.2205.
7. Marx, C. Performance of a solution of the direct geodetic problem by Taylor series of Cartesian coordinates. *J. Geod. Sci.* **2021**, *11*, 122–130. [CrossRef]
8. Bektas, S. Least squares fitting of ellipsoid using orthogonal distances. *Bol. Ciênc. Geod.* **2015**, *21*, 329–339. [CrossRef]
9. Leake, C.; Johnston, H.; Mortari, D. *The Theory of Functional Connections: A Functional Interpolation. Framework with Applications*; Lulu: Morrisville, NC, USA, 2022.
10. Mortari, D. Least-Squares Solution of Linear Differential Equations. *Mathematics* **2017**, *5*, 48. [CrossRef]
11. Mortari, D.; Johnston, H.R.; Smith, L. High accuracy least-squares solutions of nonlinear differential equations. *J. Comput. Appl. Math.* **2019**, *352*, 293–307. [CrossRef] [PubMed]
12. Hopf, H.; Rinow, W. Über den Begriff der vollständigen differentialgeometrischen Fläche. *Comment. Math. Helv.* **1931**, *3*, 209–225. [CrossRef]
13. Sun, M. *Geodesic and the Hopf-Rinow Theorem*; Technical Report; University of Chicago: Chicago, IL, USA, 2021.
14. Einstein, A. *Relativity: The Special and the General Theory*; General Press: New Delhi, India, 2016.
15. Mortari, D. The Theory of Connections: Connecting Points. *Mathematics* **2017**, *5*, 57. [CrossRef]
16. Mortari, D.; Leake, C.D. The Multivariate Theory of Connections. *Mathematics* **2019**, *7*, 296. [CrossRef] [PubMed]

17. Leake, C.D. The Multivariate Theory of Functional Connections: An n-Dimensional Constraint Embedding Technique Applied to Partial Differential Equations. Ph.D. Dissertation, Texas A&M University, College Station, TX, USA, 2021.
18. Johnston, H.R. The Theory of Functional Connections: A Journey from Theory to Application. Ph.D. Thesis, Texas A&M University, College Station, TX, USA, 2021.
19. Mortari, D.; Arnas, D. Bijective Mapping Analysis to Extend the Theory of Functional Connections to Non-Rectangular 2-Dimensional Domains. *Mathematics* **2020**, *8*, 1593. [CrossRef]
20. Johnston, H.R.; Leake, C.D.; Efendiev, Y.; Mortari, D. Selected Applications of the Theory of Connections: A Technique for Analytical Constraint Embedding. *Mathematics* **2019**, *7*, 537. [CrossRef]
21. Wang, Y.; Topputo, F. A TFC-based homotopy continuation algorithm with application to dynamics and control problems. *J. Comput. Appl. Math.* **2022**, *401*, 113777. [CrossRef]
22. Schiassi, E.; Florio, M.D.; D'Ambrosio, A.; Mortari, D.; Furfaro, R. Physics-Informed Neural Networks and Functional Interpolation for Data-Driven Parameters Discovery of Epidemiological Compartmental Models. *Mathematics* **2021**, *9*, 2069. [CrossRef]
23. Florio, M.D.; Schiassi, E.; Furfaro, R.; Ganapol, B.D.; Mostacci, D. Solutions of Chandrasekhar's basic problem in radiative transfer via theory of functional connections. *J. Quant. Spectrosc. Radiat. Transf.* **2021**, *259*, 107384. [CrossRef]
24. Florio, M.D.; Schiassi, E.; Ganapol, B.D.; Furfaro, R. Physics-informed neural networks for rarefied-gas dynamics: Thermal creep flow in the Bhatnagar–Gross–Krook approximation. *Phys. Fluids* **2021**, *33*, 047110. [CrossRef]
25. Yassopoulos, C.; Leake, C.D.; Reddy, J.; Mortari, D. Analysis of Timoshenko–Ehrenfest beam problems using the Theory of Functional Connections. *Eng. Anal. Bound. Elem.* **2021**, *132*, 271–280. [CrossRef]
26. Johnston, H.R.; Mortari, D. Least-squares solutions of boundary-value problems in hybrid systems. *J. Comput. Appl. Math.* **2021**, *393*, 113524. [CrossRef]
27. Leake, C.D.; Johnston, H.R.; Smith, L.; Mortari, D. Analytically Embedding Differential Equation Constraints into Least Squares Support Vector Machines Using the Theory of Functional Connections. *Mach. Learn. Knowl. Extr.* **2019**, *1*, 1058–1083. [CrossRef] [PubMed]
28. Leake, C.D.; Mortari, D. Deep Theory of Functional Connections: A New Method for Estimating the Solutions of Partial Differential Equations. *Mach. Learn. Knowl. Extr.* **2020**, *2*, 37–55. [CrossRef] [PubMed]
29. Schiassi, E.; Furfaro, R.; Leake, C.D.; Florio, M.D.; Johnston, H.R.; Mortari, D. Extreme theory of functional connections: A fast physics-informed neural network method for solving ordinary and partial differential equations. *Neurocomputing* **2021**, *457*, 334–356. [CrossRef]
30. D'Ambrosio, A.; Schiassi, E.; Curti, F.; Furfaro, R. Pontryagin Neural Networks with Functional Interpolation for Optimal Intercept Problems. *Mathematics* **2021**, *9*, 996. [CrossRef]
31. Mai, T.; Mortari, D. Theory of Functional Connections Applied to Quadratic and Nonlinear Programming under Equality Constraints. *J. Comput. Appl. Math.* **2022**, *406*, 113912. [CrossRef]
32. de Almeida Junior, A.K.; Johnston, H.R.; Leake, C.D.; Mortari, D. Fast 2-impulse non-Keplerian orbit transfer using the Theory of Functional Connections. *Eur. Phys. J. Plus* **2021**, *136*, 223. [CrossRef]
33. Johnston, H.R.; Lo, M.W.; Mortari, D. A Functional Interpolation Approach to Compute Periodic Orbits in the Circular-Restricted Three-Body Problem. *Mathematics* **2021**, *9*, 1210. [CrossRef]
34. Drozd, K.; Furfaro, R.; Schiassi, E.; Johnston, H.R.; Mortari, D. Energy-optimal trajectory problems in relative motion solved via Theory of Functional Connections. *Acta Astronaut.* **2021**, *182*, 361–382. [CrossRef]
35. Johnston, H.R.; Schiassi, E.; Furfaro, R.; Mortari, D. Fuel-Efficient Powered Descent Guidance on Large Planetary Bodies via Theory of Functional Connections. *J. Astron. Sci.* **2020**, *67*, 1521–1552. [CrossRef]
36. Franceschi, V. Curve Geodetiche su Superfici. Ph.D. Thesis, Department of Mathematics, University of Milan, Milan, Italy, 2006. (In Italian)

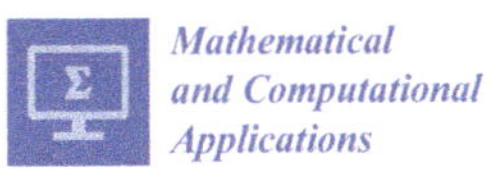
Mathematical and Computational Applications

Article

Analytical Solutions of Microplastic Particles Dispersion Using a Lotka–Volterra Predator–Prey Model with Time-Varying Intraspecies Coefficients

Lindomar Soares Dos Santos [1], José Renato Alcarás [1], Lucas Murilo Da Costa [1], Mateus Mendonça Ramos Simões [2] and Alexandre Souto Martinez [1,3,*]

[1] Faculdade de Filosofia, Ciências e Letras de Ribeirão Preto, Universidade de São Paulo, Avenida dos Bandeirantes, 3900, Ribeirão Preto 14040-901, SP, Brazil; lsoaressantos@gmail.com (L.S.D.S.); joserenato.alcaras@gmail.com (J.R.A.); lucasdacosta@usp.br (L.M.D.C.)
[2] Faculdade de Medicina de Ribeirão Preto, Universidade de São Paulo, Avenida dos Bandeirantes, 3900, Ribeirão Preto 14049-900, SP, Brazil; mateusmrsimoes@usp.br
[3] Instituto Nacional de Ciência e Tecnologia em Sistemas Complexos, Rua Dr. Xavier Sigaud 150, Urca, Rio de Janeiro 22290-180, RJ, Brazil
* Correspondence: asmartinez@usp.br; Tel.: +55-16-3315-3720

Abstract: Discarded plastic is subjected to weather effects from different ecosystems and becomes microplastic particles. Due to their small size, they have spread across the planet. Their presence in living organisms can have several harmful consequences, such as altering the interaction between prey and predator. Huang et al. successfully modeled this system presenting numerical results of ecological relevance. Here, we have rewritten their equations and solved a set of them analytically, confirming that microplastic particles accumulate faster in predators than in prey and calculating the time values from which it happens. Using these analytical solutions, we have retrieved the Lotka–Volterra predator–prey model with time-varying intraspecific coefficients, allowing us to interpret ecological quantities referring to microplastics dispersion. After validating our equations, we solved analytically particular situations of ecological interest, characterized by extreme effects on predatory performance, and proposed a second-order differential equation as a possible next step to address this model. Our results open space for further refinement in the study of predator–prey models under the effects of microplastic particles, either exploring the second-order equation that we propose or modify the Huang et al. model to reduce the number of parameters, embedding in the time-varying intraspecies coefficients all the adverse effects caused by microplastic particles.

Keywords: microplastics; Lotka–Volterra; predator; prey; predator–prey; pollution; bioaccumulation; biomagnification; predation; two-species model

Citation: Dos Santos, L.S.; Alcarás, J.R.; Da Costa, L.M.; Simões, M.M.R.; Martinez, A.S. Analytical Solutions of Microplastic Particles Dispersion Using a Lotka–Volterra Predator–Prey Model with Time-Varying Intraspecies Coefficients. *Math. Comput. Appl.* **2022**, 27, 66. https://doi.org/10.3390/mca27040066

Academic Editor: Gabriel B. Mindlin

Received: 16 June 2022
Accepted: 1 August 2022
Published: 3 August 2022

1. Introduction

The discovery of Bakelite in 1907 revolutionized modern life by introducing plastic materials to the world. The popularization of these polymers began with their commercial production around 1950 [1,2]. Plastic's versatility, stability, low weight, and low production cost leveraged its global market for this material [2]. The increase in plastic consumption has had serious repercussions on nature. It is estimated that 9.5 million tons of plastic end up in the oceans every year [3,4] and there are still no estimates for the amount of plastic deposited on land [5].

All discarded plastic is subject to weather effects of different ecosystems. This material can be slowly degraded by photo-oxidation, thermal pathways, mechanochemical interactions or biodegradation [6]. The results of all these processes are small particles known as microplastics, which have dimensions that vary between 1 μm and 5 mm [7] and the most different compositions, colors, and shapes [8].

Due to their small size, microplastics have spread across the planet. They have already been detected in all parts of the ocean [9], in the poles [10], in drinking water [11], in arable areas and pastures [12], in the habitat of terrestrial animals [13–15], and in various foods consumed by people [16]. Many studies suggest that the near-ubiquity of microplastics means that their transfer is not limited to food chain related dynamics, but also occurs through bioaccumulation from one trophic level to another [17–19].

The presence of microplastics in living organisms can have several consequences. Studies indicate the possibility of gastrointestinal inflammation [20], decreased reproductive capacity [21], reduced ability to feed [22], reduced growth rate [23], and even malformation of embryos [24]. Early research suggests that microplastics can alter the interaction between prey and predator [25], producing interesting study approaches that use Lotka–Volterra prey–predator models modified to take into account the microplastics [26].

The Lotka–Volterra model (LVM) was originally developed from the logistic equation, describing chemical reactions [27]. However, over the years, modifications of this model have allowed new and specific applications, such as the study of the interaction between prey and predators [28–30], the influence of harmful elements on population dynamics [31], and, more specifically, the effect that microplastics has on trophic relationships [18,19,26].

This paper aims to go beyond the numerical solutions of a prey–predator model by studying analytical solutions of the model proposed by Huang et al. [26], which takes into account the impact of microplastics on the population dynamics. We rewrite the Huang et al. four equations model, reducing it to two equations equivalent to one and including the time-varying intraspecies coefficients. This approach clarifies the model parameters meanings and allows us to solve analytically three special cases of ecological relationships. These special cases are based on possible extreme effects of predatory performance reduction caused by exposure to MP particles and are mathematically characterized by the decoupling of the differential equations of the model, for which we also perform numerical simulations. We also propose a second-order differential equation as a possible next step to address this model.

We organized the presentation as follows: In Section 2, we introduce Huang et al.'s work, including their model and main results. In addition, we justify our interest in their research, presenting our motivations and intentions, and rewrite four original equations of the predator–prey model, reducing them to only two ones, redefining/regrouping some quantities, including time-varying intraspecies coefficients, giving them ecological meanings analogous to the standard LVM. In Section 3, we analytically validate that microplastic (MP) particles accumulate faster in predators than in preys and calculate the characteristics times from which their concentration and changing rate of the total amount are greater in predators than in preys. After validating our model, we explore analytically and numerically three special ecological regimes characterized by extreme effects on predatory performance, which can lead these two populations to become independent. In addition, we introduce a second-order differential equation as a possible future study of our two equations system. In Section 4, we compile our results and interpretations, presenting their implications and research possibilities of refining LVM under effects of MP particles, exploring our second-order equation or modifying the standard model to reduce its number of parameters, embedding in the time varying intraspecies coefficients all the adverse effects caused by MP particles. In Section 5, we summarize our main results and point to future studies.

2. Materials and Methods

In 2020, Huang et al. [26] published a study using the LVM where they theoretically investigated predator–prey population dynamics in terms of toxicological response intensity to microplastic (MP) parts and examined the negative effects on prey feeding ability and predator performance due to MP particles. The study suggests that dynamic LVMs can be an important tool to predict the ecological impacts of MP particles on predator–prey population dynamics.

Combining a single-species model with the LVM, Huang et al. obtained the following model:

$$\begin{cases} \dot{x}_1(t) = x_1(t)[r_{10} - d_1 - r_{11}C_1(t) - (a_1 - d_3)x_2(t)] \\ \dot{x}_2(t) = x_2(t)[-r_{20} - r_{21}C_2(t) + (a_2 - d_2)x_1(t)] \\ \dot{C}_1(t) = S_1C_E - g_1 \\ \dot{C}_2(t) = S_2C_E + kC_1(t) - g_2 \end{cases}, \quad (1)$$

where $x_1(t), x_2(t), C_1(t), C_2(t) \geq 0$. This model neglects the influence of the intraspecific competition and shows the effect of the microplastics by its concentration. Considering $t \geq 0$ in those differential equations, x_1 and x_2 represent the population of prey and predator, respectively. $r_{10}x_1$ and $r_{20}x_2$ are the intrinsic growth rate of prey and mortality rate of predator without toxicity, respectively. $a_1x_1x_2$ is the lost amount of prey eaten by predators, and $a_2x_1x_2$ is the increasing number of predators due to the feeding of prey, with $a_1, a_2 > 0$. The parameters d_1, d_2, and d_3 represent the decline in the prey feeding ability, the adverse effect of reduced predatory performance, and the lost amount of prey eaten by the predator, respectively. The response intensities of MP particles on prey and predator are denoted by r_{11} and r_{21}, respectively. The microplastics egestion rates of prey and predator, $g_1, g_2 \geq 0$ ($g_1 < 0$ or $g_2 < 0$ would imply a MP particles "negative egestion", by prey or predator), are independent of the microplastics concentration in the environment $C_E \geq 0$, and the amount of microplastics concentration removed at each time step is independent of the total amount of microplastics in the organisms, C_1 and C_2. The quantity kC_1 represents the accumulated toxicity of MP particles transferred from the prey. Finally, $S_1, S_2 \geq 0$ are related to the effects of plastic particle selection of prey and predator, respectively ($S_1 < 0$ or $S_2 < 0$, which would imply an environment MP particle "negative ingestion" rate by prey or predator).

The authors estimated parameters and performed simulations that indicated that predators are more vulnerable than prey under exposure to microplastics. The effect of MP particles on both population growths can be negligible when toxicological response intensities of prey and predator are small, the system is prey-dependent for predator functional response, and the reduced feeding capacity of prey and predator induced by microplastics does not significantly affect the population dynamics. The conclusions are compatible with empirical evidence. This study indicates that this model is adequate to approach the prediction of population dynamics of the predator–prey system under toxicological effects of persistent organic pollutants.

Stimulated by the relevance of the study and the innovativeness of its approach, we decided to explore the Huang et al. model, going beyond its numerical solutions by obtaining analytical solutions that reproduce some of their results, using these solutions to validate our equations, and going even further presenting and analyzing particular cases of possible ecological interest.

After elaborating the model, Huang et al. estimated its parameters and performed numerical simulations, implementing them using MATLAB programming and its Simulink toolbox, based on differential equations such as those presented by the system (1). We have realized that it is possible to rewrite only two equations, reducing them to the derivatives $\dot{x}_1(t)$ and $\dot{x}_2(t)$. In addition, it is interesting to redefine/regroup some quantities of the model, significantly reducing its number of parameters and clarifying its understanding and ecological interpretation. Since some of the parameters in the system are related, we can regroup them into effective (net) and variable rates of decline and growth.

Initially, we define the quantities: $\alpha = a_1 - d_3$, which is the difference between the rate of prey population decline and the rate of the lost amount of prey eaten by predators (that we call effective rate of prey population decline), and $\alpha' = d_2 - a_2$, which is the difference between the rate of reduced predatory performance and the rate of predator population growth (that we call negative of the effective rate of predator population growth), and rewrite the system (1):

$$\begin{cases} \dot{x}_1(t) = x_1(t)[r_{10} - d_1 - r_{11}C_1(t) - \alpha x_2(t)] \\ \dot{x}_2(t) = x_2(t)[-r_{20} - r_{21}C_2(t) - \alpha' x_1(t)] \\ \dot{C}_1(t) = S_1 C_E - g_1 \\ \dot{C}_2(t) = S_2 C_E + kC_1(t) - g_2 \end{cases}, \tag{2}$$

where $x_1(t), x_2(t), C_1(t), C_2(t) \geq 0$.

We see that the solution to $C_1(t)$ is a first degree polynomial:

$$C_1(t) = C_1(0) + \tilde{C}_1 t\,, \tag{3}$$

with

$$\tilde{C}_1 = S_1 C_E - g_1, \tag{4}$$

which means that the total amount of microplastics in the prey population varies at a constant rate $\tilde{C}_1$. Note that the parameters S_1, C_E, and g_1 combine to form a quantity with units of $[C_1]/[t]$.

Using the solution of $C_1(t)$, we obtain the solution of $C_2(t)$, which is a second degree polynomial:

$$C_2(t) = C_2(0) + \tilde{C}_2 t + \frac{k\tilde{C}_1}{2} t^2\,, \tag{5}$$

with

$$\tilde{C}_2 = S_2 C_E - g_2 + kC_1(0), \tag{6}$$

meaning that the total amount of microplastics in the predator population, unlike the prey population, varies at a rate that is directly proportional to time, and equal to $\tilde{C}_2 + k\tilde{C}_1 t$.

Comparing the expressions for the total amount of MP particles in the prey, $C_1(t)$, and predators, $C_2(t)$, we show that, over time, the concentration of microplastics inside the predator population will eventually be larger than the concentration inside prey population, which occurs at

$$t_{C_2 > C_1} > \frac{(\tilde{C}_1 - \tilde{C}_2) + \sqrt{(\tilde{C}_1 - \tilde{C}_2)^2 - 2k\tilde{C}_1(C_2(0) - C_1(0))}}{k\tilde{C}_1}\,. \tag{7}$$

This result analytically confirms the already known fact [26] that MP particles tend to accumulate faster in predators than in prey, which occurs when the rate of change of the total amount of microplastics inside predators is greater than the rate of change of the total amount of microplastics inside prey, more specifically at

$$t_{\dot{C}_2 > \dot{C}_1} > \frac{\tilde{C}_1 - \tilde{C}_2}{k\tilde{C}_1}\,. \tag{8}$$

Replacing $C_1(t)$ and $C_2(t)$ in the first two equations of the system (2), one needs only two coupled differential equations to write the Huang et al. model:

$$\begin{cases} \dot{x}_1(t) = x_1(t)\{r_{10} - d_1 - r_{11}[C_1(0) + \tilde{C}_1 t] - \alpha x_2(t)\} \\ \dot{x}_2(t) = x_2(t)\left\{-r_{20} - r_{21}\left[C_2(0) + \tilde{C}_2 t + \frac{k\tilde{C}_1}{2}t^2\right] - \alpha' x_1(t)\right\} \end{cases}. \tag{9}$$

Regrouping further this system of equations, one sees that it is possible to rearrange some of its quantities, to obtain variable rates of growth/decline of prey/predators. Regrouping the quantities: $c_0 = r_{10} - d_1 - r_{11}C_1(0)$, $c_0' = -r_{20} - r_{21}C_2(0)$, $c_1 = r_{11}\tilde{C}_1$, $c_1' = r_{21}\tilde{C}_2$, and $c_2' = r_{21}k\tilde{C}_1/2$, we define $\beta_1(t) = c_0 - c_1 t$, which is the rate of prey population growth, and $\beta_2(t) = c_0' - c_1' t - c_2' t^2$, which is the rate of predator population decline. Note that the effect of the initial conditions appears only on the coefficients c_0 and c_0'.

In this way, for $\dot{x}_i(t)$, we have:

$$\begin{cases} \dot{x}_1(t) = x_1(t)[\beta_1(t) - \alpha x_2(t)] \\ \dot{x}_2(t) = x_2(t)[\beta_2(t) - \alpha' x_1(t)] \end{cases} . \quad (10)$$

This system is similar to the well-known standard Lotka–Volterra (predator–prey) equations, except that its prey population growth and predator population decline rates are now time-varying intraspecific coefficients, which have built-in the effects of microplastics in the organisms.

A simple steady-state analysis $\dot{x}_1(t) = \dot{x}_1(t) = 0$, as $t \to \infty$, leads to the observed regimes. In one hand, if $x_1(\infty) = 0$, then $x_2(\infty) \geq 0$ may have arbitrary values. On the other hand, if $x_2(\infty) = 0$, then $x_1(\infty) \geq 0$ may have arbitrary values. In case $x_1(\infty) \neq 0$, then $x_2(\infty) = \beta_2(\infty)/\alpha \to \infty$. Similarly, $x_2(\infty) \neq 0$, then $x_1(\infty) = \beta_1(\infty)/\alpha \to \infty$. Particular cases presented below show these asymptotic values.

Huang et al. investigate the toxicological effects of MP particles according to the value of response intensity r_{11} and r_{12} implementing the model (1) using MATLAB programming and its Simulink toolbox. The authors classified interactions into four different conditions (response intensities of prey and predator to toxicological effects induced by MP particles): (a) without the influence of MP particles ($C_1 = C_2 = r_{11} = r_{21} = 0$); (b) predator and prey have the same response strength to MP particles ($\Delta = r_{11}/r_{21} = 1.0$); (c) predator has much larger response strength than prey ($\Delta = r_{11}/r_{21} = 0.1$); (d) predator has much smaller response strength than prey ($\Delta = r_{11}/r_{21} = 10.0$). Huang et al. constructed phase-portraits and short-term population dynamics graphs of predator–prey for each of these four conditions, with the following values: $C_E = 30$, $x_1(0) = 100$, $x_2(0) = 100$, $C_1(0) = 0$, $C_2(0) = 0$, $r_{10} = 4.1$, $r_{20} = 4.0$, $d_1 = 0.1$, $d_2 = 0.002$, $d_3 = 0.002$, $a_1 = 0.052$, $a_2 = 0.052$, $g_1 = 1.2$, $g_2 = 1.3$, $S_1 = 0.042$, $S_2 = 0.039$, $a_1 = 0.052$, $k = 2.0$. In addition, they analyzed the negative effects of PM particles on prey feeding ability and predatory performance increasing the values of d_1, d_2, and d_3 to $d_1 = 0.6$, $d_2 = 0.012$, and $d_3 = 0.012$, and taking three conditions of different response strength ($r_{11} = r_{21} = 10.0$; $r_{11} = 1.0$; and $r_{21} = 10.0$; and $r_{11} = 10.0$ and $r_{21} = 1.0$).

3. Results

Implementing the model of Equation (10) using the SciPy Python library, with the function *odeint* from *Scipy.integrate* package, and considering the five scenarios described in the previous section, we successfully reproduced the phase-portraits and the population dynamics graphs of Huang et al. The system we built consists of only two first-order coupled equations, has a reduced and more comprehensible set of parameters, and reproduces all the results of Huang et al. These equations are decoupled in three situations: $\alpha = \alpha' = 0$, $\alpha = 0$, and $\alpha' = 0$, which address specific ecological regimes [32].

3.1. Maximum Reduction of Predatory Performance: $\alpha = 0$ and $\alpha' = 0$

Exposure to MP particles can cause anomalous behaviors that impair the prey feeding ability and predatory performance of organisms, which lead, for example, to a decrease in the intrinsic growth rate of prey and a reduction in the number of predators. The Huang et al. model considers some of these harmful effects. For example, it denotes d_2 to the adverse effect of reduced predatory performance, and d_3 to the consequent lost amount of prey eaten by the predator. In our study, we initially considered the extreme situation in which: 1—the presence of MP particles impairs the predatory performance to the point where there is no further increase in the predators population due to their prey feeding or, in other words, d_2 is large enough to reduce to zero the effective rate of predator population growth ($-\alpha' = a_2 - d_2 = 0$); and 2—as a consequence of the drastic reduction in the predatory performance, no more prey is eaten by predators; in other words, d_3 is large enough to reduce to zero the effective rate of prey population decline ($\alpha = a_1 - d_3 = 0$). This is a very special scenario, where the harm caused by microplastics on predators makes these two populations independent, transforming a relationship of predation into a relationship of neutralism, in a process described by the decoupling of the equations for $\dot{x_1}$ and $\dot{x_2}$.

Assuming $\alpha = \alpha' = 0$, the system (10) becomes:

$$\begin{cases} \dot{x}_1(t) = x_1(t)\beta_1(t) \\ \dot{x}_2(t) = x_2(t)\beta_2(t) \end{cases}, \tag{11}$$

which describes the population dynamics of two independent groups, prey, and predators, i.e., the variation in the number of individuals in one group does not affect the growth dynamics of the other.

Solving the system above, we get independent solutions for x_1 and x_2, which will be labeled (ind) for future reference:

$$x_1^{(ind)}(t) = x_1(t) = x_1(0)e^{\int_0^t dt'\beta_1(t')} = x_1(0)e^{c_0t - c_1t^2/2} \tag{12}$$

and

$$x_2^{(ind)}(t) = x_2(t) = x_2(0)e^{\int_0^t dt'\beta_2(t')} = x_2(0)e^{c_0't - c_1't^2/2 + c_2't^3/3}. \tag{13}$$

Analyzing the Equations (12) and (13), it is convenient to calculate $X_1^{(ind)}(t)$ and $X_2^{(ind)}(t)$:

$$\begin{aligned} X_1^{(ind)}(t) &= \int_0^t dt' x_1^{(ind)}(t') \\ &= x_1(0)\sqrt{\frac{\pi}{2c_1}}e^{\frac{c_0^2}{2c_1}}\left[\mathrm{erf}\left(\frac{c_1t - c_0}{\sqrt{2c_1}}\right) - \mathrm{erf}\left(\frac{-c_0}{\sqrt{2c_1}}\right)\right], \end{aligned} \tag{14}$$

$$\begin{aligned} X_2^{(ind)}(t) &= \int_0^t dt' x_2^{(ind)}(t') \\ &= x_2(0)\int_0^t dt'\, e^{c_0't' - c_1't'^2/2 - c_2't'^3/3}, \end{aligned} \tag{15}$$

where the error function is $\mathrm{erf}(x) = \frac{2}{\sqrt{\pi}}\int_0^x dte^{-t^2}$.

To calculate $X_1^{(ind)}(t)$, we used: $\int dt\, e^{at - bt^2} = \frac{\sqrt{\pi}e^{\frac{a^2}{4b}}}{2\sqrt{b}}\mathrm{erf}\left(\frac{2bt-a}{2\sqrt{b}}\right)$.

The model (11) was implemented using Python. We performed simulations taking the same scenarios adopted by Huang et al. and described in Section 3. Figure 1 shows the short-term population dynamics graphs of the predator–prey system—the number of organisms (No./m^3) versus time (months)—where we observe two patterns: 1—the growth of the prey population at a rate $\beta_1(t)$ accompanied by the extinction of predators and 2—the extinction of both species.

As an example of the first pattern observed when $\alpha = \alpha' = 0$, we consider the case where effects of MP particles on prey and predator are weak ($r_{11} = r_{21} = 0.1$). Since prey and predator are independent species and, according to the result of Huang et al., predators are more vulnerable than prey to the effects of MP particles, we have the growth of prey at a rate $\beta_1(t)$ accompanied by the expected extinction of predators (Figure 1a). This behavior was observed in two other scenarios: without the influence of MP particles ($C_1 = C_2 = r_{11} = r_{21} = 0$) and when predators have much larger response strength than prey to MP particles ($r_{11} = 0.01$ and $r_{21} = 0.1$), in which the effects of microplastics are not strong enough to stop the growth of the prey population. When the response intensities of MP particles on prey and predator are both increased to $r_{11} = r_{21} = 5.0$, we have an example of the second pattern, in which the effects of microplastics lead both species to extinction (Figure 1b). Prey population growth is substantially affected by this increase of r_{11}, and the number of prey rises at $t = 7.5$, peaks at $t = 13.7$, and decreases to zero. All other simulated scenarios depict this complete extinction. In these cases, the increase of the adverse effect of reduced predatory performance (d_2) and the consequent lost amount of prey eaten by the predator (d_3) led to an increasingly early extinction of prey, characterized by curves with increasingly less pronounced peaks.

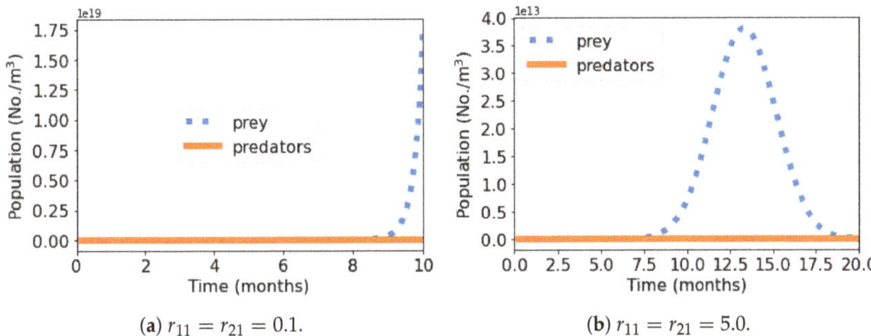

(a) $r_{11} = r_{21} = 0.1$. (b) $r_{11} = r_{21} = 5.0$.

Figure 1. Short-term population dynamics of the predator–prey for $\alpha = \alpha' = 0$. Predator and prey have the same response strength to MP particles, i.e., $\Delta = r_{11}/r_{21} = 1.0$. Abscissa measures time in months, and ordinate measures the number of organisms (No./m^3). The populations of prey (x_1) and predator (x_2) are represented by the blue dotted line and the orange continuous one, respectively.

3.2. *Reduction of Predatory Performance with No Prey Eaten by Predator: $\alpha = 0$*

Setting $\alpha = 0$ leads to the second case where the equations of the system (10) decouple. In this way, we have

$$\begin{cases} \dot{x}_1(t) = x_1(t)\beta_1(t) \\ \dot{x}_2(t) = x_2(t)[\beta_2(t) - \alpha' x_1(t)] \end{cases} . \quad (16)$$

In the original system (1), since $a_1x_1x_2$ refers to the lost amount of prey eaten by predators and $d_3x_1x_2$ refers to the decreasing of that amount, $\alpha = a_1 - d_3 = 0$ describes the regime where the prey population does not decline due to predator feeding (a "total decreasing").

For $\alpha = 0$, the population dynamics of prey are not affected by the predator population size. Nevertheless, the dynamics of predators are affected by the presence of prey. That happens because the model assumes a specific effect that affects the lost amount of prey eaten by predator (denoted by d_3) and a distinct effect that affects the number of predators due to their reduced predatory performance (denoted by d_2). Thus, even if there is not a lost amount of prey eaten by predators ($\alpha = a_1 - d_3 = 0$), the adverse effect of reduced predatory performance may not be strong enough that the number of predators does not depend on the number of prey ($\alpha' = d_2 - a_2$). In this scenario, the effects of microplastics make only the prey population independent.

In this scenario, since $x_1(t) = x_1^{(ind)}(t) = x_1(0)\exp(c_0t - c_1t^2/2)$, we have

$$\begin{aligned} x_2(t) &= x_2(0)e^{\int_0^t dt'[\beta_2(t') - \alpha' x_1^{(ind)}(t')]} \\ &= x_2^{(ind)}(t)e^{-\alpha' X_1^{(ind)}(t)} , \end{aligned}$$

where $x_2^{(ind)}(t) = x_2(0)\exp(c_0't - c_1't^2/2 - c_2't^3/3)$.

The model (16) was implemented using Python. We performed simulations taking the same scenarios adopted by Huang et al. and described in Section 3. Figure 2 shows the short-term population dynamics graphs of the predator–prey—the number of organisms (No./m^3) versus time (months)—where we observe the same pattern, characterized by the growth of the prey population at a rate $\beta_1(t)$ accompanied by the predator "population explosion".

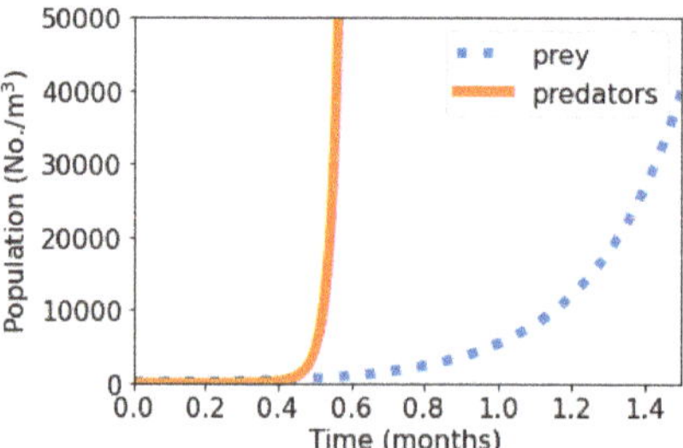

Figure 2. Short-term population dynamics of the predator–prey for $\alpha = 0$. Predator and prey have the same response strength to MP particles, $r_{11} = r_{21} = 0.1$, i.e., $\Delta = r_{11}/r_{21} = 1.0$. Abscissa measures time in months and ordinate measures the number of organisms (No./m^3). The populations of prey (x_1) and predator (x_2) are represented by the blue dotted line and the orange continuous one, respectively.

On one hand, the prey population behaves in these scenarios in the same way as in each of the situations shown in Figure 1, since, in both cases, this population is independent of the predator population and the harmful effects of MP particles are the only factor limiting its growth. On the other hand, the predator population is substantially benefited by its dependence on prey. Figure 2 shows that, in the simulation range allowed by the Python compiler, the number of predators rises very quickly at around $t = 0.4$. This "explosive growth" was also observed in all other cases where $\alpha = 0$, in which it was noted that it happens at around $t = 0.5$ when the values of d_1, d_2, and d_3 are increased.

3.3. Reduction of Predatory Performance with No Increase in the Number of Predators Due to the Feeding of Prey: $\alpha' = 0$

A third way to decouple the equations of the system (10) is to make $\alpha' = 0$:

$$\begin{cases} \dot{x}_1(t) = x_1(t)[\beta_1(t) - \alpha x_2(t)] \\ \dot{x}_2(t) = x_2(t)\beta_2(t) \end{cases} . \tag{17}$$

Since $a_2 x_1 x_2$ is the increasing number of predators due to the feeding of prey, and $d_2 x_1 x_2$ is the decreasing number of predators due to the adverse effect of reduced predatory performance, $\alpha' = d_2 - a_2 = 0$ defines a regime of reduction of predatory performance, where there is no increase in the number of predators due to the feeding of prey. Here, the exposure to MP particles makes only the predators population independent.

Since $x_2(t) = x_2^{(ind)}(t) = x_2(0)\exp(c_0' t - c_1' t^2/2 - c_2' t^3/3)$, we have

$$\begin{aligned} x_1(t) &= x_1(0)e^{\int_0^t dt'\beta_1(t') - \alpha\int_0^t dt' x_2^{(ind)}(t')} \\ &= x_1^{(ind)}(t)e^{-\alpha X_2^{(ind)}(t)} . \end{aligned} \tag{18}$$

The model (17) was implemented using Python. We performed simulations taking the same scenarios adopted by Huang et al. and described in Section 3. Figure 3 shows the short-term population dynamics graphs of the predator–prey-number of organisms (No./m^3) *versus* time (months)—where we observe the same two patterns of the Section 3.1, where $\alpha = \alpha' = 0$.

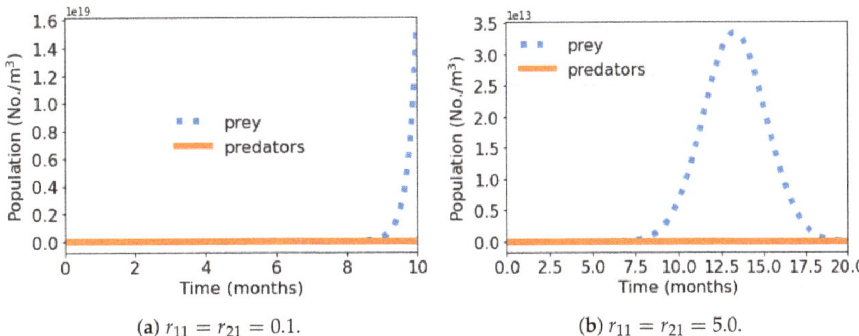

(**a**) $r_{11} = r_{21} = 0.1$. (**b**) $r_{11} = r_{21} = 5.0$.

Figure 3. Short-term population dynamics of the predator–prey for $\alpha' = 0$. Predator and prey have the same response strength to MP particles, i.e., $\Delta = r_{11}/r_{21} = 1.0$. Abscissa measures time in months and ordinate measures the number of organisms (No./m^3). The populations of prey (x_1) and predator (x_2) are represented by the blue dotted line and the orange continuous one, respectively.

When $\alpha' = 0$, the prey population is the only one affected by the presence of the other species, and predator extinction occurs early in every scenario, since there is no increase in its number due to the feeding of the prey, besides being under the harmful effects of the MP particles (Figure 3). Thus, the population dynamics of the system are similar to those in Figure 1. However, it can be seen in each of these situations that the dependence on the predator, even weak, led to an early extinction of the prey, which also presented curves with less pronounced peaks.

A possible next step for a system of ODEs such as (10) is to combine the two first-order differential equations into a single second-order one. To do so, we first write $x_2(t)$ as a function of $\beta_1(t)$ and the ratio $\dot{x}_1/x_1$ and differentiate $\dot{x}_1(t)$ in Equation (10), leading to: $\ddot{x}_1(t) = \dot{x}_1(t)[\beta_1(t) - \alpha x_2(t)] + x_1(t)[\dot{\beta}_1(t) - \alpha \dot{x}_2(t)]$, and then to: $\ddot{x}_1(t) = \dot{x}_1(t)\beta_1(t) + x_1(t)\dot{\beta}_1(t) - \alpha\{\dot{x}_1(t) + x_1(t)[\beta_2(t) - \alpha' x_1(t)]\}x_2(t)$. Using the calculated $x_2(t)$, one obtains:

$$\ddot{x}_1(t) = \beta_2 \dot{x}_1(t) + \left[\dot{\beta}_1(t) - \beta_1(t)\beta_2(t)\right]x_1(t) + \alpha' \beta_1(t) x_1^2(t) - \left[-\frac{\dot{x}_1(t)}{x(t)} + \alpha' x_1(t)\right]\dot{x}_1(t) \,,$$

which does not depend on α. Analogously, if one had written the second-order equation for $x_2(t)$, it would be independent of α'. This issue stresses that possibly Huang et al. should be reviewed.

4. Discussion

Rewriting the Huang et al. four equations model, reducing it to a two equations one and including time-varying intraspecies coefficients, allowed us to obtain an equivalent model with a format analogous to the standard LVM. This first step clarified the meanings of the model parameters and led us to explore and solve analytically special, and simpler, cases of ecological relationships. Bioaccumulation of microplastics through food web, and its consequent accelerated accumulation and magnification on predator [17–19], is a known effect that we analytically showed based on this model, calculating the time threshold from which their concentration and changing rate of the total amount are greater in predators than in prey.

The three situations that we studied are characterized by some possible behavioral abnormalities caused by the exposure to MP particles. More specifically, we considered effects of reduction of predatory performance [26], where prey is not affected by predator population size ($\alpha = 0$), and/or there is no increase in the number of predators due to the feeding of prey ($\alpha' = 0$). In our study, theses effects are considered so strong that they can make one of the species independent, or even make both independent of each other. In the cases where $\alpha = \alpha' = 0$ or only $\alpha' = 0$, results reveal basically two patterns, depending on the response strength to MP particles: 1—the growth of prey at a rate $\beta_1(t)$ accompanied by the extinction of predators, where the effects of microplastics are not strong enough to stop the growth of the prey population; and 2—the extinction of predator followed by the eventual extinction of prey. The most interesting, and counterintuitive, result of these three cases appears when $\alpha = 0$, where there is predators "population explosion", showing the predator population is substantially benefited by its dependence on prey. We also point out that the Huang et al. model is not able to address the Allee effect [33,34]. It is important to emphasize that the underlying structure of the model is exponential, and the three special situations that we studied are limit cases, which lead to very different predictions from the original study and result in the extremely large numbers of organisms shown in Figures 1–3.

5. Conclusions

We rewrite the Huang et al. model by reducing its number of equations and defining new parameters, making it analogous to the standard LVM with time varying intraspecific coefficients. The time threshold from which the MP particles concentration and changing rate of its total amount are greater in predators than in prey was calculated. We solved analytically specific and simpler ecological situations where the effect of MP particles cause severe abnormal behavior on predator and prey, leading them to become independent of each other. Our simulations reveal a counterintuitive result when toxicological effects of MP particles cause a total interruption of prey feeding ability, which can produce a predator "population explosion".

Our advance in the analytical treatment of the Huang et al. model opens space for further refinement in the study of predator–prey models under toxicological effects of MP particles, either exploring the second-order equation that we propose or modifying the original model to further reduce its number of parameters, embedding in the time-varying intraspecies coefficients all the adverse effects caused by MP particles.

Author Contributions: A.S.M. obtained the analytical solutions, designed the research, wrote the manuscript and was responsible for supervision, project administration, and funding acquisition. M.M.R.S. introduced the Huang et al. paper to the research group, and wrote and validated the first version of the numerical code. L.M.D.C. introduced the Huang et al. paper to the research group, wrote, and validated the first version of the numerical code. J.R.A. verified analytical solutions, identified the critical values of Equations (7) and (8), and wrote the manuscript. L.S.D.S. verified analytical solutions; wrote a second and independent version of the numerical code, validated it and understood the numerical overflow relative to Figure 2, interpreted the model parameters with ecological regimes, and wrote the first version of the manuscript. All authors have read and agreed to the published version of the manuscript.

Funding: José Renato Alcarás and Mateus Mendonça Ramos Simões have been funded by Coordenação de Aperfeiçoamento de Pessoal de Nível Superior—Brasil (CAPES)—Finance Code 001. José Renato Alcarás acknowledges support of Fundação de Amparo à Pesquisa do Estado de São Paulo-Brazil (FAPESP) (Grant No. 2018/21694-4). Mateus Mendonça Ramos Simões acknowledges support of Coordenação de Aperfeiçoamento de Pessoal de Nível Superior-Brazil (CAPES) (Grant No. 88887.570033/2020-00) Alexandre Souto Martinez acknowledges Conselho Nacional de Desenvolvimento Científico e Tecnológico (CNPq) (Grant No. 309851/2018-1) and NAP-FisMed for support.

Data Availability Statement: The numerical code in Python is available on demand.

Acknowledgments: The authors thank Stephany Silva Xesquevixos de Siqueira for fruitful discussions.

Conflicts of Interest: The authors declare no conflict of interest.

Abbreviations

The following abbreviations are used in this manuscript:

MP	Microplastic
LVM	Lotka–Volterra model

References

1. Shashoua, Y. *Conservation of Plastics*; Routledge: London, UK, 2012.
2. Hale, R.C.; Seeley, M.E.; La Guardia, M.J.; Mai, L.; Zeng, E.Y. A Global Perspective on Microplastics. *J. Geophys. Res. Ocean.* **2020**, *125*, e2018JC014719. [CrossRef]
3. Jambeck, J.R.; Geyer, R.; Wilcox, C.; Siegler, T.R.; Perryman, M.; Andrady, A.; Narayan, R.; Law, K.L. Plastic waste inputs from land into the ocean. *Science* **2015**, *347*, 768–771. [CrossRef] [PubMed]
4. Boucher, J.; Friot, D. *Primary Microplastics in the Oceans: A Global Evaluation of Sources*; International Union for Conservation of Nature: Gland, Switzerland, 2017; Volume 10.
5. Lau, W.W.Y.; Shiran, Y.; Bailey, R.M.; Cook, E.; Stuchtey, M.R.; Koskella, J.; Velis, C.A.; Godfrey, L.; Boucher, J.; Murphy, M.B.; et al. Evaluating scenarios toward zero plastic pollution. *Science* **2020**, *369*, 1455–1461. [CrossRef] [PubMed]
6. Singh, B.; Sharma, N. Mechanistic implications of plastic degradation. *Polym. Degrad. Stab.* **2008**, *93*, 561–584. [CrossRef]
7. Frias, J.; Nash, R. Microplastics: Finding a consensus on the definition. *Mar. Pollut. Bull.* **2019**, *138*, 145–147. [CrossRef] [PubMed]
8. Barcelo, D. Microplastics analysis. *MethodsX* **2020**, *7*, 100884. [CrossRef] [PubMed]
9. Galloway, T.; Lewis, C. Marine microplastics. *Curr. Biol.* **2017**, *27*, R445–R446. [CrossRef]
10. Tirelli, V.; Suaria, G.; Lusher, A. Microplastics in polar samples. In *Handbook of Microplastics in the Environment*; Springer: Cham, Switzerland, 2020; pp. 1–42.
11. Johnson, A.C.; Ball, H.; Cross, R.; Horton, A.A.; Jurgens, M.D.; Read, D.S.; Vollertsen, J.; Svendsen, C. Identification and quantification of microplastics in potable water and their sources within water treatment works in England and Wales. *Environ. Sci. Technol.* **2020**, *54*, 12326–12334. [CrossRef]
12. Corradini, F.; Casado, F.; Leiva, V.; Huerta-Lwanga, E.; Geissen, V. Microplastics occurrence and frequency in soils under different land uses on a regional scale. *Sci. Total Environ.* **2021**, *752*, 141917. [CrossRef]
13. Harne, R.; Rokde, A.; Jadav, K.; Chitariya, J.M.; Sengar, A. Studies on plastic bezoar ingestion in free range axis deer in summer. *J. Anim. Res.* **2019**, *9*, 383–386. [CrossRef]
14. de Souza Machado, A.A.; Kloas, W.; Zarfl, C.; Hempel, S.; Rillig, M.C. Microplastics as an emerging threat to terrestrial ecosystems. *Glob. Chang. Biol.* **2018**, *24*, 1405–1416. [CrossRef]
15. Wang, K.; Zhu, L.; Rao, L.; Zhao, L.; Wang, Y.; Wu, X.; Zheng, H.; Liao, X. Nano- and micro-polystyrene plastics disturb gut microbiota and intestinal immune system in honeybee. *Sci. Total Environ.* **2022**, *842*, 156819. [CrossRef] [PubMed]
16. Cox, K.D.; Covernton, G.A.; Davies, H.L.; Dower, J.F.; Juanes, F.; Dudas, S.E. Human consumption of microplastics. *Environ. Sci. Technol.* **2019**, *53*, 7068–7074. [CrossRef] [PubMed]
17. Yu, Q.; Hu, X.; Yang, B.; Zhang, G.; Wang, J.; Ling, W. Distribution, abundance and risks of microplastics in the environment. *Chemosphere* **2020**, *249*, 126059. [CrossRef] [PubMed]
18. Farrell, P.; Nelson, K. Trophic level transfer of microplastic: *Mytilus edulis* (L.) to *Carcinus maenas* (L.). *Environ. Pollut.* **2013**, *177*, 1–3. [CrossRef]
19. Zhao, S.; Ward, J.E.; Danley, M.; Mincer, T.J. Field-based evidence for microplastic in marine aggregates and mussels: Implications for trophic transfer. *Environ. Sci. Technol.* **2018**, *52*, 11038–11048. [CrossRef]
20. Pirsaheb, M.; Hossini, H.; Makhdoumi, P. Review of microplastic occurrence and toxicological effects in marine environment: Experimental evidence of inflammation. *Process. Saf. Environ. Prot.* **2020**, *142*, 1–14. [CrossRef]
21. Schöpfer, L.; Menzel, R.; Schnepf, U.; Ruess, L.; Marhan, S.; Brümmer, F.; Pagel, H.; Kandeler, E. Microplastics effects on reproduction and body length of the soil-dwelling nematode Caenorhabditis elegans. *Front. Environ. Sci.* **2020**, *8*, 41. [CrossRef]
22. Xu, S.; Ma, J.; Ji, R.; Pan, K.; Miao, A.J. Microplastics in aquatic environments: Occurrence, accumulation, and biological effects. *Sci. Total Environ.* **2020**, *703*, 134699. [CrossRef]
23. Besseling, E.; Wang, B.; Lürling, M.; Koelmans, A.A. Nanoplastic affects growth of S. obliquus and reproduction of D. magna. *Environ. Sci. Technol.* **2014**, *48*, 12336–12343. [CrossRef] [PubMed]
24. Tallec, K.; Huvet, A.; Di Poi, C.; González-Fernández, C.; Lambert, C.; Petton, B.; Le Goïc, N.; Berchel, M.; Soudant, P.; Paul-Pont, I. Nanoplastics impaired oyster free living stages, gametes and embryos. *Environ. Pollut.* **2018**, *242*, 1226–1235. [CrossRef]
25. Van Colen, C.; Vanhove, B.; Diem, A.; Moens, T. Does microplastic ingestion by zooplankton affect predator–prey interactions? An experimental study on larviphagy. *Environ. Pollut.* **2020**, *256*, 113479. [CrossRef] [PubMed]
26. Huang, Q.; Lin, Y.; Zhong, Q.; Ma, F.; Zhang, Y. The impact of microplastic particles on population dynamics of predator and prey: Implication of the Lotka–Volterra model. *Sci. Rep.* **2020**, *10*, 4500. [CrossRef] [PubMed]
27. Lotka, A.J. Contribution to the Theory of Periodic Reactions. *J. Phys. Chem.* **1910**, *14*, 271–274. [CrossRef]
28. Berryman, A.A. The Orgins and Evolution of Predator–Prey Theory. *Ecology* **1992**, *73*, 1530–1535. [CrossRef]
29. Ribeiro, F.; Cabella, B.C.T.; Martinez, A.S. Richards-like two species population dynamics model. *Theory Biosci.* **2014**, *133*, 135–143. [CrossRef]

30. Cabella, B.; Meloni, F.; Martinez, A.S. Inadequate Sampling Rates Can Undermine the Reliability of Ecological Interaction Estimation. *Math. Comput. Appl.* **2019**, *24*, 48. [CrossRef]
31. Hallam, T.; de Luna, J. Effects of toxicants on populations: A qualitative: Approach III. Environmental and food chain pathways. *J. Theor. Biol.* **1984**, *109*, 411–429. [CrossRef]
32. Cabella, B.C.T.; Ribeiro, F.; Martinez, A.S. Effective carrying capacity and analytical solution of a particular case of the Richards-like two-species population dynamics model. *Phys. A Stat. Mech. Appl.* **2012**, *391*, 1281–1286. [CrossRef]
33. Dos Santos, L.S.; Cabella, B.C.T.; Martinez, A.S. Generalized Allee effect model. *Theory Biosci.* **2014**, *113*, 117–124. [CrossRef]
34. Dos Santos, R.V.; Ribeiro, F.L.; Martinez, A.S. Models for Allee effect based on physical principles. *J. Theor. Biol.* **2015**, *385*, 143–152. [CrossRef] [PubMed]

Mathematical and Computational Applications

Article

Spectral Analysis of the Finite Element Matrices Approximating 3D Linearly Elastic Structures and Multigrid Proposals

Quoc Khanh Nguyen [1,*], Stefano Serra-Capizzano [2,3], Cristina Tablino-Possio [4] and Eddie Wadbro [1,5]

1 Department of Computing Science, Umeå University, SE-901 87 Umeå, Sweden
2 INdAM Unit, Department of Science and High Technology, University of Insubria, Via Valleggio 11, 22100 Como, Italy
3 Department of Information Technology, Uppsala University, Box 337, SE-751 05 Uppsala, Sweden
4 Dipartimento di Matematica e Applicazioni, Università di Milano Bicocca, Via Cozzi 53, 20125 Milano, Italy
5 Department of Mathematics and Computer Science, Karlstad University, SE-651 88 Karlstad, Sweden
* Correspondence: qukh0001@student.umu.se

Abstract: The so-called material distribution methods for topology optimization cast the governing equation as an extended or fictitious domain problem, in which a coefficient field represents the design. In practice, the finite element method is typically used to approximate that kind of governing equations by using a large number of elements to discretize the design domain, and an element-wise constant function approximates the coefficient field in that domain. This paper presents a spectral analysis of the coefficient matrices associated with the linear systems stemming from the finite element discretization of a linearly elastic problem for an arbitrary coefficient field in three spatial dimensions. The given theoretical analysis is used for designing and studying an optimal multigrid method in the sense that the (arithmetic) cost for solving the problem, up to a fixed desired accuracy, is linear in the corresponding matrix size. Few selected numerical examples are presented and discussed in connection with the theoretical findings.

Keywords: matrix sequences; spectral analysis; finite element approximations

Citation: Nguyen, K.N.; Serra-Capizzano, S.; Tablino-Possio, C.; Wadbro, E. Spectral Analysis of the Finite Element Matrices Approximating 3D Linearly Elastic Structures and Multigrid Proposals. *Math. Comput. Appl.* **2022**, *27*, 78. https://doi.org/10.3390/mca27050078

Academic Editors: Sascha Eisenträger and Gianluigi Rozza

Received: 13 April 2022
Accepted: 10 September 2022
Published: 14 September 2022

1. Introduction

In our previous paper [1], we applied the theory of generalized locally Toeplitz (GLT) sequences to compute and analyze the asymptotic spectral distribution of the sequence of stiffness matrices $\{K_n\}_n$, with K_n being the finite element (FE) approximation of the considered one spatial dimension topology optimization problem, for a given fineness parameter associated to n. In a later contribution [2], we extended the analysis to the two-dimensional setting using so-called multilevel block GLT sequences. In this paper, we further expand the theory to cover the three-dimensional case.

Since the first material distribution method for topology design was introduced in the late 1980s [3], topology optimization [4,5], a well-known computational tool for finding the optimal distribution of material within a given design domain, has been studied extensively. The material distribution topology optimization has contributed to the development of several areas, such as electromagnetic [6–8], fluid–structure interaction [9,10], acoustics [11,12], additive manufacturing [13], and especially (non-)linear elasticity [14–16]. For problems in linear elasticity, which motivates the study in this paper, the most common method to solve this type of problem is the so-called density-based or material distribution approach. In this approach, a so-called material indicator function $\alpha(x)$—typically referred to as the density or the physical design—models the presence/absence of material; $\alpha = 1$ where material is present, else $\alpha = 0$. However, the binary design problem is computationally intractable. A standard approach to make the problem computationally feasible is employing a combination of relaxation, penalization, and filtering, in which the physical density is defined as $\rho(x) = \underline{\alpha} + (1 - \underline{\alpha})g(\mathcal{F}(\alpha)(x))$ by $\underline{\rho} \geq 0$, which is a constant, the penalization operator is g, and the filtering procedure is $\mathcal{F}$. In this approach, the finite element method (FEM) is the standard choice for generating numerical solutions of a linearly elastic boundary value problem. In practice, the physical design is typically represented as an element-wise constant function. Moreover, a combination of allowing intermediate values of the design, filtering, and penalization has been crucial for the success of these methods. In this paper, we limit our attention to studying the linear system that stems from the FE discretization of the governing equation and its solution. From a high performance computing perspective, this limitation is natural since the computational effort in solving topology optimization problems is dominated by the cost of solving the discretized governing equations—a fact that is particularly prominent when studying three-dimensional problems. Over the last decades, there has been significant work in improving the efficiency of topology optimization, aiming at solving large-scale design problems [17–21]. Although our study is directly motivated by the material-distribution or density-based approach to topology optimization, the work is relevant for other design descriptions, such as level sets [22,23] or moving morphable components [24–27], provided that the physics description uses the so-called ersatz material approach, which in some cases can be justified rigorously [4].

The theory of multilevel block GLT sequences [28,29] is a generalization of the GLT sequences theory [30,31] that is typically used for computing/analyzing the spectral distribution of matrix sequences arising from, for example, the numerical discretization, such as the FE approximation of partial differential equations (PDEs) with proper boundary conditions. In the considered matrix sequences, the size of the given linear systems d_n increases with n, in which d_n tends to infinity as $n \to \infty$. In this paper, the entire sequence of linear systems with increasing size arising from the three spatial dimension problems is the primary consideration. Under mild conditions, essentially relying on regularity assumptions on the meshes, we show that the sequence of discretization matrices has an asymptotic spectral distribution. By leveraging the theory of multilevel block GLT sequences, we now extend our previous works [1,2] further to perform a detailed spectral analysis of the linear systems associated with the FE discretization of the governing equation in the three-dimensional setting. Similar to our previous work [2], we also make use of the information obtained from the spectral symbol f to design a fast, multigrid solver in three dimensions for optimizing the (arithmetic) cost to solve the related linear systems up to a fixed desired accuracy, which is proportional to the matrix–vector cost, which is linear in the corresponding matrix size. This solver is also verified and comes up with very satisfying numerical results, in terms of the linear cost and number of iterations, which are bounded independently of the matrix size and mildly depending on the desired accuracy.

In the following sections, we will go into detail on the problem description, spectral analysis, multigrid method, and a brief conclusion. The description of the continuous problem and the resulting coefficient matrices arising from our FE approximation is delivered in Sections 2 and 3 is devoted to the spectral analysis of the FE matrices from the perspective of the GLT theory. In Section 4, a brief account of multigrid methods with special attention to the block case encountered in the present context is given, and the spectral information is a core component in the development of the multigrid proposal for our specific 3D setting presented in Section 5. Eventually, the conclusions are reported in Section 6, and some relevant model information is explained in Appendice A and B.

2. Problem Description

We consider a linearly elastic structure that occupies (part of) the hyper-rectangular domain $\Omega \subset \mathbb{R}^3$. In particular, we are interested in the setting used in material distribution based topology optimization, where a function, typically denoted as the physical density, $\rho : \Omega \to [0, 1]$, describes the layout of the unloaded structure. We assume that the structure is clamped along the boundary portion $\Gamma_D \subset \partial\Omega$. Moreover, we let $b \in L^2(\Omega)^3$ be a given body load (a volume force) in Ω, $t \in L^2(\Gamma_F)^3$ be the surface traction acting on the non-clamped boundary $\Gamma_F \subset \partial\Omega$ of the solid, and u denote the resulting equilibrium displacement, which solves the following problem.

$$\text{Find } u \in \mathcal{U} \text{ such that } a(u, v; \rho) = \ell(v) \;\; \forall v \in \mathcal{U}, \tag{1}$$

where the set of all kinematically admissible displacements of the structure is

$$\mathcal{U} = \left\{ u \in H^1(\Omega)^3 \mid u|_{\Gamma_D} \equiv 0 \right\}.$$

The energy bilinear form a and the load linear form ℓ are defined as

$$a(u, v; \rho) = \int_{\Omega} \rho \left(E_c \epsilon(u) \right) : \epsilon(v),$$
$$\ell(v) = \int_{\Gamma_F} t \cdot v + \int_{\Omega} b \cdot v,$$

where $\epsilon(\boldsymbol{u}) = \left(\nabla u + \nabla u^T \right)/2$ is the strain tensor of u. The colon ":" denotes the full contraction between the two tensors; when using the standard basis, the full contraction of the two matrices is their Frobenius scalar product. E_c is a constant fourth-order elasticity tensor. In this paper, we study an FE discretization of problem (1), in which the physical density is approximated by an element-wise constant function, which is typical in material distribution based topology optimization. The domain Ω is discretized into n trilinear hexahedral elements and then applying FE approximation, the variational problem (1) is reduced to the linear system

$$\boldsymbol{K}_n(\boldsymbol{\rho})\boldsymbol{u} = \boldsymbol{f},$$

where $\boldsymbol{u}$ and $\boldsymbol{f}$ are the nodal displacement and load vector, respectively, and $\boldsymbol{K}_n(\boldsymbol{\rho})$ is the stiffness matrix of element-wise constant physical density function ρ—the entries of the vector $\boldsymbol{\rho} = [\rho_1, \rho_2, \ldots, \rho_n]$ are the element values of ρ; that is, ρ_i is the value of ρ in the ith element in the FE mesh. The stiffness matrix $\boldsymbol{K}_n$ is typically assembled by looping over each element so that

$$\boldsymbol{K}_n(\boldsymbol{\rho}) = \sum_{i=1}^{n} \rho_i \boldsymbol{K}_e^{(i)},$$

where $\boldsymbol{K}_e^{(i)}$ is the element stiffness matrix. We emphasize that the formal expression of the relevant matrices is a key ingredient for applying the multilevel block GLT theory to produce a global spectral description of the matrix sequences under consideration. More precisely, the non-zero blocks of three-dimensional element stiffness matrix can be expressed as

$$\boldsymbol{K}_e = \frac{h}{2} \frac{E_0}{1+\nu} \begin{bmatrix} K_1 & K_2 & K_3 & K_4 & K_5 & K_6 & K_7 & K_8 \\ & K_9 & K_{10} & K_{11} & K_{12} & K_{13} & K_{14} & K_7 \\ & & K_{15} & K_{16} & K_{17} & K_{18} & K_{13} & K_6 \\ & & & K_{19} & K_{20} & K_{17} & K_{12} & K_5 \\ & & & & K_{19} & K_{16} & K_{11} & K_4 \\ & & sym. & & & K_{15} & K_{10} & K_3 \\ & & & & & & K_9 & K_2 \\ & & & & & & & K_1 \end{bmatrix} \tag{2}$$

where each K_i block has size 3×3 and is associated with one node on a so-called reference element, E_0 is Young's modulus, and ν is Poisson's ratio. More details are provided in Appendix A for the explicit expressions for the element stiffness matrix and in Appendix B for the stress–strain relation and various bounds. With regard to Equation (2), the lower part of the matrix is not explicitly given because the matrix is globally real symmetric and so is every block K_l, $l = 1, \ldots, 20$, whose expressions are reported in Appendix A, Equation (A1).

3. Spectral Analysis

The current section is devoted to the spectral analysis of the FE coefficient matrices derived in the previous section and is complemented by a selection of numerical tests that confirm the theoretical analysis (more numerical experiments have been performed but here we report only a selection for the sake of brevity). We limit our focus to isotropic materials; for more information about various bounds, see the discussion in Appendix B. From a mathematics perspective, this means that we limit our attention to when the Poisson's ratio ν is in the range $[0, 0.5)$. In particular, Section 3.1 contains the minimal set of preliminary concepts and tools, while Sections 3.2 and 3.3 are focused on the specific study in 3D in the constant and variable coefficient cases, respectively.

3.1. Premises

The premises include the formal definition of the eigenvalue and singular value distribution, the notion of multi-indexing, the concepts of multilevel block Toeplitz matrices, multilevel block sampling matrices, and multilevel block GLT matrix sequences.

3.1.1. Singular Value/Eigenvalue Distributions

We first give the formal definitions, and then we briefly discuss their informal and practical meaning.

Definition 1. *Let r, t be two positive integers. Let $\{A_n\}_n$ be a sequence of matrices, with A_n of size d_n with eigenvalues $\lambda_1(A_n), \ldots, \lambda_{d_n}(A_n)$ and singular values $\sigma_1(A_n), \ldots, \sigma_{d_n}(A_n)$. Furthermore, let $f : D \subset \mathbb{R}^t \to \mathbb{C}^{r\times r}$ be a measurable function defined on a set D with $0 < \mu_t(D) < \infty$, and with $\mu_t(\cdot)$ denoting the Lebesgue measure on $\mathbb{C}^t$.*

- *We say that $\{A_n\}_n$ has a (asymptotic) singular value distribution described by f, and we write $\{A_n\}_n \sim_\sigma f$, if*

$$\lim_{n\to\infty} \frac{1}{d_n}\sum_{i=1}^{d_n} F(\sigma_i(A_n)) = \frac{1}{\mu_t(D)}\int_D \frac{\sum_{i=1}^{r} F(\sigma_i(f(\mathbf{x})))}{r}\mathbf{dx}, \quad \forall F \in C_c(\mathbb{R}), \tag{3}$$

  with $\sigma_1(f), \ldots, \sigma_r(f)$ being the singular values of f, each of them being a measurable function non-negative almost everywhere.
- *We say that $\{A_n\}_n$ has a (asymptotic) spectral (or eigenvalue) distribution described by f, and we write $\{A_n\}_n \sim_\lambda f$, if*

$$\lim_{n\to\infty} \frac{1}{d_n}\sum_{i=1}^{d_n} F(\lambda_i(A_n)) = \frac{1}{\mu_t(D)}\int_D \frac{\sum_{i=1}^{r} F(\lambda_i(f(\mathbf{x})))}{r}\mathbf{dx}, \quad \forall F \in C_c(\mathbb{C}), \tag{4}$$

  with $\lambda_1(f), \ldots, \lambda_r(f)$ being the eigenvalues of f, each of them being a complex-valued measurable function.

If $\{A_n\}_n$ has both a singular value and an eigenvalue distribution described by f, we write $\{A_n\}_n \sim_{\sigma,\lambda} f$. (In practice, in the Toeplitz setting and often in the GLT setting, the parameter r can be read at a matrix level as the size of the elementary blocks which form the global matrix A_n, as it will be clear both in our stiffness matrices in Sections 3.2 and 3.3, and in the block Toeplitz/block diagonal sampling structures in Section 3.1.2.)

The symbol f contains spectral/singular value information briefly described informally as follows. With reference to relation (4), the eigenvalues of K_n are partitioned into r subsets of the same cardinality, except possibly for a small number of outliers, such that the ith subset is approximately formed by the samples of $\lambda_i(f)$ over a uniform grid in D, $i = 1, \ldots, r$. Thus, provided that n is large enough, the symbol f provides a 'compact' and quite accurate description of the spectrum of the matrices K_n. Similarly, relation (3) has the same meaning when talking of the singular values of K_n and by replacing $\lambda_i(f)$ with $\sigma_i(f)$, $i = 1, \dots, r$.

3.1.2. Multilevel Block Toeplitz Matrices, Multilevel Block Diagonal Sampling Matrices, and Multilevel Block GLT Sequences

The present section is specifically devoted to matrix-theoretic notations and definitions, which are essential when dealing with multi-dimensional problems and with the related sequences of matrices.

Definition 2. *A multi-index $\boldsymbol{i} \in \mathbb{Z}^d$, also called a d-index, is a (row) vector in $\mathbb{Z}^d$, whose components are denoted by $i_1, \dots, i_d$.*

- $\mathbf{0}$, $\mathbf{1}$, $\mathbf{2}$, ... *are the vectors of all zeros, all ones, all twos, ... (their size will be clear from the context).*
- *For any d-index $\boldsymbol{m} \in \mathbb{N}_+^d$, we set $N(\boldsymbol{m}) = \prod_{j=1}^d m_j$ and we write $\boldsymbol{m} \to \infty$ to indicate that $\min(\boldsymbol{m}) \to \infty$.*
- *If $\boldsymbol{h}, \boldsymbol{k}$ are d-indices, $\boldsymbol{h} \le \boldsymbol{k}$ means that $h_r \le k_r$ for all $r = 1, \dots, d$.*
- *If $\boldsymbol{h}, \boldsymbol{k}$ are d-indices such that $\boldsymbol{h} \le \boldsymbol{k}$, the multi-index range $[\boldsymbol{h}, \dots, \boldsymbol{k}]$ is the set $\{\boldsymbol{j} \in \mathbb{Z}^d : \boldsymbol{h} \le \boldsymbol{j} \le \boldsymbol{k}\}$. The standard lexicographic ordering is assumed uniformly*

$$\left[\cdots \left[\left[(j_1, \dots, j_d) \right]_{j_d = h_d, \dots, k_d} \right]_{j_{d-1} = h_{d-1}, \dots, k_{d-1}} \cdots \right]_{j_1 = h_1, \dots, k_1}.$$

With regard to the previous definition, in the case $d = 2$, the lexicographic ordering is the following: (h_1, h_2), $(h_1, h_2 + 1)$, ..., (h_1, k_2), $(h_1 + 1, h_2)$,$(h_1 + 1, h_2 + 1)$, ... $(h_1 + 1, k_2)$, ... (k_1, h_1), $(k_1, h_1 + 1)$, ..., (k_1, k_2). Notice that, in general, a multi-index range $[\boldsymbol{h}, \dots, \boldsymbol{k}]$, $\boldsymbol{h} \le \boldsymbol{k}$ is used with $\boldsymbol{h} = \mathbf{0}$ or with $\boldsymbol{h} = \mathbf{1}$.

Definition 3 ((Multilevel) Block Toeplitz Matrices). *Given $\mathbf{n} \in \mathbb{N}^d$, a matrix of the form*

$$[A_{\mathbf{i}-\mathbf{j}}]_{\mathbf{i},\mathbf{j}=\mathbf{e}}^{\mathbf{n}} \in \mathbb{C}^{N(\mathbf{n})r \times N(\mathbf{n})r}$$

with $\mathbf{e}$ vector of all ones, with blocks $A_{\mathbf{k}} \in \mathbb{C}^{r \times r}$, $\mathbf{k} = -(\mathbf{n} - \mathbf{e}), \dots, \mathbf{n} - \mathbf{e}$, is called a multilevel block Toeplitz matrix, or, more precisely, a d-level r-block Toeplitz matrix. Let $\phi : [-\pi, \pi]^d \to \mathbb{C}^{r \times r}$ be a matrix-valued function in which each entry belongs to $L^1([-\pi, \pi]^d)$. We denote the Fourier coefficients of the generating function ϕ as

$$\hat{f}_{\mathbf{k}} = \frac{1}{(2\pi)^d} \int_{[-\pi,\pi]^d} \phi(\boldsymbol{\theta}) e^{-\hat{\imath}(\mathbf{k}, \boldsymbol{\theta})} d\boldsymbol{\theta} \in \mathbb{C}^{r \times r}, \quad \mathbf{k} \in \mathbb{Z}^d,$$

where the integrals are computed componentwise, $\hat{\imath}^2 = -1$, and $(\mathbf{k}, \boldsymbol{\theta}) = k_1\theta_1 + \ldots + k_d\theta_d$. For every $\mathbf{n} \in \mathbb{N}^d$, the $\mathbf{n}$th Toeplitz matrix associated with ϕ is defined as

$$T_{\mathbf{n}}(\phi) := [\hat{f}_{\mathbf{i}-\mathbf{j}}]_{\mathbf{i},\mathbf{j}=\mathbf{e}}^{\mathbf{n}}$$

or, equivalently, as

$$T_{\mathbf{n}}(\phi) = \sum_{|j_1| < n_1} \cdots \sum_{|j_d| < n_d} [J_{n_1}^{(j_1)} \otimes \ldots J_{n_d}^{(j_d)}] \otimes \hat{f}_{(j_1, \dots, j_d)},$$

where $\otimes$ denotes the (Kronecker) tensor product of matrices, while $J_m^{(l)}$ is the matrix of order m whose (i, j) entry equals 1 if $i - j = l$ and zero otherwise. We call $\{T_{\mathbf{n}}(\phi)\}_{\mathbf{n} \in \mathbb{N}^d}$ the family of (multilevel block) Toeplitz matrices associated with ϕ, which, in turn, is called the generating function of $\{T_{\mathbf{n}}(\phi)\}_{\mathbf{n} \in \mathbb{N}^d}$.

Definition 4 ((Multilevel) Block Diagonal Sampling Matrices). *For $n \in \mathbb{N}$ and $a : [0,1] \to \mathbb{C}^{r\times r}$, we define the block diagonal sampling matrix $D_n(a)$ as the diagonal matrix*

$$D_n(a) = \underset{i=1,\ldots,n}{\operatorname{diag}} a\Big(\frac{i}{n}\Big) = \begin{bmatrix} a(\frac{1}{n}) & & & \\ & a(\frac{2}{n}) & & \\ & & \ddots & \\ & & & a(1) \end{bmatrix} \in \mathbb{C}^{rn\times rn}.$$

For $\boldsymbol{n} \in \mathbb{N}^d$ and $a : [0,1]^d \to \mathbb{C}^{r\times r}$, we define the multilevel block diagonal sampling matrix $D_{\boldsymbol{n}}(a)$ as the block diagonal matrix

$$D_{\boldsymbol{n}}(a) = \underset{\boldsymbol{i}=\boldsymbol{1},\ldots,\boldsymbol{n}}{\operatorname{diag}} a\Big(\frac{\boldsymbol{i}}{\boldsymbol{n}}\Big) \in \mathbb{C}^{rN(\boldsymbol{n})\times rN(\boldsymbol{n})},$$

with the lexicographical ordering discussed at the beginning of Section 3.1.2.

Definition 5 (Zero-Distributed Sequences of Matrices). *According to Definition 1, a sequence of matrices $\{Z_n\}_n$ such that*

$$\{Z_n\}_n \sim_\sigma 0$$

is referred to as a zero-distributed sequence. Note that, for any $r \geq 1$, $\{Z_n\}_n \sim_\sigma 0$ is equivalent to $\{Z_n\}_n \sim_\sigma O_r$ (notice that O_m and I_m denote the $m \times m$ zero matrix and the $m \times m$ identity matrix, respectively).

Proposition 1 provides an important characterization of zero-distributed sequences together with a useful sufficient condition for detecting such sequences. Throughout this paper, we use the natural convention $1/\infty = 0$.

Proposition 1. [30] *Let $\{Z_n\}_n$ be a sequence of matrices, with Z_n of size d_n, and let $\|\cdot\|$ be the standard spectral matrix norm (the one induced by the Euclidean vector norm).*

- *$\{Z_n\}_n$ is zero distributed if and only if $Z_n = R_n + N_n$ with $\operatorname{rank}(R_n)/d_n \to 0$ and $\|N_n\| \to 0$ as $n \to \infty$.*
- *$\{Z_n\}_n$ is zero distributed if there exists a $p \in [1,\infty]$ such that $\|Z_n\|_p/(d_n)^{1/p} \to 0$ as $n \to \infty$.*

(Multilevel) Block GLT Matrix Sequences. Now, we give a very concise and operational description of the multilevel block GLT sequences, from which it will be clear that the multilevel block Toeplitz structures, the zero-distributed matrix sequences, and the multilevel block diagonal sampling matrices represent the basic building components. All the material is taken from the books [30,31] and from the papers [28,32].

Let $d, r \geq 1$ be fixed positive integers. A multilevel r-block GLT sequence (or simply a GLT sequence if d, r can be inferred from the context or we do not need/want to specify them) is a special r-block matrix sequence $\{A_n\}_n$ equipped with a measurable function $\kappa : [0,1]^d \times [-\pi,\pi]^d \to \mathbb{C}^{r\times r}$, the so-called symbol. We use the notation $\{A_n\}_n \sim_{\mathrm{GLT}} \kappa$ to indicate that $\{A_n\}_n$ is a GLT sequence with symbol κ. The symbol of a GLT sequence is unique in the sense that if $\{A_n\}_n \sim_{\mathrm{GLT}} \kappa$ and $\{A_n\}_n \sim_{\mathrm{GLT}} \varsigma$ then $\kappa = \varsigma$ a.e. in $[0,1]^d \times [-\pi,\pi]^d$. The main properties of r-block GLT sequences proved in [28] are listed below. If A is a matrix, we denote by $A^\dagger$ the Moore–Penrose pseudoinverse of A (recall that $A^\dagger = A^{-1}$ whenever A is invertible).

GLT 1 If $\{A_n\}_n \sim_{\mathrm{GLT}} \kappa$ then $\{A_n\}_n \sim_\sigma \kappa$. If moreover each A_n is Hermitian, then $\{A_n\}_n \sim_\lambda \kappa$.

GLT 2 We have the following:

- $\{T_n(\phi)\}_n \sim_{\mathrm{GLT}} \kappa(\mathbf{x},\boldsymbol{\theta}) = \phi(\boldsymbol{\theta})$ if $\phi : [-\pi,\pi]^d \to \mathbb{C}^{r\times r}$ is in $L^1([-\pi,\pi]^d)$;
- $\{D_n(a)\}_n \sim_{\mathrm{GLT}} \kappa(\mathbf{x},\boldsymbol{\theta}) = a(\mathbf{x})$ if $a : [0,1]^d \to \mathbb{C}^{r\times r}$ is Riemann integrable;

- $\{Z_n\}_n \sim_{\rm GLT} \kappa(\mathbf{x},\boldsymbol{\theta}) = O_r$ if and only if $\{Z_n\}_n \sim_\sigma 0$ (zero-distributed sequences coincide exactly with the GLT sequences having GLT symbol equal to O_r a.e.).

GLT 3 If $\{A_n\}_n \sim_{\rm GLT} \kappa$ and $\{B_n\}_n \sim_{\rm GLT} \varsigma$, then

- $\{A_n^*\}_n \sim_{\rm GLT} \kappa^*$;
- $\{\alpha A_n + \beta B_n\}_n \sim_{\rm GLT} \alpha\kappa + \beta\varsigma$ for all $\alpha, \beta \in \mathbb{C}$;
- $\{A_n B_n\}_n \sim_{\rm GLT} \kappa\varsigma$;
- $\{A_n^\dagger\}_n \sim_{\rm GLT} \kappa^{-1}$ provided that κ is invertible a.e.

3.2. *Constant Coefficient Case* $\rho \equiv 1$

In the current section, we report the derivation and the formal expression of the symbol for the matrix sequences in the constant coefficient setting $\rho \equiv 1$, according to the standard notion of generating function in the Toeplitz theory (see Section 3.1.2 and the paper by Garoni et al. [29] for more details) and according to the notion of symbol reported in Definition 1.

3.2.1. Symbol Definition

Recall that the computational domain Ω is a hyper-rectangular domain in 3D, and the boundary of Ω comprises six sides. Let $A_n(1, \mathtt{DN}^5)$ be the stiffness matrix obtained with a $\mathbb{Q}_1$ FE approximation with proper boundary conditions, Dirichlet 'D' on one of Ω's sides and Neumann 'N' in the other five sides of Ω (and hence the formal notation $A_n(1, \mathtt{DN}^5)$), where we have chosen a uniform meshing with n intervals in the x_1 direction, n intervals in the x_2 direction, and n intervals in the x_3 direction. According to the previously considered ordering of the nodes, the matrix $A_n(1, \mathtt{DN}^5)$ is a three-level block tridiagonal structure of size n with two-level tridiagonal blocks of size $n+1$, with tridiagonal blocks of size $n+1$, whose elements are small matrices of size 3. We notice that the size is dictated by all the meshpoints, including those in the boundaries when considering Neumann boundary conditions, while only the internal meshpoints are involved when the Dirichlet boundary conditions are enforced.

In accordance with the 2D case [2], if all the boundary conditions are of the Dirichlet type, then we obtain a matrix $A_n(1, \mathtt{D}^6)$, which is a three-level block Toeplitz structure with elementary blocks of size 3, having global dimension $3(n-1)^3$, since all the nodes on the boundaries are not unknowns. More precisely, we obtain

$$A_n(1, \mathtt{D}^6) = \begin{bmatrix} a_0 & a_{-1} & & & \\ a_1 & a_0 & a_{-1} & & \\ & \ddots & \ddots & \ddots & \\ & & a_1 & a_0 & a_{-1} \\ & & & a_1 & a_0 \end{bmatrix} = \text{tridiag}\,(a_i)_{i=-1,0,1},$$

and, for $i = -1, 0, 1$, we have

$$a_i = \begin{bmatrix} a_{i0} & a_{i-1} & & & \\ a_{i1} & a_{i0} & a_{i-1} & & \\ & \ddots & \ddots & \ddots & \\ & & a_{i1} & a_{i0} & a_{i-1} \\ & & & a_{i1} & a_{i0} \end{bmatrix} = \text{tridiag}(a_{ij})_{j=-1,0,1},$$

while, for $i, j = -1, 0, 1$, the following block tridiagonal matrices are defined:

$$a_{ij} = \begin{bmatrix} a_{ij0} & a_{ij-1} & & & \\ a_{ij1} & a_{ij0} & a_{ij-1} & & \\ & \ddots & \ddots & \ddots & \\ & & a_{ij1} & a_{ij0} & a_{ij-1} \\ & & & a_{ij1} & a_{ij0} \end{bmatrix} = \text{tridiag}(a_{ijk})_{k=-1,0,1}.$$

By reading the entries of $A_n(1, \mathrm{D}^6)$, analogously to the process in the 2D setting as in [2], we can compute explicitly the related generating function, following the rules given in Definition 3:

$$\begin{aligned}
f_{\mathbb{Q}_1}(\theta_1, \theta_2, \theta_3) &= f_{a_0}(\theta_1, \theta_2) + f_{a_{-1}}(\theta_1, \theta_2)e^{-\hat{\imath}\theta_3} + f_{a_1}(\theta_1, \theta_2)e^{\hat{\imath}\theta_3} \\
&= \left(f_{a_{00}}(\theta_1) + f_{a_{0-1}}(\theta_1)e^{-\hat{\imath}\theta_2} + f_{a_{01}}(\theta_1)e^{\hat{\imath}\theta_2}\right) \\
&\quad + \left(f_{a_{-10}}(\theta_1) + f_{a_{-1-1}}(\theta_1)e^{-\hat{\imath}\theta_2} + f_{a_{-11}}(\theta_1)e^{\hat{\imath}\theta_2}\right)e^{-\hat{\imath}\theta_3} \\
&\quad + \left(f_{a_{10}}(\theta_1) + f_{a_{1-1}}(\theta_1)e^{-\hat{\imath}\theta_2} + f_{a_{11}}(\theta_1)e^{\hat{\imath}\theta_2}\right)e^{\hat{\imath}\theta_3} \\
&= \left(a_{000} + a_{00-1}e^{-\hat{\imath}\theta_1} + a_{001}e^{\hat{\imath}\theta_1}\right) + \left(a_{0-10} + a_{0-1-1}e^{-\hat{\imath}\theta_1} + a_{0-11}e^{\hat{\imath}\theta_1}\right)e^{-\hat{\imath}\theta_2} \\
&\quad + \left(a_{010} + a_{01-1}e^{-\hat{\imath}\theta_1} + a_{011}e^{\hat{\imath}\theta_1}\right)e^{\hat{\imath}\theta_2} \\
&\quad + \left[\left(a_{-100} + a_{-10-1}e^{-\hat{\imath}\theta_1} + a_{-101}e^{\hat{\imath}\theta_1}\right)\right. \\
&\qquad + \left(a_{-1-10} + a_{-1-1-1}e^{-\hat{\imath}\theta_1} + a_{-1-11}e^{\hat{\imath}\theta_1}\right)e^{-\hat{\imath}\theta_2} \\
&\qquad \left. + \left(a_{-110} + a_{-11-1}e^{-\hat{\imath}\theta_1} + a_{-111}e^{\hat{\imath}\theta_1}\right)e^{\hat{\imath}\theta_2}\right]e^{-\hat{\imath}\theta_3} \\
&\quad + \left[\left(a_{100} + a_{10-1}e^{-\hat{\imath}\theta_1} + a_{101}e^{\hat{\imath}\theta_1}\right) + \left(a_{1-10} + a_{1-1-1}e^{-\hat{\imath}\theta_1} + a_{1-11}e^{\hat{\imath}\theta_1}\right)e^{-\hat{\imath}\theta_2}\right. \\
&\qquad \left. + \left(a_{110} + a_{11-1}e^{-\hat{\imath}\theta_1} + a_{111}e^{\hat{\imath}\theta_1}\right)e^{\hat{\imath}\theta_2}\right]e^{\hat{\imath}\theta_3},
\end{aligned}$$

where every a_{rst}, $r, s, t \in \{-1, 0, 1\}$ is a 3×3 matrix because three degrees of freedom are associated to each node of the mesh. Each a_{rst} is a sum of the building blocks of the elementary matrix (2). More precisely, in terms of the block matrices K_i, cf. definition (A1) in Appendix A, we can write

$$\begin{aligned}
a_{000} &= 2\left(K_1 + K_9 + K_{15} + K_{19}\right), & a_{001} &= a_{00-1} = 2\left(K_2 + K_{16}\right), \\
a_{010} &= a_{0-10} = 2\left(K_3 + K_{11}\right), & a_{011} &= a_{0-1-1} = 2K_4, \\
a_{01-1} &= a_{0-11} = 2K_{10}, & a_{100} &= a_{-100} = 2\left(K_5 + 2K_{13}\right), \\
a_{101} &= a_{-10-1} = 2K_6, & a_{10-1} &= a_{-101} = 2K_{12}, \\
a_{110} &= a_{-1-10} = 2K_7, & a_{1-10} &= a_{-110} = 2K_{17}, \\
a_{111} &= a_{-1-1-1} = K_8, & a_{11-1} &= a_{-1-11} = K_{14}, \\
a_{1-11} &= a_{-11-1} = K_{18}, & a_{-111} &= a_{1-1-1} = K_{20}.
\end{aligned}$$

Thus, we have

$$f_{\mathbb{Q}_1}(\theta_1,\theta_2,\theta_3)=\begin{bmatrix} f_{11}(\theta_1,\theta_2,\theta_3) & f_{12}(\theta_1,\theta_2,\theta_3) & f_{13}(\theta_1,\theta_2,\theta_3) \\ f_{12}(\theta_1,\theta_2,\theta_3) & f_{22}(\theta_1,\theta_2,\theta_3) & f_{23}(\theta_1,\theta_2,\theta_3) \\ f_{13}(\theta_1,\theta_2,\theta_3) & f_{23}(\theta_1,\theta_2,\theta_3) & f_{33}(\theta_1,\theta_2,\theta_3) \end{bmatrix}, \quad (5)$$

where

$$\begin{aligned}
f_{11}(\theta_1,\theta_2,\theta_3) &= 8k_1+8k_3\cos(\theta_1)+2\cos(\theta_2)\left(4k_3+4k_5\cos(x)\right) \\
&\quad +2\cos(\theta_3)\Big(4k_6+2\cos(\theta_2)(2k_7+2k_9\cos(\theta_1))+4k_7\cos(\theta_1)\Big), \\
f_{12}(\theta_1,\theta_2,\theta_3) &= 4k_2+4k_8+4\cos(\theta_1)(k_4+k_{10}) \\
&\quad +2e^{-i\theta_3}\Big(e^{i\theta_2}(k_2+k_4\cos(\theta_1))+e^{-i\theta_2}(k_8+k_{10}\cos(\theta_1))\Big) \\
&\quad +2e^{i\theta_3}\Big(e^{i\theta_2}(k_8+k_{10}\cos(\theta_1))+e^{-i\theta_2}(k_2+k_4\cos(\theta_1))\Big), \\
f_{13}(\theta_1,\theta_2,\theta_3) &= 4k_2+4k_8+8k_0\cos(\theta_1)+4\cos(\theta_2)(k_4+k_{10}+2k_0\cos(\theta_1)) \\
&\quad +2e^{-i\theta_3}\Big(k_2e^{i\theta_1}+k_8e^{-i\theta_1}+\left(k_4e^{i\theta_1}+k_{10}e^{-i\theta_1}\right)\cos(\theta_2)\Big) \\
&\quad +2e^{i\theta_3}\Big(k_2e^{-i\theta_1}+k_8e^{i\theta_1}+\left(k_4e^{-i\theta_1}+k_{10}e^{i\theta_1}\right)\cos(\theta_2)\Big), \\
f_{22}(\theta_1,\theta_2,\theta_3) &= 8k_1+8k_3\cos(\theta_1)+8\cos(\theta_2)(k_6+k_7\cos(\theta_1)) \\
&\quad +8\cos(\theta_3)\Big(k_3+k_5\cos(\theta_1)+\cos(\theta_2)(k_7+k_9\cos(\theta_1))\Big), \\
f_{23}(\theta_1,\theta_2,\theta_3) &= 4k_2+4k_8+2e^{-i\theta_2}\left(k_2e^{i\theta_1}+k_8e^{-i\theta_1}\right)+2e^{i\theta_2}\left(k_2e^{-i\theta_1}+k_8e^{i\theta_1}\right) \\
&\quad +2\cos(\theta_3)\Big(2k_4+2k_{10}+4k_0\cos(\theta_1)+4k_0\cos(\theta_2) \\
&\qquad +e^{-i\theta_2}\left(k_4e^{i\theta_1}+k_{10}e^{-i\theta_1}\right)+e^{i\theta_2}\left(k_4e^{-i\theta_1}+k_{10}e^{i\theta_1}\right)\Big), \\
f_{33}(\theta_1,\theta_2,\theta_3) &= 8k_1+8k_6\cos(\theta_1)+8\cos(\theta_2)(k_3+k_7\cos(\theta_1)) \\
&\quad +8\cos(\theta_3)\Big(k_3+k_7\cos(\theta_1)+\cos(\theta_2)(k_5+k_9\cos(\theta_1))\Big).
\end{aligned}$$

Finally, by expanding the k_is according to expressions (A2), we deduce the formal expression of the generating functions

$$\begin{aligned}
f_{11}(\theta_1,\theta_2,\theta_3) &= 2\cos(\theta_3)\Bigg[\cos(\theta_2)\left(\frac{2\nu}{3(2\nu-1)}+2\cos(\theta_1)\left(\frac{\nu}{3(2\nu-1)}-\frac{1}{9}\right)-\frac{4}{9}\right) \\
&\qquad +\cos(\theta_1)\left(\frac{2\nu}{3(2\nu-1}-\frac{4}{9}\right)-\frac{8}{9}\Bigg] \\
&\quad +\frac{8}{9}\cos(\theta_1)-\frac{16\nu}{3(2\nu-1)} \\
&\quad +\cos(\theta_2)\left(2\cos(\theta_1)\left(\frac{2\nu}{3(2\nu-1)}+\frac{2}{9}\right)+\frac{8}{9}\right)+\frac{16}{9}, \\
f_{22}(\theta_1,\theta_2,\theta_3) &= 8\cos(\theta_3)\Bigg[\cos(\theta_2)\left(\frac{\nu}{6(2\nu-1)}+\cos(\theta_1)\left(\frac{\nu}{6(2\nu-1)}-\frac{1}{18}\right)-\frac{1}{9}\right) \\
&\qquad +\cos(\theta_1)\left(\frac{\nu}{6(2\nu-1)}+\frac{1}{18}\right)+\frac{1}{9}\Bigg] \\
&\quad +\frac{8}{9}\cos(\theta_1)-\frac{16\nu}{3(2\nu-1)} \\
&\quad +8\cos(\theta_2)\left(\cos(\theta_1)\left(\frac{\nu}{6(2\nu-1)}-\frac{1}{9}\right)-\frac{2}{9}\right)+\frac{16}{9},
\end{aligned}$$

$$\begin{aligned}
f_{33}(\theta_1,\theta_2,\theta_3) &= 8\cos(\theta_3)\left[\cos(\theta_2)\left(\frac{\nu}{6(2\nu-1)}+\cos(\theta_1)\left(\frac{\nu}{6(2\nu-1)}-\frac{1}{18}\right)+\frac{1}{18}\right)\right.\\
&\quad\left.+\cos(\theta_1)\left(\frac{\nu}{6(2\nu-1)}-\frac{1}{9}\right)+\frac{1}{9}\right]\\
&\quad-\frac{16\nu}{3(2\nu-1)}-\frac{16}{9}\cos(\theta_1)+\frac{16}{9}\\
&\quad+8\cos(\theta_2)\left(\cos(\theta_1)\left(\frac{\nu}{6(2\nu-1)}-\frac{1}{9}\right)+\frac{1}{9}\right),\\
f_{12}(\theta_1,\theta_2,\theta_3) &= -\frac{4}{3(2\nu-1)}\Big(\nu\sin(\theta_2)\sin(\theta_3)\big(\cos(\theta_1)+2\big)\Big),\\
f_{13}(\theta_1,\theta_2,\theta_3) &= -\frac{4}{3(2\nu-1)}\Big(\nu\sin(\theta_1)\sin(\theta_3)\big(\cos(\theta_2)+2\big)\Big),\\
f_{23}(\theta_1,\theta_2,\theta_3) &= -\frac{4}{3(2\nu-1)}\Big(\nu\sin(\theta_1)\sin(\theta_2)\big(\cos(\theta_3)+2\big)\Big).
\end{aligned}$$

The following proposition links in a precise way the involved Toeplitz structures and matrices $A_n(1,\mathtt{D}^6)$ and $A_n(1,\mathtt{DN}^5)$.

Proposition 2. *Let $f_{\mathbb{Q}_1}(\theta_1,\theta_2,\theta_3)$ be the symbol defined in the previous lines; see (5). Let us consider a uniform meshing in all the three directions with n subintervals. Then we have the following relationships:*

$$A_n(1,\mathrm{D}^6) = T_{\boldsymbol{n}}(f_{\mathbb{Q}_1}),\quad \boldsymbol{n}=(n_1,n_2,n_3),\ n_1=n_2=n_3=n-1,$$

$$A_n(1,\mathrm{DN}^5) = T_{\boldsymbol{n}}(f_{\mathbb{Q}_1})+R_{\boldsymbol{n}},\quad \boldsymbol{n}=(n_1,n_2),\ n_1=n,n_2=n_3=n+1,$$

where $R_{\boldsymbol{n}}$ has rank of order $n^2=o(\mathrm{size}(R_{\boldsymbol{n}}))$, with $\mathrm{size}(R_{\boldsymbol{n}})=\mathrm{size}(A_n(1,\mathrm{DN}^5))=n_1n_2n_3$.

If the more general setting is considered with n_1 subintervals in the x_1 direction, n_2 subintervals in the x_2 direction, and n_3 intervals in the x_3 direction, then the analogous is true:

$$A_{\boldsymbol{n}}(1,\mathrm{D}^6) = T_{\boldsymbol{n}}(f_{\mathbb{Q}_1}),\quad \boldsymbol{n}=(n_1-1,n_2-1,n_3-1),$$

$$A_{\boldsymbol{n}}(1,\mathrm{DN}^5) = T_{\boldsymbol{n}}(f_{\mathbb{Q}_1})+R_{\boldsymbol{n}},\quad \boldsymbol{n}=(n_1,n_2+1,n_3+1),$$

where $R_{\boldsymbol{n}}$ has a rank proportional to $n_1n_2+n_1n_3+n_2n_3 = o(\mathrm{size}(R_{\boldsymbol{n}}))$, with $\mathrm{size}(R_{\boldsymbol{n}}) = \mathrm{size}(A_n(1,\mathrm{DN}^5))=n_1n_2n_3$.

Proof. Since the grid points in a cube are $n_1n_2n_3$, if we consider Dirichlet boundary conditions in a facet orthogonal, for example, to the direction x_1, then the number of points and equations which are affected by the Neumann boundary conditions on the other 5 facets are $2l_3n_1n_2+2l_2n_1n_3+l_1n_2n_3$, where typically $l_3=l_2=l_1=2$ if the standard linear FEs are employed. Hence, the rank correction induced by the matrix $R_{\boldsymbol{n}}$ is proportional, as claimed to $n_1n_2+n_1n_3+n_2n_3=o(\mathrm{size}(R_{\boldsymbol{n}}))$. □

In the following section, we prove that the function $f_{\mathbb{Q}_1}$ is the eigenvalue symbol, in the sense of Definition 1, of the matrix sequences $\{A_n(1,\mathtt{D}^6)\}_n$ and $\{A_n(1,\mathtt{DN}^5)\}_n$.

3.2.2. Symbol Spectral Analysis in 3D: Distribution, Extremal Eigenvalues, and Conditioning

We study the spectral distribution, extremal eigenvalues, and conditioning of our matrix sequences in 3D. All the results are based on the symbol and on its analytical features.

Proposition 3. *Let $f_{\mathbb{Q}_1}(\theta_1, \theta_2, \theta_3)$ be the symbol defined in Section 3.2.1 According to the notation in Proposition 2, all the matrix sequences $\{T_{\boldsymbol{n}}(f_{\mathbb{Q}_1})\}_{\boldsymbol{n}}$, $\{A_n(1, \mathrm{D}^6)\}_n$, $\{A_{\boldsymbol{n}}(1, \mathrm{D}^6)\}_{\boldsymbol{n}}$, $\{A_n(1, \mathrm{DN}^5)\}_n$, $\{A_{\boldsymbol{n}}(1, \mathrm{DN}^5)\}_{\boldsymbol{n}}$ are spectrally distributed as $f_{\mathbb{Q}_1}$ in the sense of Definition 1.*

Proof. For the multilevel block Toeplitz sequences $\{T_{\boldsymbol{n}}(f_{\mathbb{Q}_1})\}_{\boldsymbol{n}}$, $\{A_n(1, \mathtt{D}^6)\}_n$, and $\{A_{\boldsymbol{n}}(1, \mathtt{D}^6)\}_{\boldsymbol{n}}$ refer to Item **GLT 2.**, part 1, and Item **GLT 1.**, part 2 in Section 3.1.2. For the sequences $\{A_n(1, \mathtt{DN}^5)\}_n$, $\{A_{\boldsymbol{n}}(1, \mathtt{DN}^5)\}_{\boldsymbol{n}}$ first observe that the matrix sequence $\{R_{\boldsymbol{n}}\}_{\boldsymbol{n}}$ defined in Proposition 2 is zero distributed, thanks to Proposition 1 and Proposition 2 since $\mathrm{rank}(R_{\boldsymbol{n}})/\mathrm{size}(R_{\boldsymbol{n}}) \to 0$, as $\boldsymbol{n} \to \infty$. Then the claim follows thanks to the $*$-algebra structure of the GLT sequences and more specifically thanks to Item **GLT 3.**, part 2, Item **GLT 2.**, part 1, and Item **GLT 1.**, part 2. □

Now, we identify a few key analytical features of the spectral symbol which are shared by all the matrix sequences mentioned in Proposition 3. Theorem 1 is crucial for the study of the extremal eigenvalues and of the conditioning of the same matrix sequences and, in Section 5, it is the main ingredient for designing ad hoc multigrid solvers when dealing with the associated large linear systems.

Theorem 1. *Let $f_{\mathbb{Q}_1}(\theta_1, \theta_2, \theta_3)$ be the symbol defined in (5). The following statements hold true:*

1. $f_{\mathbb{Q}_1}(0,0,0)\mathbf{e} = 0$, $\mathbf{e} = [1,\ 1,\ 1]^T$;
2. *All three eigenvalues of $f_{\mathbb{Q}_1}$ have a zero of order 2 at $(0,0,0)$.*

Proof. Claim 1. The function $f_{\mathbb{Q}_1}$ evaluated at $(0,0,0)$ equals

$$f_{\mathbb{Q}_1}(0,0,0) = \begin{bmatrix} \alpha & \beta & \beta \\ \beta & \alpha & \beta \\ \beta & \beta & \alpha \end{bmatrix}$$

with $\alpha = 8(k_1 + 2k_3 + k_5 + k_6 + 2k_7 + k_9)$ and $\beta = 8(4k_0 + k_2 + k_4 + k_8 + k_{10})$, whose row-sum $\alpha + 2\beta = 0$ according to (A2).

Claim 2. It is just a direct check. Indeed, it is enough to check that the quantities

$$\mathrm{tr}(f_{\mathbb{Q}_1}(\theta_1, \theta_2, \theta_3)) \text{ and } \det(f_{\mathbb{Q}_1}(\theta_1, \theta_2, \theta_3))$$

have a zero of orders two and six, respectively, by considering the Taylor expansion centered at $(0,0,0)$. □

In accordance to what was proven in the 2D setting, the minimal eigenvalue of the symbol $f_{\mathbb{Q}_1}(\theta_1, \theta_2, \theta_3)$ has a unique zero of order two at $(\theta_1, \theta_2, \theta_3) = (0,0,0)$. In fact, the symbol $f_{\mathbb{Q}_1}$ is positive semi-definite, and all the eigenvalues of the symbol $f_{\mathbb{Q}_1}(\theta_1, \theta_2, \theta_3)$ have a unique zero of order two at $(\theta_1, \theta_2, \theta_3) = (0,0,0)$, that is, there exist positive constants $c(j), C(j), j = 1,2,3$ such that

$$c(j)\|(\theta_1, \theta_2, \theta_3)^T\|^2 \le \lambda_j(f_{\mathbb{Q}_1}(\theta_1, \theta_2, \theta_3)) \le C(j)\|(\theta_1, \theta_2, \theta_3)^T\|^2$$

uniformly in a proper neighborhood of $(0,0,0)$.

Therefore, we can infer important information on the extremal eigenvalues and on the conditioning of the corresponding matrix sequences.

Proposition 4. *Let $f_{\mathbb{Q}_1}(\theta_1, \theta_2, \theta_3)$ be the symbol defined in (5). Let us consider a uniform meshing in all the directions with n subintervals. Then we have the following relationships:*

$$\begin{aligned}
&\lambda_{\min}(A_n(1, \mathrm{D}^6)) \sim n^{-2}, \\
&\max f_{\mathbb{Q}_1} - \lambda_{\max}(A_n(1, \mathrm{D}^6)) \sim n^{-2}, \\
&\mu(A_n(1, \mathrm{D}^6)) \sim n^2, \\
&A_n(1, \mathrm{D}^6) = T_{\boldsymbol{n}}(f_{\mathbb{Q}_1}), \ \boldsymbol{n} = (n_1, n_2, n_3), \ n_1 = n_2 = n_3 = n - 1
\end{aligned}$$

and

$$\begin{aligned}
&\lambda_{\min}(A_n(1, \mathrm{DN}^5)) \sim n^{-2}, \\
&\max f_{\mathbb{Q}_1} - \lambda_{\max}(A_n(1, \mathrm{DN}^5)) \sim n^{-2}, \\
&\mu(A_n(1, \mathrm{DN}^5)) \sim n^2, \\
&A_n(1, \mathrm{DN}^5) = T_{\boldsymbol{n}}(f_{\mathbb{Q}_1}) + R_{\boldsymbol{n}}, \quad \boldsymbol{n} = (n_1, n_2, n_3), \ n_1 = n, n_2 = n_3 = n + 1,
\end{aligned}$$

with $R_{\boldsymbol{n}}$ as in Proposition 2 and $\mu(\cdot)$ denoting the condition number in spectral norm (the one induced by the Euclidean vector norm).

If the more general setting is considered with n_1 subintervals in the x_1 direction, n_2 subintervals in the x_2 direction, and n_3 subintervals in the x_3 direction, then analogous relations are true:

$$\begin{aligned}
&\lambda_{\min}(A_{\boldsymbol{n}}(1, \mathrm{D}^6)) \sim [\min n_j]^{-2}, \\
&\max f_{\mathbb{Q}_1} - \lambda_{\max}(A_{\boldsymbol{n}}(1, \mathrm{D}^6)) \sim [\min n_j]^{-2}, \\
&\mu(A_{\boldsymbol{n}}(1, \mathrm{D}^6)) \sim n^2, \\
&A_{\boldsymbol{n}}(1, \mathrm{D}^6) = T_{\boldsymbol{n}}(f_{\mathbb{Q}_1}), \ \boldsymbol{n} = (n_1 - 1, n_2 - 1, n_3 - 1)
\end{aligned}$$

and

$$\begin{aligned}
&\lambda_{\min}(A_{\boldsymbol{n}}(1, \mathrm{DN}^5)) \sim [\min n_j]^{-2}, \\
&\max f_{\mathbb{Q}_1} - \lambda_{\max}(A_{\boldsymbol{n}}(1, \mathrm{DN}^5)) \sim [\min n_j]^{-2}, \\
&\mu(A_{\boldsymbol{n}}(1, \mathrm{DN}^5)) \sim [\min n_j]^2, \\
&A_{\boldsymbol{n}}(1, \mathrm{DN}^5) = T_{\boldsymbol{n}}(f_{\mathbb{Q}_1}) + R_{\boldsymbol{n}}, \quad \boldsymbol{n} = (n_1, n_2 + 1, n_3 + 1),
\end{aligned}$$

with $R_{\boldsymbol{n}}$ as in Proposition 2.

3.3. Non-Constant Coefficient ρ Case—3D

It should be observed that the natural extension of the previous analysis refers to the case of a non-constant coefficient ρ. According to a standard assembling procedure in FEs, given the triangulation $\mathcal{T}$ made by all the basic elements τ, we write the stiffness matrix as

$$A_n(\rho) = \sum_{\tau \in \mathcal{T}} \rho_\tau A^{El}_{n,\tau}, \tag{6}$$

where $A^{El}_{n,\tau}$ is the elementary matrix K_n in (2) (possibly properly cut when nodes on the boundary are involved), but widened to size $N(\boldsymbol{n})$ according to the chosen global ordering of nodes. Here, $\boldsymbol{n} = (n-1, n-1, n-1)$ in the case of the Dirichlet boundary conditions and n subintervals in all the directions such that $A_n(\rho) = A_n(\rho, \mathrm{D}^6)$, while $\boldsymbol{n} = (n, n+1, n+1)$ in the case of Dirichlet boundary conditions on one-side Neumann boundary conditions in the remaining ones with n subintervals in all the directions so that $A_n(\rho) = A_n(\rho, \mathrm{DN}^5)$.

Clearly, the elementary matrix K_e is positive semi-definite for $0 \leq \nu < 1/2$. More precisely, the not zero eigenvalues are

$$1 \text{ (double)}, \quad \frac{(1-\nu)}{3(1-2\nu)} \text{ (triple)}, \quad \frac{2\nu}{(1-2\nu)} \text{ (triple)}, \quad \frac{\nu}{3(1-2\nu)} \text{ (double)},$$
$$\frac{4\nu}{3(1-2\nu)} \text{ (simple)}, \quad \frac{(\nu+1)}{(1-2\nu)} \text{ (simple)}, \quad \frac{(\nu+1)}{3(1-2\nu)} \text{ (triple)}, \quad \frac{(\nu+1)}{9(1-2\nu)} \text{ (triple)}.$$

In addition, every elementary matrix, which has been cut due to nodes on the boundary, is positive semi-definite as well since it is a principal submatrix of K_e in (2).

Thus, on the basis of (6) and by applying the Courant–Fisher theorem, we can claim again

$$\rho_{\min}\lambda_{\min}(A_n(1)) \leq \lambda_{\min}(A_n(\rho)) \leq \rho_{\max}\lambda_{\min}(A_n(1)),$$
$$\rho_{\min}\lambda_{\max}(A_n(1)) \leq \lambda_{\max}(A_n(\rho)) \leq \rho_{\max}\lambda_{\max}(A_n(1)).$$

with $\rho_{\min}$ and $\rho_{\max}$ minimum and maximum of ρ, respectively (see Figure 1 for an upper-bound of the maximal eigenvalues of $A_n(1)$ as a function of ν).

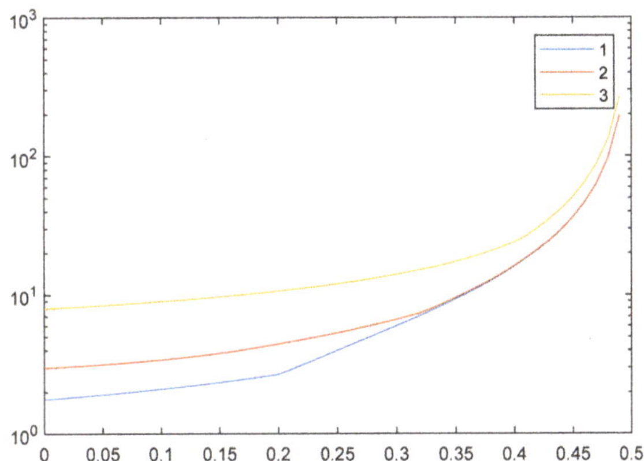

Figure 1. Maxima of eigenvalue surfaces of the symbol $f_{\mathbb{Q}_1}$ vs. ν.

By combining the above inequalities and Proposition 4, we infer that the extremal eigenvalues and the conditioning have the same asymptotical behavior as in the constant coefficient setting.

The whole eigenvalues distribution is sketched below by referring to the basics of the GLT theory [29] reported in Section 3.1.2. Let $D_n(\rho)$ be a multilevel block diagonal sampling matrix according to the notions introduced in Section 3.1.2 and let $A_n(1)$ be the multilevel block Toeplitz matrix $T_{\boldsymbol{n}}(f_{\mathbb{Q}_1})$ if the Dirichlet boundary conditions are used or $T_{\boldsymbol{n}}(f_{\mathbb{Q}_1}) + R_{\boldsymbol{n}}$ in the other case. Then the following facts hold:

Fact 1 $\{D_n(\rho)\}_n \sim_{\rm GLT} \rho$ according to Item **GLT 2.**

Fact 2 $\{R_{\boldsymbol{n}}\}_n \sim_{\rm GLT} 0$ according to Proposition 2, Proposition 1, and Item **GLT 2.**

Fact 3 $\{T_{\boldsymbol{n}}(f_{\mathbb{Q}_1})\}_n \sim_{\rm GLT} f_{\mathbb{Q}_1}$ according to Item **GLT 2.**

Fact 4 $\{A_n(1)\}_n \sim_{\rm GLT} f_{\mathbb{Q}_1}$ according to Fact 2, Fact 3, and to the $*$-algebra structure of GLT sequences that is Item **GLT 3.**

Fact 5 given $\Delta_n = A_n(\rho) - D_n(\rho)A_n(1)$ a simple check shows that $\{\Delta_n\}_n \sim_{\rm GLT} 0$, using Proposition 1.

Fact 6 $\{A_n(\rho)\}_n \sim_{\rm GLT} \rho\, f_{\mathbb{Q}_1}$ as a consequence of Fact5, Fact 1, Fact 4, and of the $*$-algebra structure of GLT sequences that is Item **GLT 3.**; moreover since the matrix sequence $\{A_n(\rho)\}_n$ is made up by Hermitian matrices, by Item **GLT 1.** it follows that $\{A_n(\rho)\}_n \sim_{\sigma,\lambda} \rho\, f_{\mathbb{Q}_1}$.

We stress that a unique notation has been employed for the sake of notational compactness. However, the main statement, Fact 6, holds for all the matrix sequences reported in Proposition 3 in the case of a variable ρ.

Finally, it is worth noticing that non-square domains can be treated as well, using the reduced GLT theory (see pages 398–399 in [33], Section 3.1.4 in [34], and the recent analysis [35]): here, we do not go in the details and this will be the subject of future investigations.

4. Two-Grid and Multigrid Methods

In this section, we concisely report few relevant results concerning the convergence theory of algebraic multigrid methods with special attention of the case of multilevel block Toeplitz structures generated by a matrix-valued symbol f.

We start by taking into consideration the generic linear system $A_m x_m = b_m$ with large dimension m, where $A_m \in \mathbb{C}^{m\times m}$ is a Hermitian positive definite matrix and $x_m, b_m \in \mathbb{C}^m$. Let $m_0 = m > m_1 > \ldots > m_s > \ldots > m_{s_{\min}}$ and let $P_s^{s+1} \in \mathbb{C}^{m_{s+1}\times m_s}$ be a full-rank matrix for any s. At last, let us denote by $\mathcal{V}_s$ a class of stationary iterative methods for given linear systems of dimension m_s.

In accordance with [36], the algebraic two-grid method (TGM) can be easily seen as a stationary iterative method whose generic steps are reported below.

$$x_s^{\mathrm{out}} = \mathrm{TGM}(s, x_s^{\mathrm{in}}, b_s)$$

Step	Description
$x_s^{\mathrm{pre}} = \mathcal{V}_{s,\mathrm{pre}}^{\nu_{\mathrm{pre}}}(x_s^{\mathrm{in}}, b_s)$	Pre-smoothing iterations
$r_s = A_s x_s^{\mathrm{pre}} - b_s$ $r_{s+1} = P_{m_s}^{m_{s+1}} r_s$ $A_{s+1} = P_{m_s}^{m_{s+1}} A_s (P_{m_s}^{m_{s+1}})^H$ Solve $A_{s+1} y_{s+1} = r_{s+1}$ $\hat{x}_s = x_s^{\mathrm{pre}} - (P_{m_s}^{m_{s+1}})^H y_{s+1}$	Exact coarse grid correction (CGC)
$x_s^{\mathrm{out}} = \mathcal{V}_{s,\mathrm{post}}^{\nu_{\mathrm{post}}}(\hat{x}_s, b_s)$	Post-smoothing iterations

Here, we refer to the dimension m_s by means of its subscript s.

In the first and last steps, a *pre-smoothing iteration* and a *post-smoothing iteration* are applied ν_{pre} times and ν_{post} times, respectively, taking into account the considered stationary iterative method in the class $\mathcal{V}_s$. Furthermore, the intermediate steps define the *exact coarse grid correction operator*, which is dependent on the considered projector operator P_{s+1}^s. The resulting iteration matrix of the TGM is then defined as

$$\begin{aligned} TGM_s &= V_{s,\mathrm{post}}^{\nu_{\mathrm{post}}} CGC_s V_{s,\mathrm{pre}}^{\nu_{\mathrm{pre}}}, \\ CGC_s &= I^{(s)} - (P_{m_s}^{m_{s+1}})^H A_{s+1}^{-1} P_{m_s}^{m_{s+1}} A_s \\ A_{s+1} &= P_{m_s}^{m_{s+1}} A_s (P_{m_s}^{m_{s+1}})^H, \end{aligned}$$

where $V_{s,\mathrm{pre}}$ and $V_{s,\mathrm{post}}$ represent the pre-smoothing and post-smoothing iteration matrices, respectively, and $I^{(s)}$ is the identity matrix at the sth level.

By employing a recursive procedure, the TGM leads to a multigrid method (MGM). Indeed, the standard V-cycle can be expressed in the following way:

$$x_s^{\text{out}} = \text{MGM}(s, x_s^{\text{in}}, b_s)$$

```
if s ≤ s_min then
```

Solve $A_s x_s^{\text{out}} = b_s$	Exact solution

```
else
```

$x_s^{\text{pre}} = \mathcal{V}_{s,\text{pre}}^{\nu_{\text{pre}}}(x_s^{\text{in}}, b_s)$	Pre-smoothing iterations
$r_s = A_s x_s^{\text{pre}} - b_s$ $r_{s+1} = P_{m_s}^{m_{s+1}} r_s$ $y_{s+1} = \text{MGM}(s+1, \mathbf{0}_{s+1}, r_{s+1})$ $\hat{x}_s = x_s^{\text{pre}} - (P_{m_s}^{m_{s+1}})^H y_{s+1}$	Coarse grid correction
$x_s^{\text{out}} = \mathcal{V}_{s,\text{post}}^{\nu_{\text{post}}}(\hat{x}_s, b_s)$	Post-smoothing iterations

From a computational viewpoint, to reduce the related costs, it is more efficient that the matrices $A_{s+1} = P_s^{s+1} A_s (P_s^{s+1})^H$ are computed in the so-called *setup phase*.

According to the previous setting, the global iteration matrix of the MGM is recursively defined as

$$\begin{aligned} MGM_{s_{\min}} &= O \in \mathbb{C}^{s_{\min} \times s_{\min}}, \\ MGM_s &= V_{s,\text{post}}^{\nu_{\text{post}}} \left[I^{(s)} - (P_{m_s}^{m_{s+1}})^H \left(I^{(s+1)} - MGM_{s+1} \right) A_{s+1}^{-1} P_{m_s}^{m_{s+1}} A_s \right] V_{s,\text{pre}}^{\nu_{\text{pre}}}, \\ & \qquad s = s_{\min} - 1, \ldots, 0. \end{aligned}$$

Lastly, the W-cycle is just a variation of the previous V-cycle considering two recursive calls in the coarse grid correction as follows:

$$x_s^{\text{out}} = \text{MGMW}(s, x_s^{\text{in}}, b_s)$$

```
if s ≤ s_min then
```

Solve $A_s x_s^{\text{out}} = b_s$	Exact solution

```
else
```

$x_s^{\text{pre}} = \mathcal{V}_{s,\text{pre}}^{\nu_{\text{pre}}}(x_s^{\text{in}}, b_s)$	Pre-smoothing iterations
$r_s = A_s x_s^{\text{pre}} - b_s$ $r_{s+1} = P_{m_s}^{m_{s+1}} r_s$ $y_{s+1}^{(0)} = \mathbf{0}_{s+1}$ for $\mu = 1:2$ $y_{s+1}^{(\mu)} = \text{MGMW}(s+1, y_{s+1}^{(\mu-1)}, r_{s+1})$ $\hat{x}_s = x_s^{\text{pre}} - (P_{m_s}^{m_{s+1}})^H y_{s+1}^{(2)}$	Coarse grid correction
$x_s^{\text{out}} = \mathcal{V}_{s,\text{post}}^{\nu_{\text{post}}}(\hat{x}_s, b_s)$	Post-smoothing iterations

In the following remark, we emphasize the relevant computational properties of the considered methods.

Remark 1. *The first part of the current remark concerns the computational cost. In the V-cycle, there is just one recursive call, while in the W-cycle, there are two recursive calls, which in principle, is more expensive. Let us analyze the related costs. Since our matrices are sparse and they remain* *sparse at the lower levels, the number of arithmetic operations in the V-cycle and W-cycle algorithms is of the type*

$$C_V(m) \le C_V(m/q) + \alpha m, \qquad C_W(m) \le 2C_W(m/q) + \alpha m,$$

where $m = m_0$ *and* $m_{s+1} = m_s/q$*. Now in the one-dimensional setting* $q = 2$ *and* $C_V(m) = O(m)$*, while* $C_W(m) = O(m \log m)$ *and thus the W-cycle is asymptotically more expensive. Conversely, in a d-dimensional setting with* $d \ge 2$*, we have* $q = 2^d$ *and, as a consequence of the recursive relations, both* $C_V(m)$ *and* $C_W(m)$ *are linear in the matrix-size m. To be precise, we have*

$$C_V(m) \le \frac{2^d}{2^d - 1}\alpha m, \qquad C_W(m) \le \frac{2^{d-1}}{2^{d-1} - 1}\alpha m,$$

so that, in the current three-dimensional setting, the bounds of the V-cycle and W-cycle complexity are very close, i.e.,

$$C_V(m) \le \frac{8}{7}\alpha m, \qquad C_W(m) \le \frac{4}{3}\alpha m.$$

Another issue to be considered is the case where the discretization matrices appear in checkerboard fashion (see, for example, [37] and the references therein). We emphasize that our analysis, which is based on the Toeplitz generating function, is not directly applicable since we lose the Toeplitz character of the approximation matrices when a checkerboard (called also red–black) ordering is used. However, this is just a matter of a similarity transformation by a permutation matrix. Hence, with a careful work and without increasing the complexity analyzed in the previous lines, the algorithm discussed here and the related theoretical analysis can be adapted to the checkerboard context.

Remark 2. *In the relevant literature (see, for instance, [38]), the convergence analysis of the TGM splits into checking two separate conditions: the smoothing property and the approximation property. Regarding the latter and regarding scalar structured matrices [38,39], the TGM optimality is given in terms of choosing the proper conditions that the symbol p of a family of projection operators has to fulfill. Indeed, let* $T_n(f)$ *with* $n = (2^t - 1)$*, where f is a non-negative trigonometric polynomial. Let* θ^0 *be the unique zero of f. Then the TGM optimality applied to* $T_n(f)$ *is guaranteed if we choose the symbol p of the family of projection operators such that*

$$\begin{aligned} &\limsup_{\theta \to \theta^0} \frac{|p(\eta)|^2}{f(\theta)} < \infty, \ \eta \in \mathcal{M}(\theta), \\ &\sum_{\eta \in \Omega(\theta)} |p(\eta)|^2 > 0, \end{aligned} \tag{7}$$

where, for $d = 1$*, the sets* $\Omega(\theta)$ *and* $\mathcal{M}(\theta)$ *are the following corner and mirror points*

$$\Omega(\theta) = \{\eta \in \{\theta, \theta + \pi\}\}, \qquad \mathcal{M}(\theta) = \Omega(\theta) \setminus \{\theta\},$$

respectively. In the general case of $d > 1$*, we have*

$$\Omega(\boldsymbol{\theta}) = \{\boldsymbol{\eta} \in \{\boldsymbol{\theta} + \pi \mathbf{s}\}, \mathbf{s} = (s_1, \ldots, s_d), s_j \in \{0,1\}, j = 1, \ldots, d\}$$

with $\mathcal{M}(\boldsymbol{\theta}) = \Omega(\boldsymbol{\theta}) \setminus \{\boldsymbol{\theta}\}$*, so that the cardinality of* $\Omega(\boldsymbol{\theta})$ *and* $\mathcal{M}(\boldsymbol{\theta})$ *is* 2^d *and* $2^d - 1$*, respectively.*

Informally, for $d = 1$, it means that the optimality of the two-grid method is obtained by choosing the family of projection operators associated to a symbol p such that $|p|^2(\vartheta) + |p|^2(\vartheta + \pi)$ does not have zeros and $|p|^2(\vartheta + \pi)/f(\vartheta)$ is bounded (if we require the optimality of the V-cycle, then the second condition is a bit stronger); see [38]. For approximation of an elliptic differential operator of order 2α by local methods (e.g., finite differences, FEs, and isogeometric analysis), the previous conditions mean that p has a unique zero of order at least α at $\vartheta = \pi$ whenever f has a unique zero at $\theta^0 = 0$ of order 2α.

In our specific block setting, by interpreting the analysis given in [40], all the involved symbols are matrix valued, and the conditions which generalize (7) and are sufficient for the TGM convergence and optimality are the following:

(A) Zero of order 2 at all the mirror points of the eigenvalue functions of the symbol of the projector for our matrix sequences having common symbol $f_{\mathbb{Q}_1}$ (mirror point theory [38,39]);

(B) Positive definiteness of $\sum_{\eta \in \Omega(\theta)} pp^H(\eta)$;

(C) Commutativity of all $p(\eta)$ for η varying in the corner points.

Even if the theoretical extension to the V-cycle and W-cycle convergence and optimality is not given, in the subsequent section, we propose specific choices of the projection operators numerically showing how this leads to two-grid, V-cycle, and W-cycle procedures converging optimally or quasi-optimally with respect to all the relevant parameters (size, dimensionality, and polynomial degree k).

Our choices are in agreement with the mathematical conditions set in items **(A)**, **(B)**, and **(C)**. We remark that the violation of condition **C)** is discussed by Donatelli et al. [40], in their conclusion section.

5. Multigrid Proposals

Based on the theory in Section 4, we define the prolongation operator as

$$P_{2h}^{h} = P \otimes P \otimes P \otimes I_3/\sqrt{2}$$

in the case of matrices $A_n(1, \mathtt{D}^6)$, and

$$P_{2h}^{h} = P_c \otimes P_t \otimes P_t \otimes I_3/\sqrt{2}$$

in the case of matrices $A_n(1, \mathtt{DN}^5)$. In the expressions above,

$$P = \begin{bmatrix} \frac{1}{2} & & & \\ 1 & & & \\ \frac{1}{2} & \frac{1}{2} & & \\ & 1 & & \\ & \frac{1}{2} & \ddots & \frac{1}{2} \\ & & & 1 \\ & & & \frac{1}{2} \end{bmatrix}, \quad P_t = \begin{bmatrix} 1 & & & & \\ \frac{1}{2} & \frac{1}{2} & & & \\ & 1 & & & \\ & \frac{1}{2} & \frac{1}{2} & & \\ & & 1 & & \\ & & \frac{1}{2} & \ddots & \frac{1}{2} & \\ & & & & 1 & \\ & & & & \frac{1}{2} & \frac{1}{2} \\ & & & & & 1 \end{bmatrix}, \text{ and } P_c = \begin{bmatrix} \frac{1}{2} & & & \\ 1 & & & \\ \frac{1}{2} & \frac{1}{2} & & \\ & 1 & & \\ & \frac{1}{2} & \ddots & \frac{1}{2} \\ & & & 1 \\ & & & \frac{1}{2} & \frac{1}{2} \\ & & & & 1 \end{bmatrix}.$$

Note that $A_{2h} = P^T A_h P$ equals the same FEM approximation on the coarse mesh in both cases, independently of the boundary conditions.

Independently of the boundary conditions, we observe that the symbol of the prolongation operator is a 3×3 matrix-valued function as the symbol of the coefficient matrix sequence of the linear systems to be solved: more precisely we define

$$p(\theta_1, \theta_2, \theta_3) = \sqrt{2^{-1}}(1 + \cos(\theta_1)(1 + \cos(\theta_2)(1 + \cos(\theta_3)I_3.$$

Since the matrix-valued function $p(\theta_1, \theta_2, \theta_3)$ is a scalar function times the identity, it automatically follows that the difficult condition **(C)** is satisfied. Moreover, taking into account Theorem 1, since the scalar function $1 + \cos(\theta)$ has a zero of order 2 at $\theta = \pi$, also condition **(A)** holds, while the positive definiteness of $\sum_{\eta \in \Omega(\theta)} pp^H(\eta)$ is verified as well, so that also condition **(B)** is met.

Finally, we choose one iteration of Gauss–Seidel as the pre and post smoother. Given the analysis reported in the previous section, we expect that our multigrid method is convergent in an optimal way, that is, with a convergence speed independent of the matrix size. In Table 1, we find a plain confirmation of our theoretical expectations, which holds not only for a constant ρ, but also in the variable coefficient setting as long as ρ remains positive.

Table 1. Number of iteration for matrices $A_n(1, \mathtt{D}^6)$ and $A_n(1, \mathtt{DN}^5)$ of increasing dimension $N(n)$, with the numeric precision $\varepsilon = 10^{-4}, 10^{-6}, 10^{-8}$, respectively.

$A_n(1,\mathtt{D}^6)$										$A_n(1,\mathtt{DN}^5)$									
$\nu = 0.1$																			
$N(n)$	Twogrid			Vcycle			Wcycle			$N(n)$	Twogrid			Vcycle			Wcycle		
81	5	7	10	-	-	-	-	-	-	300	7	13	18	-	-	-	-	-	-
1029	7	10	15	7	10	15	7	10	15	1944	8	12	17	8	12	17	8	12	17
10125	8	13	18	8	13	18	8	13	18	13872	8	13	18	9	14	19	8	13	18
$\nu = 0.2$																			
$N(n)$	Twogrid			Vcycle			Wcycle			$N(n)$	Twogrid			Vcycle			Wcycle		
81	4	5	7	-	-	-	-	-	-	300	5	8	11	-	-	-	-	-	-
1029	5	7	9	5	7	9	5	7	9	1944	6	8	11	6	8	11	6	8	11
10125	6	8	11	6	8	11	6	8	11	13872	8	10	12	8	10	12	8	10	12
$\nu = 0.4$																			
$N(n)$	Twogrid			Vcycle			Wcycle			$N(n)$	Twogrid			Vcycle			Wcycle		
81	3	4	6	-	-	-	-	-	-	300	6	10	14	-	-	-	-	-	-
1029	4	7	9	4	7	9	4	7	9	1944	7	11	15	7	11	15	7	11	15
10125	4	7	11	5	8	12	4	7	11	13872	6	11	15	8	14	19	7	11	15

6. Conclusions

We provided a quite complete spectral analysis of the (large) coefficient matrices associated with the linear systems stemming from the FE discretization of a linearly elastic problem for an arbitrary element-wise constant coefficient field. Our interest in this problem stems from the fact that in the material distribution method for topology optimization, such a problem is solved at each iteration. The solution for these linear systems typically dominates the computational effort required to solve the topology optimization problems. Based on the spectral information, we proposed a specialized multigrid method, which turned out to be optimal in the sense that the (arithmetic) cost for solving the related linear systems, up to a fixed desired accuracy, is proportional to the computational cost of matrix–vector products, which is linear in the corresponding matrix size and mildly depending on the given accuracy. The method was tested, and the preliminary numerical results are very satisfactory and promising, in terms of having a linear cost and a number of iterations that is bounded by a constant independent of the matrix size and only lightly influenced by the desired accuracy.

Finally, we mention future lines of research:

- The present analysis was performed also with variable coefficients, but with the restriction that the domain is Cartesian: this limitation can be easily overcome, but the computations are not trivial using the notion of reduced GLT sequences of matrices (see [35] for a recent very complete work and [33,34] for examples of applications and for the initial idea).
- The use of higher order FEs is another direction to explore: for more standard problems, we refer to [41,42] whose analysis can be the starting point for adapting the techniques to the present context.
- In the direction of more general approaches, we remind that the GLT machinery was used also in connection with finite differences, finite volumes, isogeometric analysis

of high order and intermediate regularity (see [30,31] and references therein). Hence, we are convinced that such extensions can be formally obtained, also in the case of the current topology optimization problem.

Author Contributions: Conceptualization, Q.K.N., S.S.-C., C.T.-P. and E.W.; Investigation, Q.K.N., S.S.-C., C.T.-P. and E.W.; Methodology, Q.K.N., S.S.-C., C.T.-P. and E.W.; Writing—original draft, Q.K.N., S.S.-C., C.T.-P. and E.W.; Writing—review and editing, Q.K.N., S.S.-C., C.T.-P. and E.W. All authors have read and agreed to the published version of the manuscript.

Funding: This work is financially supported by the Swedish strategic research program eSSENCE and the Italian Institution for High Mathematics (INdAM).

Data Availability Statement: The data in this paper are results of numerical computations and not experimental measurements. The datasets generated and analyzed during the current study are available from the corresponding author on reasonable request.

Conflicts of Interest: The authors declare that they have no known competing financial interests or personal relationships that could have appeared to influence the work reported in this paper.

Appendix A. Explicit Expressions for the Element Stiffness Matrix

Recall that for the studied problem, the three-dimensional element stiffness matrix is

$$\boldsymbol{K}_n = \frac{h}{2}\frac{E_0}{1+\nu}\begin{bmatrix} K_1 & K_2 & K_3 & K_4 & K_5 & K_6 & K_7 & K_8 \\ & K_9 & K_{10} & K_{11} & K_{12} & K_{13} & K_{14} & K_7 \\ & & K_{15} & K_{16} & K_{17} & K_{18} & K_{13} & K_6 \\ & & & K_{19} & K_{20} & K_{17} & K_{12} & K_5 \\ & & & & K_{19} & K_{16} & K_{11} & K_4 \\ & & & & & K_{15} & K_{10} & K_3 \\ & & & & & & K_9 & K_2 \\ & & & & & & & K_1 \end{bmatrix}$$

where

$$\begin{aligned}
K_1 &= \begin{bmatrix} k_1 & k_2 & k_2 \\ & k_1 & k_2 \\ & & k_1 \end{bmatrix}, \quad K_2 = \begin{bmatrix} k_3 & k_4 & 0 \\ & k_3 & 0 \\ & & k_6 \end{bmatrix}, \quad K_3 = \begin{bmatrix} k_3 & 0 & k_4 \\ & k_6 & 0 \\ & & k_3 \end{bmatrix}, \quad K_4 = \begin{bmatrix} k_5 & 0 & 0 \\ & k_7 & k_8 \\ & & k_7 \end{bmatrix}, \\
K_5 &= \begin{bmatrix} k_6 & 0 & 0 \\ & k_3 & k_4 \\ & & k_3 \end{bmatrix}, \quad K_6 = \begin{bmatrix} k_7 & 0 & k_8 \\ & k_5 & 0 \\ & & k_7 \end{bmatrix}, \quad K_7 = \begin{bmatrix} k_7 & k_8 & 0 \\ & k_7 & 0 \\ & & k_5 \end{bmatrix}, \quad K_8 = \begin{bmatrix} k_9 & k_{10} & k_{10} \\ & k_9 & k_{10} \\ & & k_9 \end{bmatrix}, \\
K_9 &= \begin{bmatrix} k_1 & k_2 & k_8 \\ & k_1 & k_8 \\ & & k_1 \end{bmatrix}, \quad K_{10} = \begin{bmatrix} k_5 & 0 & 0 \\ & k_7 & k_2 \\ & & k_7 \end{bmatrix}, \quad K_{11} = \begin{bmatrix} k_3 & 0 & k_{10} \\ & k_6 & 0 \\ & & k_3 \end{bmatrix}, \quad K_{12} = \begin{bmatrix} k_7 & 0 & k_2 \\ & k_5 & 0 \\ & & k_7 \end{bmatrix}, \\
K_{13} &= \begin{bmatrix} k_6 & 0 & 0 \\ & k_3 & k_{10} \\ & & k_3 \end{bmatrix}, \quad K_{14} = \begin{bmatrix} k_9 & k_{10} & k_4 \\ & k_9 & k_4 \\ & & k_9 \end{bmatrix}, \quad K_{15} = \begin{bmatrix} k_1 & k_8 & k_2 \\ & k_1 & k_8 \\ & & k_1 \end{bmatrix}, \quad K_{16} = \begin{bmatrix} k_3 & k_{10} & 0 \\ & k_3 & 0 \\ & & k_6 \end{bmatrix}, \\
K_{17} &= \begin{bmatrix} k_7 & k_2 & 0 \\ & k_7 & 0 \\ & & k_5 \end{bmatrix}, \quad K_{18} = \begin{bmatrix} k_9 & k_4 & k_{10} \\ & k_9 & k_4 \\ & & k_9 \end{bmatrix}, \quad K_{19} = \begin{bmatrix} k_1 & k_8 & k_8 \\ & k_1 & k_2 \\ & & k_1 \end{bmatrix}, \quad K_{20} = \begin{bmatrix} k_9 & k_4 & k_4 \\ & k_3 & k_{10} \\ & & k_9 \end{bmatrix}.
\end{aligned} \tag{A1}$$

We stress that all the 3×3 blocks mentioned before are real symmetric and hence the lower part is defined accordingly; see Section 3.2.1 for the use of these blocks in the definition of the GLT symbol. We observe that the quantity $hE_0/2(1+\nu)$ is a multiplicative term that will be neglected in the spectral analysis since it will be simplified and included in the right-hand vector when solving the related large linear systems). An analytical evaluation of the entries of the stiffness matrix is reported in the following lines:

$$\begin{aligned} &k_1 = \frac{2\hat{\lambda}}{3} + \frac{4\hat{\mu}}{9}, \quad k_2 = \frac{\hat{\lambda}}{3}, \quad k_3 = \frac{2\hat{\mu}}{9}, \quad k_4 = \frac{\hat{\lambda}}{6}, \quad k_5 = \frac{\hat{\mu}}{9} - \frac{\hat{\lambda}}{6}, \\ &k_6 = \frac{-4\hat{\mu}}{9}, \quad k_7 = \frac{-\hat{\lambda}}{6} - \frac{2\hat{\mu}}{9}, \quad k_8 = -\frac{\hat{\lambda}}{3}, \quad k_9 = -\frac{\hat{\lambda}}{6} - \frac{\hat{\mu}}{9}, \quad k_{10} = -\frac{\hat{\lambda}}{6}, \end{aligned} \tag{A2}$$

where

$$\hat{\lambda} = \frac{\nu}{(1-2\nu)}, \quad \hat{\mu} = \frac{1}{2}.$$

Note that $\hat{\lambda}$ and $\hat{\mu}$ in (A2) do not represent the Lamé parameters because they have been modified to facilitate the spectral analysis conducted in Section 3. In fact, in this article, $\lambda = \hat{\lambda}\frac{E_0}{1+\nu}$ and $\mu = \hat{\mu}\frac{E_0}{1+\nu}$ are the actual Lamé parameters.

Appendix B. Stress–Strain Relation and Various Bounds

In the three-dimensional setting, there are six independent strain components in total at a point in an element, and they are written as a vector

$$\epsilon = [\epsilon_{11}\ \epsilon_{22}\ \epsilon_{33}\ 2\epsilon_{12}\ 2\epsilon_{23}\ 2\epsilon_{31}]^T.$$

Similarly, corresponding to the six strain components above, there are also six independent stress components written in vector form as

$$\sigma = [\sigma_{11}\ \sigma_{22}\ \sigma_{33}\ \sigma_{12}\ \sigma_{23}\ \sigma_{31}]^T.$$

By the generalized Hooke's law, the most general linear relation among components of the stress and strain tensor can then be written as

$$\sigma = \boldsymbol{E}\epsilon, \tag{A3}$$

where $\boldsymbol{E}$ is a matrix that corresponds to the constant fourth-order elasticity tensor E_c. The relationship between stresses and strains is

$$\begin{aligned} &\epsilon_{11} = \frac{1}{E_0}(\sigma_{11} - \nu(\sigma_{22} + \sigma_{33})), \quad \epsilon_{12} = \frac{\sigma_{12}}{2G}, \quad \epsilon_{13} = \frac{\sigma_{13}}{2G}, \\ &\epsilon_{22} = \frac{1}{E_0}(\sigma_{22} - \nu(\sigma_{11} + \sigma_{33})), \quad \epsilon_{23} = \frac{\sigma_{23}}{2G}, \quad \epsilon_{33} = \frac{1}{E_0}(\sigma_{33} - \nu(\sigma_{11} + \sigma_{22})), \end{aligned} \tag{A4}$$

where ν is Poisson's ratio, E_0 is Young's modulus, and the shear modulus G is

$$G = \frac{E_0}{2+2\nu}. \tag{A5}$$

The relationships above can be rewritten in matrix form as

$$\epsilon = \frac{1}{E_0}\begin{bmatrix} 1 & -\nu & -\nu & 0 & 0 & 0 \\ -\nu & 1 & -\nu & 0 & 0 & 0 \\ -\nu & -\nu & 1 & 0 & 0 & 0 \\ 0 & 0 & 0 & 2(1+\nu) & 0 & 0 \\ 0 & 0 & 0 & 0 & 2(1+\nu) & 0 \\ 0 & 0 & 0 & 0 & 0 & 2(1+\nu) \end{bmatrix}\sigma. \tag{A6}$$

Furthermore, the equation system (A6) can be inverted to obtain Hooke's law (A3) with

$$\boldsymbol{E}=\begin{bmatrix}\lambda+2\mu & \lambda & \lambda & 0 & 0 & 0\\ \lambda & \lambda+2\mu & \lambda & 0 & 0 & 0\\ \lambda & \lambda & \lambda+2\mu & 0 & 0 & 0\\ 0 & 0 & 0 & \lambda & 0 & 0\\ 0 & 0 & 0 & 0 & \lambda & 0\\ 0 & 0 & 0 & 0 & 0 & \lambda\end{bmatrix} \tag{A7}$$

where

$$\begin{aligned}\lambda &= \frac{E_0\nu}{(1+\nu)(1-2\nu)},\\ \mu &= \frac{E_0}{2(1+\nu)}.\end{aligned} \tag{A8}$$

In this case, the bulk modulus K can be expressed as

$$K=\frac{(\sigma_{11}+\sigma_{22}+\sigma_{33})/3}{\epsilon_{11}+\epsilon_{22}+\epsilon_{33}}=\frac{E_0}{3(1-2\nu)}. \tag{A9}$$

The Young's (E_0), shear (G), and bulk (K) moduli need to be positive. Thus, Equations (A5) and (A9) imply that the Poisson's ratio in three dimensions must satisfy $-1<\nu<0.5$.

Remark A1. *Physically, there are no known isotropic materials with a negative Poisson's ratio. Therefore, from a practical perspective [43], we can limit our study to* ν*, satisfying the inequalities* $0\leq\nu<0.5$*. Furthermore, in some more delicate circumstances, the Poisson ratio lies in the interval* $0.2\leq\nu<0.5$*, as proved experimentally in view of the elastic properties of real materials [44].*

References

1. Nguyen, Q.K.; Serra-Capizzano, S.; Wadbro, E. On using a zero lower bound on the physical density in material distribution topology optimization. *Comput. Methods Appl. Mech. Eng.* **2020**, *359*, 112669. [CrossRef]
2. Nguyen, Q.K.; Serra-Capizzano, S.; Tablino-Possio, C.; Wadbro, E. Spectral Analysis of the Finite Element Matrices Approximating 2D Linearly Elastic Structures and Multigrid Proposals. *Numer. Linear Algebra Appl.* **2022**, *29*, e2433. [CrossRef]
3. Bendsøe, M.P.; Kikuchi, N. Generating optimal topologies in structural design using a homogenization method. *Comput. Methods Appl. Mech. Eng.* **1988**, *71*, 197–224. [CrossRef]
4. Allarire, G. *Shape Optimization by the Homogenization Method*; Springer: New York, NY, USA, 2002.
5. Bendsøe, M.P.; Sigmund, O. *Topology Optimization. Theory, Methods, and Applications*; Springer: Berlin/Heidelberg, Germany; New York, NY, USA, 2003.
6. Elesin, Y.; Lazarov, B.S.; Jensen, J.S.; Sigmund, O. Time domain topology optimization of 3D nanophotonic devices. *Photonics Nanostruct. Appl.* **2014**, *12*, 23–33. [CrossRef]
7. Erentok, A.; Sigmund, O. Topology optimization of sub-wavelength antennas. *IEEE Trans. Antennas Propag.* **2011**, *59*, 58–69. [CrossRef]
8. Wadbro, E.; Engström, C. Topology and shape optimization of plasmonic nano-antennas. *Comput. Methods Appl. Mech. Eng.* **2015**, *293*, 155–169. [CrossRef]
9. Andreasen, C.; Sigmund, O. Topology optimization of fluid–structure-interaction problems in poroelasticity. *Comput. Methods Appl. Mech. Eng.* **2013**, *258*, 55–62. [CrossRef]
10. Yoon, G.H. Topology optimization for stationary fluid–structure interaction problems using a new monolithic formulation. *Int. J. Numer. Methods Eng.* **2010**, *82*, 591–616. [CrossRef]
11. Christiansen, R.E.; Lazarov, B.S.; Jensen, J.S.; Sigmund, O. Creating geometrically robust designs for highly sensitive problems using topology optimization: Acoustic cavity design. *Struct. Multidiscip. Optim.* **2015**, *52*, 737–754. [CrossRef]
12. Kook, J.; Koo, K.; Hyun, J.; Jensen, J.S.; Wang, S. Acoustical topology optimization for Zwicker's loudness model—Application to noise barriers. *Comput. Methods Appl. Mech. Eng.* **2012**, *237–240*, 130–151. [CrossRef]
13. Wadbro, E.; Niu, B. Multiscale design for additive manufactured structures with solid coating and periodic infill pattern. *Comput. Methods Appl. Mech. Eng.* **2019**, *357*, 112605. [CrossRef]
14. Clausen, A.; Aage, N.; Sigmund, O. Topology optimization of coated structures and material interface problems. *Comput. Methods Appl. Mech. Eng.* **2015**, *290*, 524–541. [CrossRef]
15. Klarbring, A.; Strömberg, N. Topology optimization of hyperelastic bodies including non-zero prescribed displacements. *Struct. Multidiscip. Optim.* **2013**, *47*, 37–48. [CrossRef]

16. Park, J.; Sutradhar, A. A multi-resolution method for 3D multi-material topology optimization. *Comput. Methods Appl. Mech. Eng.* **2015**, *285*, 571–586. [CrossRef]
17. Aage, N.; Andreassen, E.; Lazarov, B.S.; Sigmund, O. Giga-voxel computational morphogenesis for structural design. *Nature* **2017**, *550*, 84–86. [CrossRef]
18. Amir, O.; Stolpe, M.; Sigmund, O. Efficient use of iterative solvers in nested topology optimization. *Struct. Multidiscip. Optim.* **2010**, *42*, 55–72. [CrossRef]
19. Baandrup, M.; Sigmund, O.; Polk, H.; Aage, N. Closing the gap towards super-long suspension bridges using computational morphogenesis. *Nat. Commun.* **2020**, *11*, 2735. [CrossRef]
20. Borrvall, T.; Petersson, J. Large-scale topology optimization in 3D using parallel computing. *Comput. Methods Appl. Mech. Eng.* **2001**, *190*, 6201–6229. [CrossRef]
21. Wadbro, E.; Berggren, M. Megapixel topology optimization using a graphics processing unit. *SIAM Rev.* **2009**, *51*, 707–721. [CrossRef]
22. Allaire, G.; Jouve, F.; Toader, A.M. Structural optimization using sensitivity analysis and a level-set method. *J. Comput. Phys.* **2004**, *194*, 363–393. [CrossRef]
23. Dijk, N.P.; Maute, K.; Langelaar, M.; van Keulen, F. Level-set methods for structural topology optimization: A review. *Struct. Multidiscip. Optim.* **2013**, *48*, 437–472. [CrossRef]
24. Du, Z.; Cui, T.; Liu, C.; Zhang, W.; Guo, Y.; Guo, X. An efficient and easy-to-extend Matlab code of the Moving Morphable Component (MMC) method for three-dimensional topology optimization. *Struct. Multidiscip. Optim.* **2022**, *65*, 158. [CrossRef]
25. Guo, X.; Zhang, W.; Zhong, W. Doing topology optimization explicitly and geometrically—A new moving morphable components based framework. *J. Appl. Mech.* **2014**, *81*, 081009 [CrossRef]
26. Niu, B.; Wadbro, E. On equal-width length-scale control in topology optimization. *Struct. Multidiscip. Optim.* **2019**, *59*, 1321–1334. [CrossRef]
27. Zhang, W.; Yuan, J.; Zhang, J.; Guo, X. A new topology optimization approach based on Moving Morphable Components (MMC) and the ersatz material model. *Struct. Multidiscip. Optim.* **2016**, *53*, 1243–1260. [CrossRef]
28. Barbarino, G.; Garoni, C.; Serra-Capizzano, S. Block Generalized Locally Toeplitz Sequences: Theory and applications in the multidimensional case. *Electron. Trans. Numer. Anal.* **2020**, *53*, 113–216. [CrossRef]
29. Garoni, C.; Mazza, M.; Serra-Capizzano, S. Block Generalized Locally Toeplitz Sequences: From the Theory to the Applications. *Axioms* **2018**, *7*, 49. [CrossRef]
30. Garoni, C.; Serra-Capizzano, S. *Generalized Locally Toeplitz Sequences: Theory and Applications*; Springer: Cham, Switzerland, 2017; Volume I.
31. Garoni, C.; Serra-Capizzano, S. *Generalized Multilevel Locally Toeplitz Sequences: Theory and Applications*; Springer: Cham, Switzerland, 2018; Volume II.
32. Barbarino, G.; Garoni, C.; Serra-Capizzano, S. Block generalized locally Toeplitz sequences: Theory and applications in the unidimensional case. *Electron. Trans. Numer. Anal.* **2020**, *53*, 28–112. [CrossRef]
33. Serra-Capizzano, S. Generalized locally Toeplitz sequences: Spectral analysis and applications to discretized partial differential equations. *Linear Algebra Appl.* **2003**, *366*, 371–402. [CrossRef]
34. Serra-Capizzano, S. The GLT class as a generalized Fourier analysis and applications. *Linear Algebra Appl.* **2006**, *419*, 180–233. [CrossRef]
35. Barbarino, G. A systematic approach to reduced GLT. *BIT Numer. Math.* **2022**, *62*, 681–743. [CrossRef]
36. Hackbusch, W. *Multigrid Methods and Applications*; Springer: Berlin/Heidelberg, Germany, 1985.
37. Saye, R.I. Efficient multigrid solution of elliptic interface problems using viscosity-upwinded local discontinuous Galerkin methods. *Commun. Appl. Math. Comput. Sci.* **2019**, *14*, 247–283. [CrossRef]
38. Aricò, A.; Donatelli, M.; Serra-Capizzano, S. V-cycle optimal convergence for certain (multilevel) structured linear systems. *SIAM J. Matrix Anal. Appl.* **2004**, *26*, 186–214. [CrossRef]
39. Fiorentino, G.; Serra, S. Multigrid methods for Toeplitz matrices. *Calcolo* **1991**, *28*, 283–305. [CrossRef]
40. Donatelli, M.; Ferrari, P.; Furci, I.; Serra-Capizzano, S.; Sesana, D. Multigrid methods for block-Toeplitz linear systems: Convergence analysis and applications. *Numer. Linear Algebra Appl.* **2021**, *28*, e2356. [CrossRef]
41. Garoni, C.; Serra-Capizzano, S.; Sesana, D. Spectral analysis and spectral symbol of d-variate $\mathbb{Q}_p$ Lagrangian FEM stiffness matrices. *SIAM J. Matrix Anal. Appl.* **2015**, *36*, 1100–1128. [CrossRef]
42. Rahla, R.; Serra-Capizzano, S.; Tablino-Possio, C. Spectral analysis of $\mathbb{P}_k$ finite element matrices in the case of Friedrichs-Keller triangulations via generalized locally Toeplitz technology. *Numer. Linear Algebra Appl.* **2020**, *27*, e2302. [CrossRef]
43. Love, A.E.H.; Goldstine, H.H. *A Treatise on the Mathematical Theory of Elasticity*; Dover Publications: New York, NY, USA, 1944.
44. Mott, P.H.; Roland, C.M. Limits to Poisson's ratio in isotropic materials. *Phys. Rev. B* **2009**, *80*, 132104. [CrossRef]

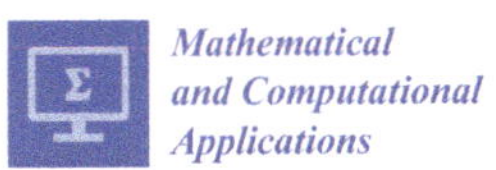
Mathematical and Computational Applications

Article

Entropy Analysis for Hydromagnetic Darcy–Forchheimer Flow Subject to Soret and Dufour Effects

Sohail A. Khan * and Tasawar Hayat

Department of Mathematics, Quaid-I-Azam University, Islamabad 45320, Pakistan
* Correspondence: sohailahmadkhan93@gmail.com

Abstract: Here, our main aim is to examine the impacts of Dufour and Soret in a radiative Darcy–Forchheimer flow. Ohmic heating and the dissipative features are outlined. The characteristics of the thermo-diffusion and diffusion-thermo effects are addressed. A binary chemical reaction is deliberated. To examine the thermodynamical system performance, we discuss entropy generation. A non-linear differential system is computed by the finite difference technique. Variations in the velocity, concentration, thermal field and entropy rate for the emerging parameters are scrutinized. A decay in velocity is observed for the Forchheimer number. Higher estimation of the magnetic number has the opposite influence for the velocity and temperature. The velocity, concentration and thermal field have a similar effect on the suction variable. The temperature against the Dufour number is augmented. A decay in the concentration is found against the Soret number. A similar trend holds for the entropy rate through the radiation and diffusion variables. An augmentation in the entropy rate is observed for the diffusion variable.

Keywords: Darcy–Forchheimer model; thermal radiation; finite difference technique; viscous dissipation; Soret and Dufour impacts; chemical reaction; entropy generation

Citation: Khan, S.A.; Hayat, T. Entropy Analysis for Hydromagnetic Darcy–Forchheimer Flow Subject to Soret and Dufour Effects. *Math. Comput. Appl.* **2022**, 27, 80. https://doi.org/10.3390/mca27050080

Academic Editor: Zhuojia Fu

Received: 26 June 2022
Accepted: 14 September 2022
Published: 19 September 2022

1. Introduction

Henry Darcy established the framework of homogeneous liquid flow through a permeable medium throughout his work on the progression of water over saturated sand [1]. At higher flow rates, when inertial and boundary impacts arise, then the Darcy law cannot work appropriately. To overcome such an issue, Forchheimer gave the concept of the non-Darcy model through the insertion of a quadratic velocity term in a momentum expression [2]. Later on, the Forchheimer term was so named by Muskat [3]. The moment of fluid flowing through a permeable surface is of keen interest due to its significance in technical, biological and scientific fields such as artificial dialysis, gas turbines, atherosclerosis, catalytic converters, geo energy production and many others. Hayat et al. [4] explored the 2-D Darcy–Forchheimer flow of non-Newtonian liquid with variable properties. Pal and Mondal [5] addressed the convective flow of Darcy–Forchheimer liquid subject to a variable heat sink or source and viscosity. Non-uniform heat conductivity analysis in the reactive flow of Darcy–Forchheimer Carreau nanomaterial subject to a magnetic dipole was presented by Mallawi and Ullah [6]. Alshomrani and Ullah [7] studied the convective flow of Darcy–Forchheimer hybrid nanomaterial subject to a cubic autocatalysis chemical reaction. Seth and Mandal [8] discussed the hydromagnetic effect in rotating the flow of Darcy–Forchheimer Casson liquid toward a permeable space. Little analysis concerning a porous medium is discussed in [9–17].

Thermal and solutal transportation in a permeable surface has been a significant consideration of researchers during the last two decades. This is due to its usefulness in geothermal systems, catalytic reactors, nuclear waste repositories, areas of geosciences, chemical engineering, energy storage units, drying technology, heat insulation, heat exchangers for packed beds and many others. Initially, the Dufour effect in liquid was explored by Rastogi and Madan [18]. After that, the diffusion-thermo impact in a homogeneous mixture was investigated in [19,20]. Moorthy and Senthilvalivu [21] studied the Dufour and Soret outcomes in the convective flow of liquid of a non-uniform viscosity subject to a permeable medium. Non-uniform temperature in a convective fluid flow subject to thermal-diffusion and diffusion-thermo effects was illustrated by El-Arabawy [22]. Few reviews with reference to relevant titles have been discussed through certain studies [23–27]. The important reason for entropy production is the conversion of thermal energy in the occurrence of numerous examples of processes, such as fluid friction, kinetic energy, rotational moments, molecular resistance, the Joule Thomson effect, mass transport rate, and molecular vibration. The concept of entropy optimization in liquid flow was initially given by Bejan [28,29]. Entropy analysis in a water-based hybrid nanoliquid subject to mixed convection was discussed by Buonomo [30]. Irreversibility exploration in the dissipative flow of nanomaterial with melting and radiation over a stretching sheet was addressed by Khan et al. [31]. Some investigations of the entropy rate are mentioned in [32–39].

To our knowledge, no study has reported about entropy-optimized radiative Darcy–Forchheimer flows with Soret and Dufour features yet. A porous medium through a Darcy–Forchheimer relation is discussed. Dissipation, Ohmic heating and radiation are scrutinized in energy equations. The physical characteristics for the Soret and Dufour impacts are addressed. A first-order chemical reaction is deliberated. To examine the thermodynamical system performance, we discuss entropy optimization. Nonlinear differential systems are obtained through appropriate transformations. Non-dimensional differential systems are solved through the finite difference method. The performance of appropriate variables concerning the velocity, entropy generation, concentration and thermal field have been scrutinized.

2. Formulation

An unsteady radiative hydromagnetic Darcy–Forchheimer flow saturating a porous medium is discussed. Joule heating, viscous dissipation and thermal radiation in energy expression have been scrutinized. The Soret and Dufour effects are inspected. The impact of the entropy rate is addressed. Additionally, the flow is subject to a chemical reaction of the first order. A constant magnetic field with a strength (B_0) is applied. Consider $u = u_w = ax$ as the stretching velocity, with $a > 0$. The chosen magnetic Reynolds number is small. Figure 1 shows a flow sketch [31].

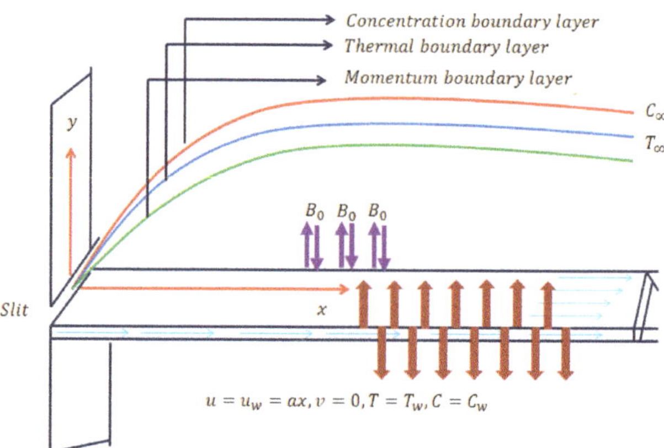

Figure 1. Flow sketch.

By taking into account the infinite plate, the term $\frac{\partial u}{\partial x}$ becomes zero. Here, the continuity equation becomes $\frac{\partial v}{\partial y} = 0$.

Under the above discussion, the related expression for constant suction becomes

$$v = -v_0 = \text{constant} \tag{1}$$

$$\frac{\partial u}{\partial t} - v_0\frac{\partial u}{\partial y} = \nu\frac{\partial^2 u}{\partial y^2} - \frac{\sigma B_0^2}{\rho}u - \frac{\nu}{k_p}u - Fu^2, \tag{2}$$

$$\left.\begin{array}{c} \frac{\partial T}{\partial t} - v_0\frac{\partial T}{\partial y} = \alpha\frac{\partial^2 T}{\partial y^2} + \frac{16}{3}\frac{\sigma^* T_\infty^3}{k^*(\rho c_p)}\frac{\partial^2 T}{\partial y^2} + \frac{\mu}{(\rho c_p)}(\frac{\partial u}{\partial y})^2 \\ +\frac{\sigma B_0^2}{(\rho c_p)}u^2 + \frac{D_B K_T}{C_s c_p}\frac{\partial^2 C}{\partial y^2} \end{array}\right\}, \tag{3}$$

$$\frac{\partial C}{\partial t} - v_0\frac{\partial C}{\partial y} = D_B\frac{\partial^2 C}{\partial y^2} + \frac{D_B K_T}{T_m}\frac{\partial^2 T}{\partial y^2} - k_r(C - C_\infty), \tag{4}$$

with

$$\left.\begin{array}{c} u = 0,\ T = T_\infty,\ C = C_\infty, \text{ at } t = 0 \\ u = ax,\ T = T_w,\ C = C_w, \text{ at } y = 0 \\ u \to 0,\ T \to T_\infty,\ C \to C_\infty, \text{ as } y \to \infty \end{array}\right\}. \tag{5}$$

Consider the following formula:

$$\left.\begin{array}{c} \tau = \frac{v}{L_1^2}t,\ \xi = \frac{x}{L_1},\ \eta = \frac{y}{L_1},\ U(\tau,\eta) = \frac{L_1}{v}u, \\ \theta(\tau,\eta) = \frac{(T-T_\infty)}{(T_w-T_\infty)},\ \phi(\tau,\eta) = \frac{C-C_\infty}{C_w-C_\infty}, \end{array}\right\}, \tag{6}$$

Then, we have

$$\frac{\partial U}{\partial \tau} - S\frac{\partial U}{\partial \eta} = \frac{\partial^2 U}{\partial \eta^2} - MU - \lambda U - FrU^2, \tag{7}$$

$$\frac{\partial \theta}{\partial \tau} - S\frac{\partial \theta}{\partial \eta} = \frac{1}{\Pr}(1 + Rd)\frac{\partial^2 \theta}{\partial \eta^2} + Ec(\frac{\partial U}{\partial \eta})^2 + MEcU^2 + Du\frac{\partial^2 \phi}{\partial \eta^2}, \tag{8}$$

$$\frac{\partial \phi}{\partial \tau} - S\frac{\partial \phi}{\partial \eta} = \frac{1}{Sc}\frac{\partial^2 \phi}{\partial \eta^2} + Sr\frac{\partial^2 \theta}{\partial \eta^2} - \gamma\phi, \tag{9}$$

with

$$\left.\begin{array}{c} U = 0,\ \theta = 0,\ \phi = 0,\ \text{at } \tau = 0 \\ U = \xi\,\mathrm{Re},\ \theta = 1, \phi = 1,\ \text{at } \eta = 0 \\ U \to 0,\ \theta \to 0,\ \phi \to 0,\ \text{as } \eta \to \infty \end{array}\right\} \tag{10}$$

Here the non-dimensional variables are $S(= \frac{v_0}{\nu}L_1)$, $M\left(= \frac{\sigma B_0^2 L_1^2}{\nu\rho}\right)$, $\mathrm{Re}\left(= \frac{aL_1^2}{\nu}\right)$, $Sr\left(= \frac{DK_T(T_w-T_\infty)}{\nu T_m(C_w-C_\infty)}\right)$, $F_r\left(= \frac{C_b}{\sqrt{k_p}}L_1\right)$, $\Pr(= \frac{\nu}{\alpha})$, $\lambda\left(= \frac{L_1^2}{k_p}\right)$, $E_C\left(= \frac{\nu^2}{c_p L_1^2(T_w-T_\infty)}\right)$, $\gamma\left(= \frac{k_r L_1^2}{\nu}\right)$, $Du\left(= \frac{D_B K_T(C_w-C_\infty)}{\nu C_s c_p(T_w-T_\infty)}\right)$, $Rd\left(= \frac{16\sigma^* T_\infty^3}{3k^* k}\right)$, $Sc\left(= \frac{\nu}{D_B}\right)$ and $Br(= \Pr Ec)$.

3. Engineering Contents of Interest

3.1. Nusselt Number

Here, we have

$$Nu_x = \frac{xq_w}{k(T_w - T_\infty)}, \tag{11}$$

with the heat flux q_w given by

$$q_w = -\left(k + \frac{16\sigma^* T_\infty^3}{3k^*}\right)\left(\frac{\partial T}{\partial y}\right)\Big|_{y=0} \tag{12}$$

We finally have

$$Nu_x = -\xi(1+Rd)\left(\frac{\partial\theta}{\partial\eta}\right)_{\eta=0}. \tag{13}$$

3.2. Sherwood Number

This is given as

$$Sh_x = \frac{xj_w}{D_B(C_w - C_\infty)}, \tag{14}$$

in which the mass flux j_w is defined as

$$j_w = -D_B\left(\frac{\partial C}{\partial y}\right)\Big|_{y=0}, \tag{15}$$

We can write

$$Sh_x = -\xi\left(\frac{\partial\phi}{\partial\eta}\right)_{\eta=0}. \tag{16}$$

4. Entropy

The important reason for entropy production is the conversion of thermal energy in the occurrence of numerous processes, such as fluid friction, kinetic energy, rotational moment, molecular resistance, the Joule–Thomson effect, mass transport rate and molecular vibration. We have the following [30–33]:

$$\left.\begin{array}{l} S_G = \frac{k}{T_\infty^2}\left(1+\frac{16\sigma^* T_\infty^3}{3k^* k}\right)\left(\frac{\partial T}{\partial y}\right)^2 + \frac{\mu_f}{T_\infty}\left(\frac{\partial u}{\partial y}\right)^2 + \frac{\mu}{k_p T_\infty}u^2 \\ +\frac{\sigma B_0^2}{T_\infty}u^2 + \frac{RD_B}{T_\infty}\left(\frac{\partial T}{\partial y}\frac{\partial C}{\partial y}\right) + \frac{RD_B}{C_\infty}\left(\frac{\partial C}{\partial y}\right)^2 \end{array}\right\}, \tag{17}$$

The dimensionless expression is

$$\left.\begin{array}{l} N_G(\tau,\eta) = \alpha_1(1+Rd)\left(\frac{\partial\theta}{\partial\eta}\right)^2 + Br\left(\frac{\partial U}{\partial\eta}\right)^2 + Br\lambda U^2 \\ +MBrU^2 + L\left(\frac{\partial\theta}{\partial\eta}\frac{\partial\phi}{\partial\eta}\right) + L\frac{\alpha_2}{\alpha_1}\left(\frac{\partial\phi}{\partial\eta}\right)^2 \end{array}\right\} \tag{18}$$

In the above expression, the dimensionless parameters are $\alpha_1\left(=\frac{(T_w - T_\infty)}{T_\infty}\right)$, $N_G\left(=\frac{S_G T_\infty L_1^2}{k(T_w - T_\infty)}\right)$, $L\left(=\frac{RD_B(C_w - C_\infty)}{k}\right)$ and $\alpha_2\left(=\frac{(C_w - C_\infty)}{C_\infty}\right)$.

5. Solution Methodology

Using the finite difference method, we can solve the nonlinear differential system [40–43] by writing

$$\left.\begin{array}{c} \frac{\partial U}{\partial\tau} = \frac{U_a^{n+1} - U_a^n}{\Delta\tau}, \frac{\partial U}{\partial\eta} = \frac{U_{a+1}^n - U_a^n}{\Delta\eta} \\ \frac{\partial\theta}{\partial\tau} = \frac{\theta_a^{n+1} - \theta_s^k}{\Delta\tau}, \frac{\partial\theta}{\partial\eta} = \frac{\theta_{a+1}^n - \theta_a^n}{\Delta\eta} \\ \frac{\partial\phi}{\partial\tau} = \frac{\phi_a^{n+1} - \phi_a^n}{\Delta\tau}, \frac{\partial\phi}{\partial\eta} = \frac{\phi_{a+1}^n - \phi_a^n}{\Delta\eta} \\ \frac{\partial^2 U}{\partial\eta^2} = \frac{U_{a+1}^n - 2U_a^n + U_{a-1}^n}{(\Delta\eta)^2}, \frac{\partial^2\theta}{\partial\eta^2} = \frac{\theta_{a+1}^n - 2\theta_a^n + \theta_{a-1}^n}{(\Delta\eta)^2} \\ \frac{\partial^2\phi}{\partial\eta^2} = \frac{\phi_{a+1}^n - 2\phi_a^n + \phi_{a-1}^n}{(\Delta\eta)^2} \end{array}\right\}, \tag{19}$$

By employing Equation (23) in Equations (8)–(10), we obtain

$$\frac{U_a^{n+1} - U_a^n}{\Delta\tau} - S\frac{U_{a+1}^n - U_a^n}{\Delta\eta} = \frac{U_{a+1}^n - 2U_a^n + U_{a-1}^n}{(\Delta\eta)^2} - MU_a^n - \lambda U_a^n - Fr(U_a^n)^2, \tag{20}$$

$$\left.\begin{array}{c}\frac{\theta_a^{n+1}-\theta_a^n}{\Delta\tau}-S\frac{\theta_{a+1}^n-\theta_a^n}{\Delta\eta}=\frac{(1+Rd)}{P_r}\frac{\theta_{a+1}^n-2\theta_a^n+\theta_{a-1}^n}{(\Delta\eta)^2}+Ec\left(\frac{U_{a+1}^n-U_a^n}{\Delta\eta}\right)^2\\+MEc(U_a^n)^2+Du\left(\frac{\phi_{a+1}^n-2\phi_a^n+\phi_{a-1}^n}{(\Delta\eta)^2}\right)\end{array}\right\}\tag{21}$$

$$\frac{\phi_a^{n+1}-\phi_a^n}{\Delta\tau}-S\frac{\phi_{a+1}^n-\phi_a^n}{\Delta\eta}=\frac{1}{Sc}\frac{\phi_{a+1}^n-2\phi_a^n+\phi_{a-1}^n}{(\Delta\eta)^2}+Sr\frac{\theta_{a+1}^n-2\theta_a^n+\theta_{a-1}^n}{(\Delta\eta)^2}-\gamma\phi_a^n,\tag{22}$$

with

$$\left.\begin{array}{c}U_a^0=0,\ \theta_a^0=0,\ \phi_a^0=0,\\U_0^n=1,\ \theta_0^n=1,\ \phi_0^n=1,\\U_\infty^n\to 0,\ \theta_\infty^n\to 0,\ \phi_\infty^n\to 0\end{array}\right\}\tag{23}$$

The entropy generation expression yields

$$\left.\begin{array}{c}N_G(\tau,\xi,\eta)=\alpha_1(1+Rd)(\frac{\theta_{a+1}^n-\theta_a^n}{\Delta\eta})^2+Br\left(\frac{U_{a+1}^n-U_a^n}{\Delta\eta}\right)^2+Br\lambda(U_a^n)^2\\+MBr(U_a^n)^2+L(\frac{\theta_{a+1}^n-\theta_a^n}{\Delta\eta}.\frac{\phi_{a+1}^n-\phi_a^n}{\Delta\eta})+L\frac{\alpha_2}{\alpha_1}(\frac{\phi_{a+1}^n-\phi_a^n}{\Delta\eta})^2\end{array}\right\}\tag{24}$$

6. Graphical Results and Review

The physical interpretation of the parameters of the concentration, entropy rate, velocity and temperature have been investigated. The present observations are compared with previous published results in Table 1, and excellent agreement is noticed.

Table 1. Comparison of Nusselt numbers with [44].

Pr	Bidin and Nazar [44]	Recent Outcomes
1.0	0.9547	0.954710
2.0	1.4714	1.471409
3.0	1.8961	1.896115

6.1. Velocity

The influence of suction (S) upon the velocity $(U(\tau,\eta))$ is sketched in Figure 2. Obviously, a higher estimation of the suction parameter (S) decays the velocity. As expected, this is in accordance to the physical facts. Figure 3 displays the velocity against a magnetic field. Actually, reduction occurs in the velocity for (M). A physically higher (M) value corresponds to amplifying the Lorentz force which the flow opposes. Hence, velocity decay is guaranteed. Figure 4 was developed in order to recognize the velocity $(U(\tau,\eta))$ design with variation in the Forchheimer number (Fr). A larger estimation for the Forchheimer number decays the velocity $(U(\tau,\eta))$. The influence of (λ) on $(U(\tau,\eta))$ is illustrated in Figure 5. Clearly, $U(\tau,\eta)$ decays against higher λ values.

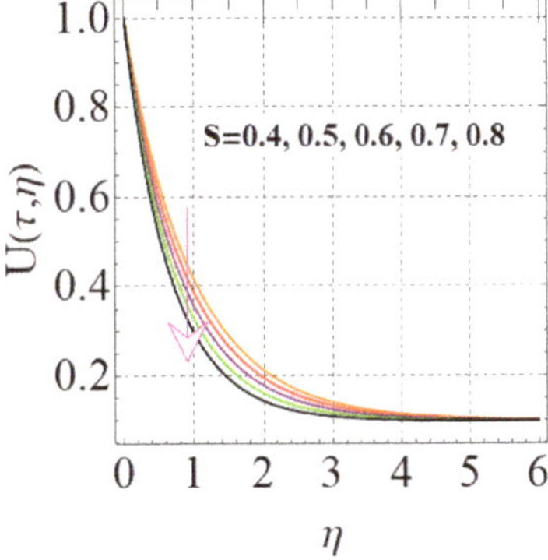

Figure 2. $U(\tau,\eta)$ via S.

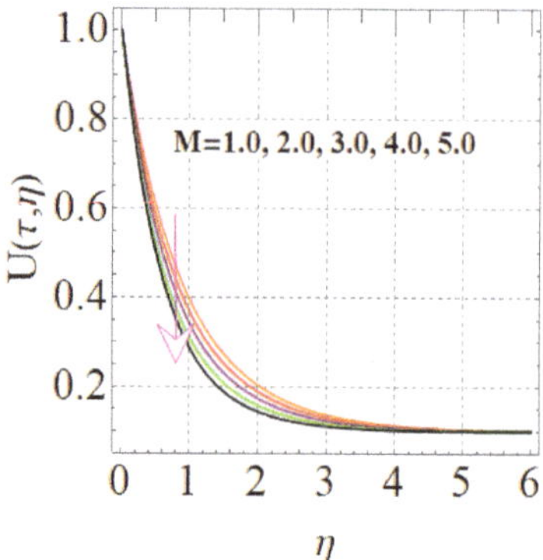

Figure 3. $U(\tau, \eta)$ via M.

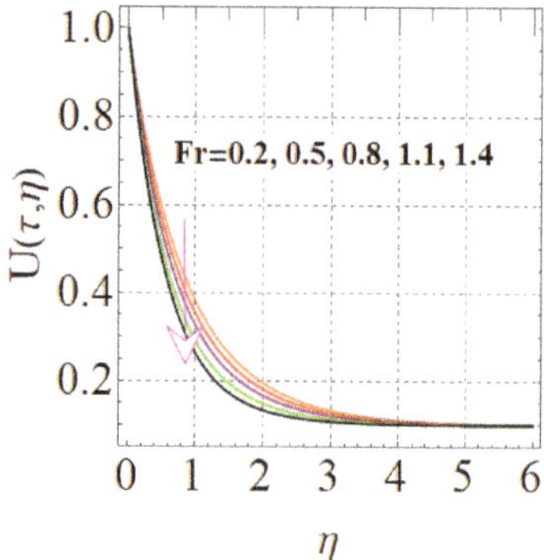

Figure 4. $U(\tau, \eta)$ via Fr.

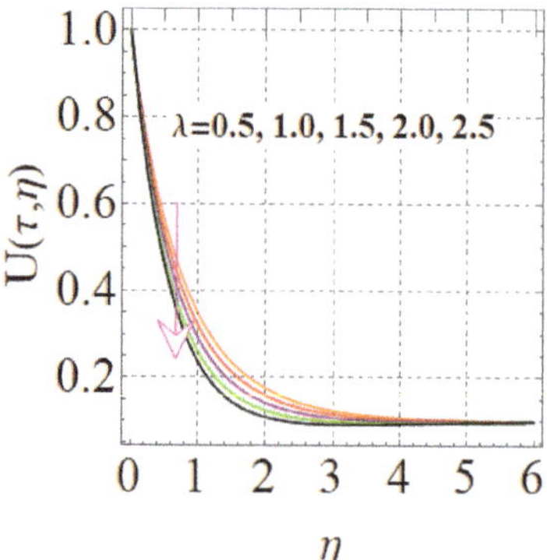

Figure 5. $U(\tau, \eta)$ via λ.

6.2. Temperature

Figures 6 and 7 are for the thermal field against the suction and magnetic variables (S and M). A similar scenario holds for the thermal field ($\theta(\tau, \eta)$) through the suction and magnetic variables. Figure 8 portrays the performance of the radiation against the temperature ($\theta(\tau, \eta)$). Larger radiation values lead to the temperature ($\theta(\tau, \eta)$) increasing. Figure 9 displays the performance of the thermal field against the Prandtl number. A larger approximation of (Pr) corresponds to the decay of the thermal diffusivity, and consequently, the temperature ($\theta(\tau, \eta)$) decreases. Figure 10 displays the impact of the thermal field $\theta(\tau, \eta)$ on the Eckert number (Ec). A higher estimation for (Ec) corresponds to the temperature being higher.

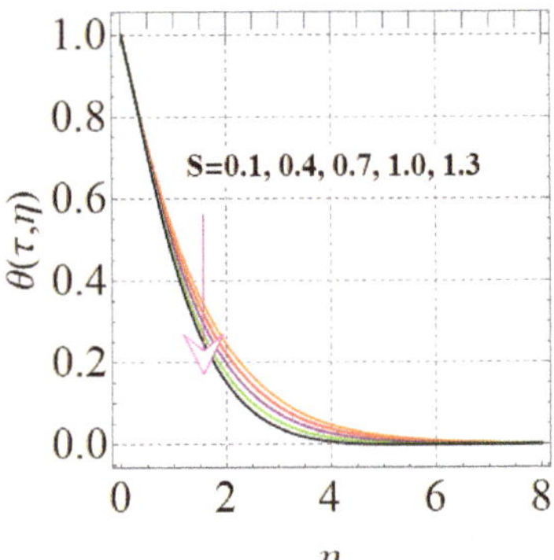

Figure 6. $\theta(\tau, \eta)$ via S.

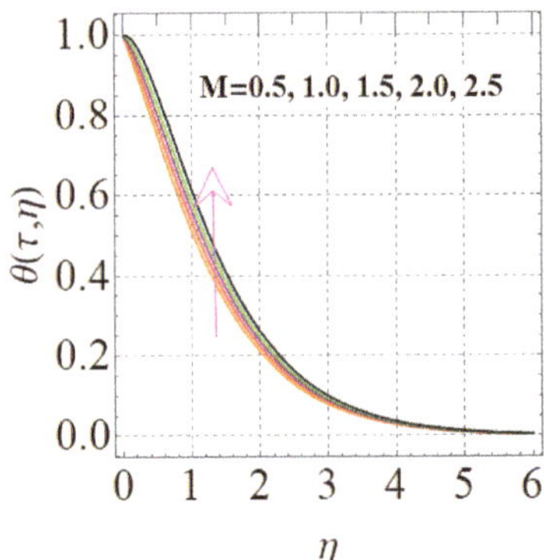

Figure 7. $\theta(\tau, \eta)$ via M.

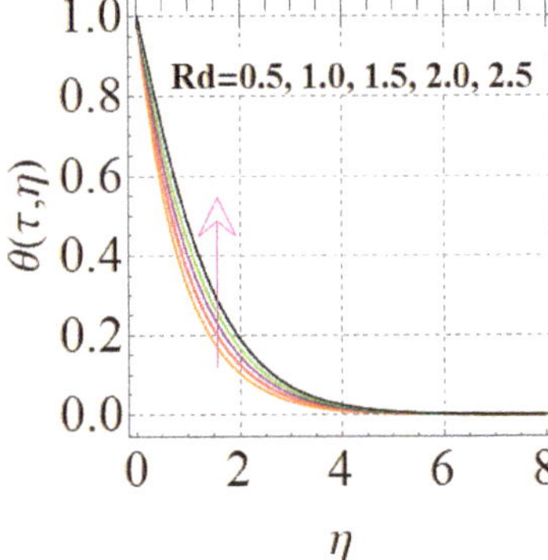

Figure 8. $\theta(\tau, \eta)$ via Rd.

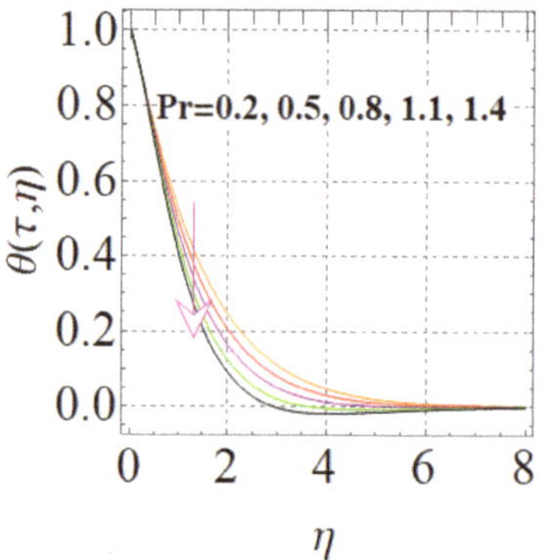

Figure 9. $\theta(\tau, \eta)$ via Pr.

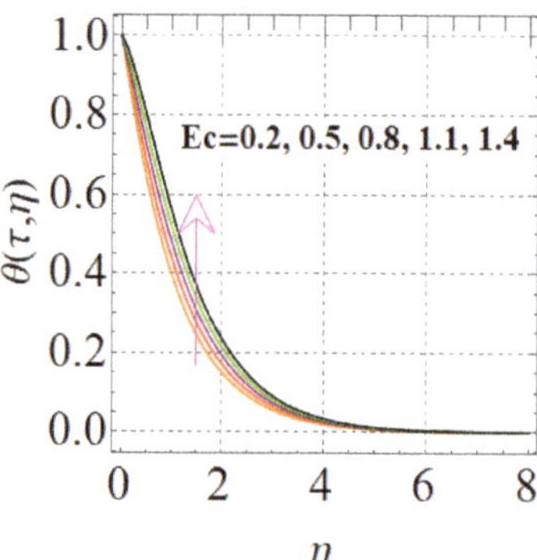

Figure 10. $\theta(\tau, \eta)$ via *Ec*.

6.3. Concentration

Figure 11 exhibits the concentration performance against the suction variable (S). Clearly, the concentration $(\phi(\tau, \eta))$ was reduced against larger (S) values. Figure 12 shows the performance of the concentration $(\phi(\tau, \eta))$ versus (Sc). An increment (Sc) decayed the mass diffusivity, and thus the concentration $(\phi(\tau, \eta))$ diminished. An amplification of the Soret number (Sr) led to a decaying value for $\phi(\tau, \eta)$ (see Figure 13). Figure 14 comprises the impact of $\phi(\tau, \eta)$ on γ. Here, $\phi(\tau, \eta)$ decreased against γ.

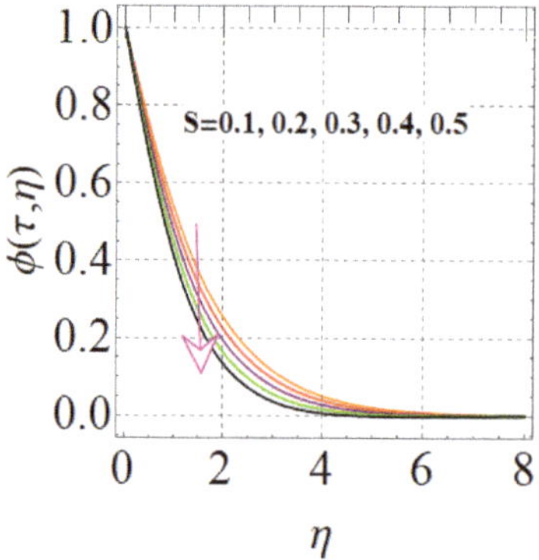

Figure 11. $\phi(\tau, \eta)$ via S.

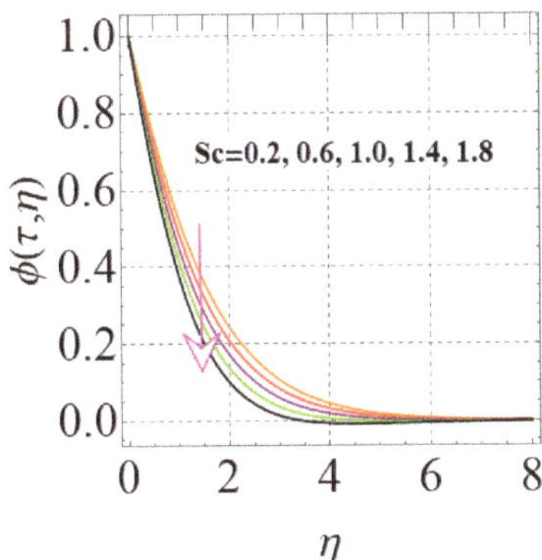

Figure 12. $\phi(\tau, \eta)$ via Sc.

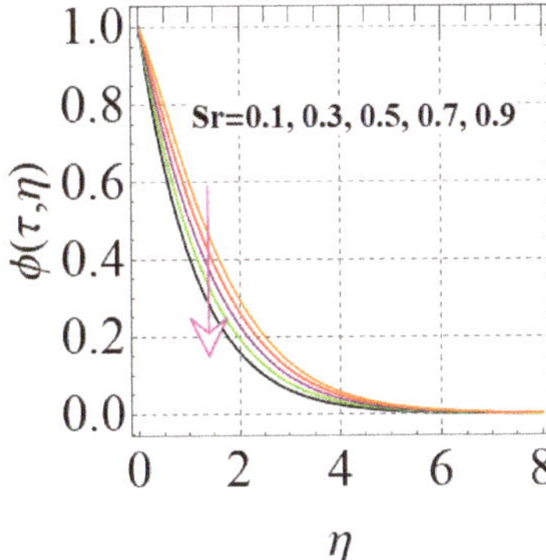

Figure 13. $\phi(\tau, \eta)$ via Sr.

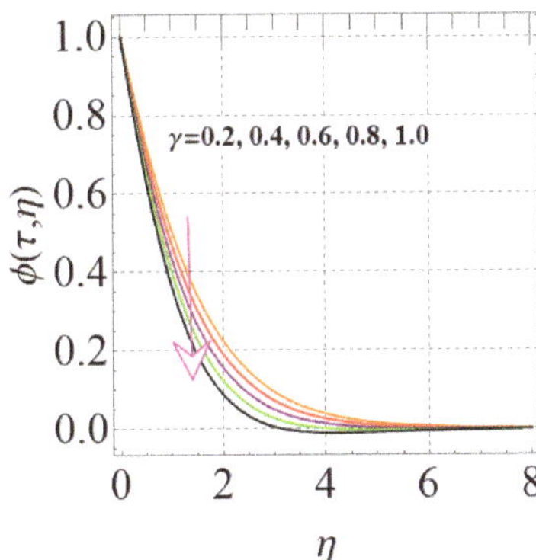

Figure 14. $\phi(\tau, \eta)$ via γ.

6.4. Entropy Generation Rate

The influence of the entropy rate ($N_G(\tau, \eta)$) via the radiation variable is disclosed in Figure 15. Clearly, a greater Rd value improved the radiation emission, which boosted the collision between the fluid particles, and so $N_G(\tau, \eta)$ was enhanced. Figure 16 discloses the impact of L on $N_G(\tau, \eta)$). As predicted, the entropy generation ($S_G(\eta)$) was greater via the higher approximation of L. A larger approximation of the Brinkman (Br) number enhanced the entropy generation ($S_G(\eta)$) (see Figure 17). This is because of augmentation through a higher Br value causing the viscous features to improve. As a result, the entropy rate rose. Figure 18 demonstrates the entropy rate for the magnetic parameter. A larger approximation of the magnetic variable led to an increase in the entropy rate.

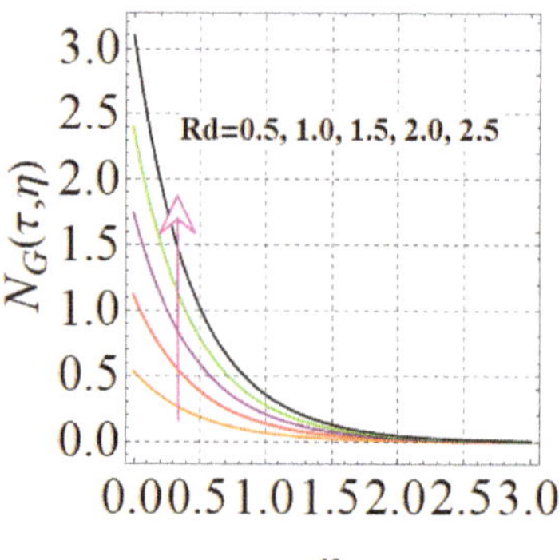

Figure 15. $N_G(\tau, \eta)$ via Rd.

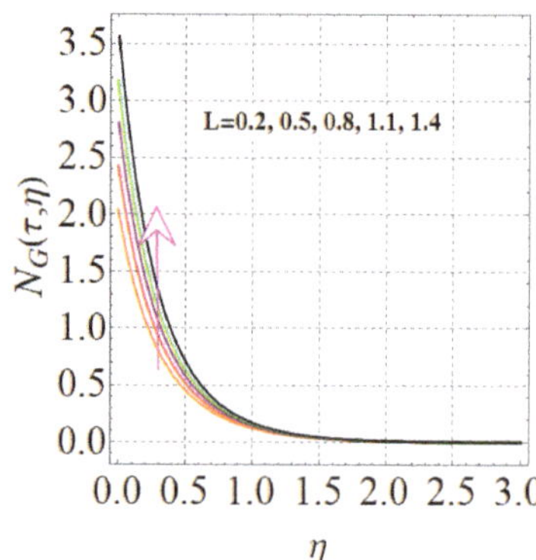

Figure 16. $N_G(\tau, \eta)$ via L.

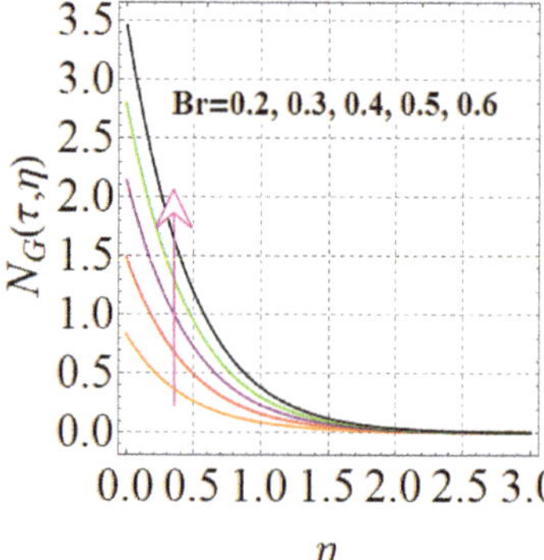

Figure 17. $N_G(\tau, \eta)$ via Br.

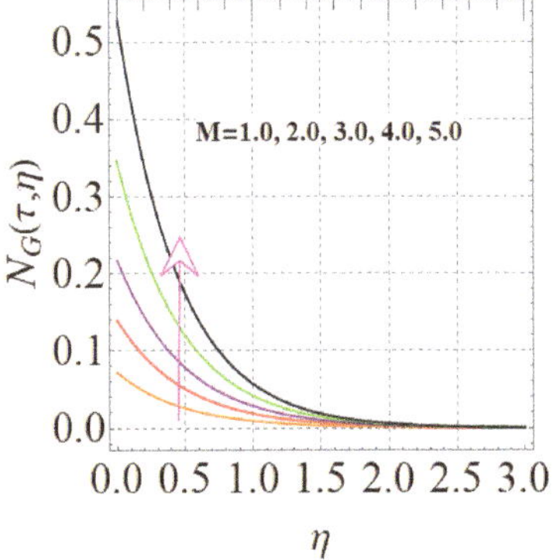

Figure 18. $N_G(\tau, \eta)$ via M.

7. Closing Points

- The theermal field and velocity for the magnetic field had opposing trends.
- A decrease in velocity was noted for the Forchheimer number and suction variable.
- The velocity versus the porosity parameter was decreased.
- Similar behavior for the concentration and temperature against suction was noticed.
- The temperatures for the Eckert and Prandtl numbers were dissimilar.
- Radiation for the entropy and temperature had a similar role.
- The concentration decayed via larger approximation of the Soret number and reaction parameter.
- A decay in concentration against the Schmidt number held.
- Entropy generation enhancement against the Brinkman number and diffusion variable was noticed.
- The entropy rate was boosted with variation in the diffusion variable.

Author Contributions: Conceptualization, S.A.K. and T.H.; Formal analysis, S.A.K. and T.H.; Investigation, S.A.K. and T.H.; Methodology, S.A.K. and T.H.; Supervision, T.H.; Writing—original draft, S.A.K.; Writing—review & editing, S.A.K. All authors have read and agreed to the published version of the manuscript.

Funding: This research received no external funding.

Conflicts of Interest: The authors declare no conflict of interest.

Nomenclature

u, v	Velocity components (ms^{-1})	x, y	Cartesian coordinates (m)
t	Time (s)	$v_0 > 0$	Suction velocity (ms^{-1})
ρ	Density (kgm^{-3})	σ	Electrical conductivity (Sm^{-1})
T	Temperature (K)	c_p	Specific heat ($Jkg^{-1}K^{-1}$)
k_p	Porous medium permeability (m^2)	C_b	Drag coefficient
T_w	Wall temperature (K)	α	Thermal diffusivity ($m^2\ s^{-1}$)
k	Thermal conductivity ($Wm^{-1}K^{-1}$)	T_∞	Ambient temperature (K)
σ^*	Stefan–Boltzman constant ($Wm^{-2}K^{-4}$)	K_T	Thermal diffusion ratio
C_s	Concentration susceptibility	k^*	Mean absorption coefficient (cm^{-1})
C	Concentration	k_r	Reaction rate (s)
C_w	Wall concentration	D_B	Mass diffusivity ($m^2\ s^{-1}$)
L_1	Reference length (m)	C_∞	Ambient concentration
u_w	Stretching velocity (ms^{-1})	a	Stretching rate constant (s^{-1})
Nu_x	Nusselt number	q_w	Heat flux (Wm^2)

Sh_x	Sherwood number	j_w	Mass flux
R	Molar gas constant ($\text{kgm}^2\ \text{s}^{-2}\text{K}^{-1}\text{mol}^{-1}$)	M	Magnetic variable
λ	Porosity variable	Fr	Forchheimer number
S	Suction parameter	Pr	Prandtl number
Rd	Radiation variable	Du	Dufour number
Ec	Eckert number	γ	Reaction variable
Sr	Soret number	Re	Reynold number
Sc	Schmidt number	N_G	Entropy rate
α_1	Temperature ratio variable	Br	Brinkman number
α_2	Concentration ratio variable	L	Diffusion variable
T_m	Mean fluid temperature (K)	B_0	Magnetic field strength

References

1. Darcy, H. *Les Fontaines Publiques de la Ville dr Dijion*; Dalmont, V., Ed.; Typ. Hennuyer: Paris, France, 1856; pp. 647–658.
2. Forchheimer, P. Wasserbewegung durch boden. *Z. Vereins Dtsch. Ingenieure* **1901**, *45*, 1782–1788.
3. Muskat, M. *The Flow of Homogeneous Fluids through Porous Media*; JW Edwards, Inc.: Ann Arbor, MI, USA, 1946.
4. Hayat, T.; Muhammad, T.; Al-Mezal, S.; Liao, S.J. Darcy-Forchheimer flow with variable thermal conductivity and Cattaneo-Christov heat flux. *Int. J. Numer. Methods Heat Fluid Flow* **2016**, *26*, 2355–2369. [CrossRef]
5. Pal, D.; Mondal, H. Hydromagnetic convective diffusion of species in Darcy-Forchheimer porous medium with non-uniform heat source/sink and variable viscosity. *Int. Commun. Heat Mass Transf.* **2012**, *39*, 913–917. [CrossRef]
6. Mallawi, F.; Ullah, M.Z. Conductivity and energy change in Carreau nanofluid flow along with magnetic dipole and Darcy-Forchheimer relation. *Alex. Eng. J.* **2021**, *60*, 3565–3575. [CrossRef]
7. Alshomrani, A.S.; Ullah, M.Z. Effects of homogeneous-heterogeneous reactions and convective condition in Darcy-Forchheimer flow of carbon nanotubes. *J. Heat Transf.* **2019**, *141*, 012405. [CrossRef]
8. Seth, G.S.; Mandal, P.K. Hydromagnetic rotating flow of Casson fluid in Darcy-Forchheimer porous medium. *MATEC Web Conf.* **2018**, *192*, 02059. [CrossRef]
9. Khan, S.A.; Hayat, T.; Alsaedi, A. Irreversibility analysis in Darcy-Forchheimer flow of viscous fluid with Dufour and Soret effects via finite difference method. *Case Stud. Therm. Eng.* **2021**, *26*, 101065. [CrossRef]
10. Azam, M.; Xu, T.; Khan, M. Numerical simulation for variable thermal properties and heat source/sink in flow of Cross nanofluid over a moving cylinder. *Int. Commun. Heat Mass Transf.* **2020**, *118*, 104832. [CrossRef]
11. Wu, Y.; Kou, J.; Sun, S. Matrix acidization in fractured porous media with the continuum fracture model and thermal Darcy-Brinkman-Forchheimer framework. *J. Pet. Sci. Eng.* **2022**, *211*, 110210. [CrossRef]
12. Haider, F.; Hayat, T.; Alsaedi, A. Flow of hybrid nanofluid through Darcy-Forchheimer porous space with variable characteristics. *Alex. Eng. J.* **2021**, *60*, 3047–3056. [CrossRef]
13. Tayyab, M.; Siddique, I.; Jarad, F.; Ashraf, M.A.; Ali, B. Numerical solution of 3D rotating nanofluid flow subject to Darcy-Forchheimer law, bio-convection and activation energy. *S. Afr. J. Chem. Eng.* **2022**, *40*, 48–56. [CrossRef]
14. Nawaz, M.; Sadiq, M.A. Unsteady heat transfer enhancement in Williamson fluid in Darcy-Forchheimer porous medium under non-Fourier condition of heat flux. *Case Stud. Therm. Eng.* **2021**, *28*, 101647. [CrossRef]
15. Ali, L.; Wang, Y.; Ali, B.; Liu, X.; Din, A.; Mdallal, Q.A. The function of nanoparticle's diameter and Darcy-Forchheimer flow over a cylinder with effect of magnetic field and thermal radiation. *Case Stud. Therm. Eng.* **2021**, *28*, 101392. [CrossRef]
16. Bejawada, S.G.; Reddy, Y.D.; Jamshed, W.; Eid, M.R.; Safdar, R.; Nisar, K.S.; Isa, S.S.P.M.; Alam, M.M.; Parvin, S. 2D mixed convection non-Darcy model with radiation effect in a nanofluid over an inclined wavy surface. *Alex. Eng. J.* **2022**, *61*, 9965–9976. [CrossRef]
17. Eid, M.R.; Mahny, K.L.; Al-Hossainy, A.F. Homogeneous-heterogeneous catalysis on electromagnetic radiative Prandtl fluid flow: Darcy-Forchheimer substance scheme. *Surf. Interfaces* **2021**, *24*, 101119. [CrossRef]
18. Rastogi, R.P.; Madan, G.L. Dufour Effect in Liquids. *J. Chem. Phys.* **1965**, *43*, 4179–4180. [CrossRef]
19. Rastogi, R.P.; Nigam, R.K. Cross-phenomenological coefficients. Part 6—Dufour effect in gases. *Trans. Faraday Soc.* **1966**, *62*, 3325–3330. [CrossRef]
20. Rastogi, R.P.; Yadava, B.L.S. Dufour effect in liquid mixtures. *J. Chem. Phys.* **1969**, *51*, 2826–2830. [CrossRef]
21. Moorthy, M.B.K.; Senthilvadivu, K. Soret and Dufour effects on natural convection flow past a vertical surface in a porous medium with variable viscosity. *J. Math. Phys.* **2012**, *2012*, 634806. [CrossRef]
22. El-Arabawy, H.A.M. Soret and dufour effects on natural convection flow past a vertical surface in a porous medium with variable surface temperature. *J. Math. Stat.* **2009**, *5*, 190–198. [CrossRef]
23. Reddy, G.J.; Raju, R.S.; Manideep, C.; Rao, J.A. Thermal diffusion and diffusion thermo effects on unsteady MHD fluid flow past a moving vertical plate embedded in porous medium in the presence of Hall current and rotating system. *Trans. A. Razmadze Math. Inst.* **2016**, *170*, 243–265. [CrossRef]

24. Dursunkaya, Z.; Worek, W.M. Diffusion-thermo and thermal-diffusion effects in transient and steady natural convection from vertical surface. *Int. J. Heat Mass Transf.* **1992**, *35*, 2060–2067. [CrossRef]
25. Khan, S.A.; Hayat, T.; Khan, M.I.; Alsaedi, A. Salient features of Dufour and Soret effect in radiative MHD flow of viscous fluid by a rotating cone with entropy generation. *Int. J. Hydrogen Energy* **2020**, *45*, 4552–14564. [CrossRef]
26. Bekezhanova, V.B.; Goncharova, O.N. Influence of the Dufour and Soret effects on the characteristics of evaporating liquid flows. *Int. J. Heat Mass Transf.* **2020**, *154*, 119696. [CrossRef]
27. Jiang, N.; Studer, E.; Podvin, B. Physical modeling of simultaneous heat and mass transfer: Species interdiffusion, Soret effect and Dufour effect. *Int. J. Heat Mass Transf.* **2020**, *156*, 119758. [CrossRef]
28. Bejan, A. Second law analysis in heat transfer. *Energy Int. J.* **1980**, *5*, 721–732. [CrossRef]
29. Bejan, A. *Entropy Generation Minimization*; CRC Press: New York, NY, USA, 1996.
30. Buonomo, B.; Pasqua, A.; Manca, O.; Nappo, S.; Nardini, S. Entropy generation analysis of laminar forced convection with nanofluids at pore length scale in porous structures with Kelvin cells. *Int. Commun. Heat Mass Transf.* **2022**, *132*, 105883. [CrossRef]
31. Khan, S.A.; Hayat, T.; Alsaedi, A.; Ahmad, B. Melting heat transportation in radiative flow of nanomaterials with irreversibility analysis. *Renew. Sustain. Energy Rev.* **2021**, *140*, 110739. [CrossRef]
32. Tayebi, T.; Öztop, H.F.; Chamkha, A.J. Natural convection and entropy production in hybrid nanofluid filled-annular elliptical cavity with internal heat generation or absorption. *Therm. Sci. Eng. Prog.* **2020**, *19*, 100605. [CrossRef]
33. Abbas, Z.; Naveed, M.; Hussain, M.; Salamat, N. Analysis of entropy generation for MHD flow of viscous fluid embedded in a vertical porous channel with thermal radiation. *Alex. Eng. J.* **2020**, *59*, 3395–3405. [CrossRef]
34. Rahmanian, S.; Koushkaki, H.R.; Shahsavar, A. Numerical assessment on the hydrothermal behaviour and entropy generation characteristics of boehmite alumina nanofluid flow through a concentrating photovoltaic/thermal system considering various shapes for nanoparticle. *Sustain. Energy Technol. Assess.* **2022**, *52*, 102143. [CrossRef]
35. Nayak, M.K.; Mabood, F.; Dogonchi, A.S.; Khan, W.A. Electromagnetic flow of SWCNT/MWCNT suspensions with optimized entropy generation and cubic auto catalysis chemical reaction. *Int. Commun. Heat Mass Transf.* **2020**, *2020*, 104996. [CrossRef]
36. Kumawat, C.; Sharma, B.K.; Al-Mdallal, Q.M.; Gorji, M.R. Entropy generation for MHD two phase blood flow through a curved permeable artery having variable viscosity with heat and mass transfer. *Int. Commun. Heat Mass Transf.* **2022**, *133*, 105954. [CrossRef]
37. Liu, Y.; Jian, Y.; Tan, W. Entropy generation of electromagnetohydrodynamic (EMHD) flow in a curved rectangular microchannel. *Int. J. Heat Mass Transf.* **2018**, *127*, 901–913. [CrossRef]
38. Alotaibi, H.; Eid, M.R. Thermal analysis of 3D electromagnetic radiative nanofluid flow with suction/blowing: Darcy–Forchheimer scheme. *Micromachines* **2021**, *12*, 1395. [CrossRef]
39. Eid, M.R.; Mabood, F. Entropy analysis of a hydromagnetic micropolar dusty carbon NTs-kerosene nanofluid with heat generation: Darcy–Forchheimer scheme. *J. Therm. Anal. Calorim.* **2021**, *143*, 2419–2436. [CrossRef]
40. Swain, I.; Pattanayak, H.; Das, M.; Singh, T. Finite difference solution of free convective heat transfer of non-Newtonian power law fluids from a vertical plate. *Glob. J. Pure Appl. Math.* **2015**, *11*, 339–348.
41. Adekanye, O.; Washington, T. Nonstandard finite difference scheme for a Tacoma narrows bridge model. *Appl. Math. Model.* **2018**, *62*, 223–236. [CrossRef]
42. Hayat, T.; Ullah, H.; Ahmad, B.; Alhodaly, M.S. Heat transfer analysis in convective flow of Jeffrey nanofluid by vertical stretchable cylinder. *Int. Commun. Heat Mass Transf.* **2021**, *120*, 104965. [CrossRef]
43. Khan, Z.H.; Makinde, O.D.; Ahmad, R.; Khan, W.A. Numerical study of unsteady MHD flow and entropy generation in a rotating permeable channel with slip and Hall effects. *Commun. Theor. Phys.* **2018**, *70*, 641–650. [CrossRef]
44. Bidin, B.; Nazar, R. Numerical solution of the boundary layer flow over an exponentially stretching sheet with thermal radiation. *Eur. J. Sci. Res.* **2009**, *33*, 710–717.

Mathematical and Computational Applications

Article

A New Material Model for Agglomerated Cork

Gabriel Thomaz de Aquino Pereira [1], Ricardo J. Alves de Sousa [2,3], I-Shih Liu [4], Marcello Goulart Teixeira [1,*] and Fábio A. O. Fernandes [2,3]

[1] Intitute of Computing, Federal University of Rio de Janeiro, Rio de Janeiro 21941-853, Brazil
[2] TEMA—Centre for Mechanical Technology and Automation, Department of Mechanical Engineering, University of Aveiro, 3810-193 Aveiro, Portugal
[3] LASI—Intelligent Systems Associate Laboratory, 4800-058 Guimarães, Portugal
[4] Intitute of Mathematics, Federal University of Rio de Janeiro, Rio de Janeiro 21941-853, Brazil
* Correspondence: marcellogt@ic.ufrj.br

Abstract: It is increasingly necessary to promote means of production that are less polluting and less harmful to the environment following the UN 2030 agenda for sustainable development. Using natural cellular materials in structural applications can be essential for enabling a future in this direction. Cork is a natural cellular material with an excellent energy absorption capacity. Its use in engineering applications and products has grown over time, so predicting its mechanical response through numerical tools is crucial. Classical cork modeling uses a model developed for foam material, including an adjustment function that does not have a clear physical interpretation. This work presents a new material model for an agglomerated cork based solely on well-known hypotheses of continuum mechanics using fewer parameters than the classical model and further a finite element framework to validate the new model against experimental data.

Keywords: agglomerated cork; material modeling; successive linear approximation; finite element

Citation: Pereira, G.T.d.A.; Sousa, R.J.A.d.; Liu, I.-S.; Teixeira, M.G.; Fernandes, F.A.O. A New Material Model for Agglomerated Cork. *Math. Comput. Appl.* **2022**, *27*, 92. https://doi.org/10.3390/mca27060092

Academic Editor: Gianluigi Rozza

Received: 3 September 2022
Accepted: 7 November 2022
Published: 9 November 2022

1. Introduction

The use of cellular materials in engineering applications has been established worldwide. This kind of material has excellent crashworthiness and insulation properties. Styrofoam (expanded polystyrene), derived from oil, is widely present in packaging and safety apparel, with excellent cost/performance ratios but with well-known recyclability and biodegradability issues after usage. Cork, the outer bark of *Quercus suber* L. tree, is a natural cellular material by excellence, allying crashworthiness and insulation properties. Compared with its synthetic counterparts, it is a sustainable and eco-friendly solution.

Dart and Eugene [1] made one of the first attempts to characterize cork mechanical behavior under compression loading. They found that each stress–time curve for various compressions can be obtained from each other by scaling. Such scalability was justified by the shape of the stress–strain compression curve. A highly non-linear stress–strain response curve characterizes agglomerated cork under compression. This curve is divided into three parts: a small elastic linear behavior at the beginning, followed by a plateau, and finished with a highly non-linear densification part.

All types of cork, natural, agglomerated, and expanded, are characterized by this compression behavior. The material properties vary with density, cellular dimensions, and porosity, as seen in [2,3]. The Poisson effect in natural cork compression was studied by [4]. They found that for compression in axial and tangential directions, the Poisson's ratio is almost null, while in the radial direction, cork presents a Poisson's ratio of 0.3. The cell geometry of the cork explained this effect.

Unlike natural cork, which presents the natural material anisotropy [5], cork agglomerates are obtained by compressing together randomly oriented cork grains [6]. This manufacturing procedure promotes a fairly regular isotropic mechanical behavior in cork agglomerates.

The authors performed experimental tests to obtain the Young and shear modulus in [7,8]. To determine the Young modulus, a tensile test was conducted, and, in the case of the shear modulus, they performed a torsion test on a cork cylinder. Ref. [8] also concluded that the amount of strain necessary to fracture in the radial direction is much larger than in the other directions. More recently, agglomerated cork is being used as an ideal core material for sandwich components of lightweight structures [9–14]. This structure is interesting because cork is used as an energy dissipator.

Usually, a material characterized by a mechanical behavior similar to cork's under compression is numerically modeled as a foam material model [15]. The Ogden-Hill hyperelastic model, [16,17], is a well-known numerical model for elastomeric polymer foams and hence a good model for agglomerated cork. More recently, in [18], an investigation was carried out to determine the influence of the variability related to the material properties of natural cork based on a numerical homogenization approach to predict the temperature-dependent equivalent elastic properties. These approaches consider only the elastic behavior of the material since the plastic period is only reached through substantial deformation and high strain energies.

This work focuses on the modeling of the mechanical response of cork. We present a new material model for cork agglomerates based on an extension of the Mooney–Rivlin material. The material parameterization will be made through an optimization problem using uniaxial and equibiaxial experimental results and the analytical solutions for these compression tests. To the authors' best knowledge, it is also the first time that experimental data from biaxial compressions of cork has been reported. After the parameterization, a finite element framework is used to validate this new material model against experimental data.

Due to the non-linear nature of both the material model and the large deformation, we use the successive linear approximation method for numerical simulation purposes to deal with the problem. This method calculates the constitutive equations at each state, with the reference configuration updated for each time step. The new reference configuration is the current configuration of the body. Assuming that in each time step that occurs, a small deformation is added, both the constitutive equations and the PDE system are linearized.

In Section 2, we briefly present the Successive Linear Approximation method used in this work, and in Section 3, the hyperelastic material model is considered. Section 4 presents the parameter fitting for the cork agglomerates model, and finally, in Section 5, we show the comparison between numerical results, obtained by a finite element code against analytical and experimental results.

2. Successive Linear Approximation Method

The Successive Linear Approximation (SLA) Method [19], a relative Lagrangian formulation based on the "small-on-large" idea, allows implementing the solution to large deformation in a successive incremental manner. In other words, at each time step, the constitutive function is calculated at the present state of deformation, which will be regarded as the reference configuration for the next state. From this point of view, it is considered a relative motion description; see, for example [20]. Assuming that the deformation to the next state is small, the constitutive function and the partial differential equation can be linearized. This procedure for large deformation problems was presented in [21].

This approach has significant fields of application, such as salt tectonics [22]. In [23], the influence of temperature in salt domes, a thermoviscoelastic material, was studied.

2.1. *Relative Motion Description*

Let κ_0 be the preferred reference configuration of an elastic body $\mathcal{B}$, $\mathcal{B}_0 = \kappa_0(\mathcal{B})$, and $\mathbf{x} = \chi(X,t)$, with $X \in \mathcal{B}_0$ be the motion of the body, see Figure 1. The configuration with specific material symmetries, such as isotropy, is usually chosen as the reference configuration. Such material symmetries may be lost in the deformed configuration in general.

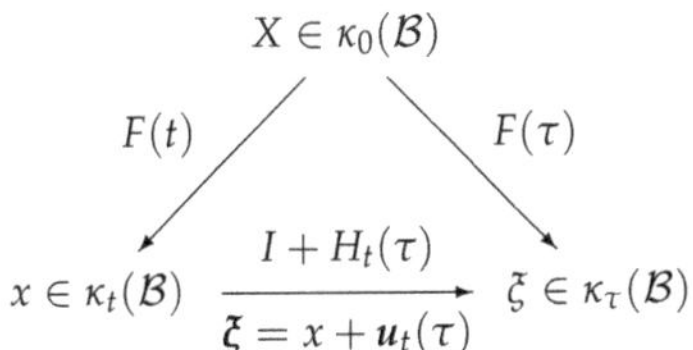

Figure 1. The deformation and deformation gradient diagram.

In this paper, we consider the time t as the present time. Thus, κ_t is the deformed configuration at time t, $\mathcal{B}_t = \kappa_t(\mathcal{B})$. Now, we can calculate the deformation gradient with respect to configuration κ_0 as

$$\mathbf{F}(X,t) = \nabla_X \chi(\mathbf{X},\mathbf{t}). \tag{1}$$

Now, let κ_τ be the deformed configuration at time τ. The deformation $\xi = \chi(X,\tau)$ from κ_0 can be described in the current configuration at the present time t by

$$\xi = \chi(X,\tau) := \chi_t(\mathbf{x},\tau) \quad \text{for} \quad \mathbf{x} = \chi(X,t), \tag{2}$$

where $\chi_t(\cdot,\tau) : \mathcal{B}_t \longrightarrow \mathcal{B}_\tau$ is called relative deformation. With these in mind we can define the relative displacement vector $\mathbf{u}$ as

$$\mathbf{u} = \xi - \mathbf{x} = \mathbf{u}_t(\mathbf{x},\tau) = \chi_t(\mathbf{x},\tau) - \mathbf{x}, \qquad \mathbf{x} \in \mathcal{B}_t. \tag{3}$$

Taking the gradient relative to X and $\mathbf{x}$, respectively, we have

$$\begin{aligned} \mathbf{H}_t(\mathbf{x},\tau)\mathbf{F}(X,t) &= \nabla_{\mathbf{x}}\mathbf{u}_t(\mathbf{x},\tau)\mathbf{F}(X,t) = \mathbf{F}(X,\tau) - \mathbf{F}(X,t), \\ \mathbf{H}_t(\mathbf{x},\tau) &= \nabla_{\mathbf{x}}\chi_t(\mathbf{x},\tau) - \mathbf{I} = \mathbf{F}_t(\mathbf{x},\tau) - \mathbf{I}, \end{aligned} \tag{4}$$

where $\mathbf{I}$, $\mathbf{H}_t(\mathbf{x},\tau)$ and $\mathbf{F}_t(\mathbf{x},\tau)$ are the identity tensor, relative displacement gradient and relative deformation gradient, respectively. We can rewrite the above equations as

$$\mathbf{F}_t(\mathbf{x},\tau) = \mathbf{F}(X,\tau)\mathbf{F}(X,t)^{-1}, \qquad \mathbf{F}(X,\tau) = (\mathbf{I} - \mathbf{H}_t(\mathbf{x},\tau))\mathbf{F}(X,t). \tag{5}$$

With all these equations, we can define the motions of a body, and it is called relative description formulation.

2.2. Linearized Constitutive Equation

Consider $\tau = t + \Delta t$. By taking Δt small enough, we can assume that the gradient of the displacement is small, and for simplicity, we denote

$$\mathbf{H}(\tau) := \mathbf{H}_t(\mathbf{x},\tau)), \text{ with } \parallel \mathbf{H}(\tau) \parallel \ll 1, \tag{6}$$

moreover, from (4), we have

$$\mathbf{F}(\tau) - \mathbf{F}(t) = \mathbf{H}(\tau)\mathbf{F}(t), \qquad \mathbf{F}_t(\tau) = \mathbf{I} + \mathbf{H}(\tau). \tag{7}$$

Without loss of generality, the Cauchy stress tensor of an elastic body in the preferred reference configuration is given by

$$\mathbf{T}(X,t) = -p\mathbf{I} + \mathcal{F}_{\kappa_0}(\mathbf{F}(X,t)), \tag{8}$$

where, for compressible bodies, $p = p(\mathbf{F})$. We shall assume that this dependence is only through the determinant, or equivalently, from the mass balance, depending only on the mass density, i.e., $p = p(\rho)$ where $\rho = \frac{\rho_0}{\det\mathbf{F}}$.

For time $\tau = t + \Delta t$, we have

$$\begin{aligned}\rho(\tau) - \rho(t) &= \rho_0(\det\mathbf{F}(\tau)^{-1} - \det\mathbf{F}(t)^{-1}) = \rho(t)(\det(\mathbf{F}(t)\mathbf{F}(\tau)^{-1}) - 1)\\ &= \rho(t)(\det(\mathbf{I} + \mathbf{H}(\tau))^{-1} - 1) = -\rho(t)\mathrm{tr}\mathbf{H}(\tau) + o(2),\end{aligned} \tag{9}$$

where in the passages, (5) and (7) were used. Using Taylor series expansion in (8) and (9), it follows that

$$\begin{aligned}\mathbf{T}(\tau) &= \mathbf{T}(t) - \left(\frac{\partial p}{\partial \rho}\right)_t (\rho(\tau) - \rho(t))\mathbf{I} + \nabla_{\mathbf{F}}\mathcal{F}(\mathbf{F}(t))(\mathbf{F}(\tau) - \mathbf{F}(t))\\ &= \mathbf{T}(t) - \beta(\mathrm{tr}\mathbf{H}(\tau))\mathbf{I} + \nabla_{\mathbf{F}}\mathcal{F}(\mathbf{F}(t))(\mathbf{H}(\tau)\mathbf{F}(t),\end{aligned} \tag{10}$$

or

$$\mathbf{T}(\tau) = \mathbf{T}(t) + \mathcal{L}(\mathbf{F}(t))[\mathbf{H}(\tau)], \tag{11}$$

where $\beta = \rho\frac{\partial p}{\partial \rho}$ is a material parameter and

$$\mathcal{L}(\mathbf{F}(t))[\mathbf{H}(\tau)] := -\beta(\mathrm{tr}\mathbf{H}(\tau))\mathbf{I} + \nabla_{\mathbf{F}}\mathcal{F}(\mathbf{F}(t))[\mathbf{H}(\tau)\mathbf{F}(t)] \tag{12}$$

defines the fourth-order elasticity tensor, relative to the current configuration κ_t. Furthermore, the first Piola–Kirchhoff stress tensor at time τ relative to the current configuration, is given by

$$\begin{aligned}\mathbf{T}_{\kappa_t}(\tau) &= \det\mathbf{F}_t(\tau)\mathbf{T}(\tau)\mathbf{F}_t(\tau)^{-T} = \det(\mathbf{I} + \mathbf{H})\mathbf{T}(\tau)(\mathbf{I} + \mathbf{H})^{-T}\\ &= \det(\mathbf{I} + \mathbf{H})(\mathbf{T}(t) + \mathcal{L}(\mathbf{F})[\mathbf{H}] + o(2))(\mathbf{I} + \mathbf{H})^{-T}\\ &= (\mathbf{I} + \mathrm{tr}\mathbf{H})(\mathbf{T}(t) + \mathcal{L}(\mathbf{F})[\mathbf{H}])(\mathbf{I} - \mathbf{H}^T) + o(2)\\ &= \mathbf{T}(t) + (\mathrm{tr}\mathbf{H})\mathbf{T}(t) - \mathbf{T}(t)\mathbf{H}^T + \mathcal{L}(\mathbf{F})[\mathbf{H}] + o(2),\end{aligned} \tag{13}$$

and we write the linearized first Piola–Kirchhoff as

$$\mathbf{T}_{\kappa_t}(\tau) = \mathbf{T}(t) + \mathcal{K}(\mathbf{F}(t), \mathbf{T}(t))[\mathbf{H}(\tau)], \tag{14}$$

where

$$\mathcal{K}(\mathbf{F}, \mathbf{T})[\mathbf{H}] := (\mathrm{tr}\mathbf{H})\mathbf{T}(t) - \mathbf{T}(t)\mathbf{H}^T + \mathcal{L}(\mathbf{F})[\mathbf{H}] \tag{15}$$

is the fourth-order elasticity tensor for the first Piola–Kirchhoff stress tensor.

To define the fourth-order elasticity tensor we have to define the material model. In Section 3 we will propose one based on Mooney–Rivlin hyperelastic material model.

3. Hyperelastic Material Model

To obtain a constitutive model for large strains of an isotropic elastic solid, we shall start with the free energy function $\psi = \psi(\mathrm{I}_\mathbf{B}, \mathrm{II}_\mathbf{B}, \mathrm{III}_\mathbf{B})$, and the Cauchy stress given by

$$\mathbf{T} = 2\rho\frac{\partial\psi}{\partial\mathbf{B}}\mathbf{B}, \tag{16}$$

where $\mathbf{B}$ is the left Cauchy–Green strain tensor and $\mathrm{I}_\mathbf{B}, \mathrm{II}_\mathbf{B}, \mathrm{III}_\mathbf{B}$ are the first, second and third invariants of $\mathbf{B}$. By Taylor series expansion, we can write

$$\begin{aligned}\psi(\mathrm{I}_\mathbf{B}, \mathrm{II}_\mathbf{B}, \mathrm{III}_\mathbf{B}) &= \psi_0(3, 3, \mathrm{III}_\mathbf{B}) + \psi_1(\mathrm{I}_\mathbf{B} - 3) + \psi_2(\mathrm{II}_\mathbf{B} - 3)\\ &\quad + \frac{1}{2}\psi_3(\mathrm{I}_\mathbf{B} - 3)^2 + \psi_4(\mathrm{I}_\mathbf{B} - 3)(\mathrm{II}_\mathbf{B} - 3) + \frac{1}{2}\psi_5(\mathrm{II}_\mathbf{B} - 3)^2,\end{aligned} \tag{17}$$

where $\psi_k = \psi_k(\mathrm{III}_\mathbf{B})$ for $k = 0, 1 \ldots, 5$, which stands for the expansion up to the second order for moderate strains.

From (16), by the use of the relations

$$\frac{\partial \mathrm{I}_\mathbf{B}}{\partial \mathbf{B}} = \mathbf{I}, \qquad \frac{\partial \mathrm{II}_\mathbf{B}}{\partial \mathbf{B}} = \mathrm{I}_\mathbf{B}\mathbf{I} - \mathbf{B}, \qquad \frac{\partial \mathrm{III}_\mathbf{B}}{\partial \mathbf{B}} = \mathrm{III}_\mathbf{B}\mathbf{B}^{-1}, \tag{18}$$

we obtain

$$\begin{aligned} \mathbf{T} + p\mathbf{I} = {} & 2\rho[\psi_1 + \psi_3(\mathrm{I}_\mathbf{B} - 3) + \psi_4(\mathrm{II}_\mathbf{B} - 3)]\mathbf{B} \\ & - 2\rho[\psi_2 + \psi_4(\mathrm{I}_\mathbf{B} - 3) + \psi_5(\mathrm{II}_\mathbf{B} - 3)](\mathbf{B}^2 - \mathrm{I}_\mathbf{B}\mathbf{B}), \end{aligned} \tag{19}$$

where terms with the identity tensor are lumped into the pressure $p(\mathrm{I}_\mathbf{B}, \mathrm{II}_\mathbf{B}, \mathrm{III}_\mathbf{B})$.

By the use of the Cayley–Hamilton theorem,

$$\mathbf{B}^2 - \mathrm{I}_\mathbf{B}\mathbf{B} = \mathrm{III}_\mathbf{B}\mathbf{B}^{-1} - \mathrm{II}_\mathbf{B}\mathbf{I}, \tag{20}$$

it becomes

$$\begin{aligned} \mathbf{T} + p\mathbf{I} = {} & 2\rho[\psi_1 + \psi_3(\mathrm{I}_\mathbf{B} - 3) + \psi_4(\mathrm{II}_\mathbf{B} - 3)]\mathbf{B} \\ & - 2\rho[\psi_2 + \psi_4(\mathrm{I}_\mathbf{B} - 3) + \psi_5(\mathrm{II}_\mathbf{B} - 3)]\mathrm{III}_\mathbf{B}\mathbf{B}^{-1}, \end{aligned} \tag{21}$$

in which the term $\mathrm{II}_\mathbf{B}\mathbf{I}$ is absorbed into the indeterminate pressure $p\mathbf{I}$.

We can rewrite the above equation with the definition of the parameters,

$$\begin{aligned} & s_1 = 2\rho(\psi_1 - 3\psi_3 - 3\psi_4), \\ & s_2 = 2\rho\mathrm{III}_\mathbf{B}(\psi_2 - 3\psi_4 - 3\psi_5), \\ & s_3 = 2\rho\psi_3, \quad s_4 = 2\rho\mathrm{III}_\mathbf{B}\psi_4, \\ & s_5 = 2\rho\psi_4, \quad s_6 = 2\rho\mathrm{III}_\mathbf{B}\psi_5, \end{aligned} \tag{22}$$

and propose an isotropic elastic model as:

An extended Mooney–Rivlin material. *The constitutive equation*

$$\mathbf{T} = -p\mathbf{I} + (s_1 + s_3\mathrm{I}_\mathbf{B} + s_5\mathrm{II}_\mathbf{B})(\mathbf{B} - \mathbf{I}) - (s_2 + s_4\mathrm{I}_\mathbf{B} + s_6\mathrm{II}_\mathbf{B})(\mathbf{B}^{-1} - \mathbf{I}), \tag{23}$$

is an extended version of Mooney–Rivlin model for isotropic solids at large strain. We shall assume that six parameters $s_1, \ldots, s_6$ are material constants, and the pressure is a function of mass density only, $p(\rho)$ so that

$$\beta = \rho\frac{\partial p}{\partial \rho}, \tag{24}$$

is a material parameter.

With the proposed constitutive equation, from (12) we can derive the fourth-order elastic tensor. Therefore, in Einstein notation, it follows that

$$\begin{aligned} \mathcal{L}_{ijkl} = {} & \beta(\delta_{ij}\delta_{kl}) \\ & + (s_1 + s_3\mathrm{I}_\mathbf{B} + s_5\mathrm{II}_\mathbf{B})(B_{lj}\delta_{ki} + B_{il}\delta_{kj}) \\ & + (s_2 + s_4\mathrm{I}_\mathbf{B} + s_6\mathrm{II}_\mathbf{B})(B^{-1}_{ik}\delta_{jl} + B^{-1}_{kj}\delta_{il}) \\ & + 2(s_3B_{kl} + s_5(\mathrm{II}_\mathbf{B}\delta_{kl} - \mathrm{III}_\mathbf{B}B^{-1}_{kl}))(B_{ij} - \delta_{ij}) \\ & - 2(s_4B_{kl} + s_6(\mathrm{II}_\mathbf{B}\delta_{kl} - \mathrm{III}_\mathbf{B}B^{-1}_{kl}))(B^{-1}_{ij} - \delta_{ij}). \end{aligned} \tag{25}$$

Remark on β value for cork: Since, in this paper, we are considering the cork agglomerated as a material, from the definition of β and the compressibility of cork, it follows that $\beta \approx 0$.

Remarks on linear model: Let $\mathbf{B} = \mathbf{I} + 2\mathbf{E}$, hence $\mathbf{B}^{-1} = \mathbf{I} - 2\mathbf{E}$ and $\mathrm{tr}\mathbf{H} = \mathrm{tr}\mathbf{E}$, for small linear strain $\mathbf{E}$, then

$$\mathrm{I}_{\mathbf{B}} = \mathrm{tr}(\mathbf{I} + 2\mathbf{E}) = 3 + 2\mathrm{tr}\mathbf{E}, \tag{26}$$

$$\mathrm{II}_{\mathbf{B}} = (\det\mathbf{B})\mathrm{I}_{\mathbf{B}^{-1}} = (1 + \mathrm{tr}\mathbf{E})\mathrm{tr}(\mathbf{I} - 2\mathbf{E}) \tag{27}$$

$$= (1 + \mathrm{tr}\mathbf{E})(3 - 2\mathrm{tr}\mathbf{E}) = 3 + \mathrm{tr}\mathbf{E}, \tag{28}$$

and, using a Taylor expansion of $p(\rho)$, the stress becomes

$$\mathbf{T} = -(p_0 - \beta_0\mathrm{tr}\mathbf{E})\mathbf{I} + 2[s_1 + s_3(3 + 2\mathrm{tr}\mathbf{E}) + s_5(3 + \mathrm{tr}\mathbf{E})]\mathbf{E} \tag{29}$$

$$- 2[s_2 + s_4(3 + 2\mathrm{tr}\mathbf{E}) + s_6(3 + \mathrm{tr}\mathbf{E})]\mathbf{E} \tag{30}$$

$$= -p_0\mathbf{I} + \lambda(\mathrm{tr}\mathbf{E})\mathbf{I} + 2\mu\mathbf{E} + o(2), \tag{31}$$

where

$$\lambda = \beta_0, \tag{32}$$

$$\mu = s_1 - s_2 + 3(s_3 + s_5 - s_4 - s_6) \tag{33}$$

are the elastic Lamé constants. Note that we have the relation from linear elasticity,

$$\beta_0 = \lambda = \frac{2\nu\mu}{1 - 2\nu}, \tag{34}$$

where ν is the Poisson ratio.

4. Parameter Fitting for Cork Agglomerates

Cork agglomerate can be modeled as a hyperelastic material. As any other hyperelastic material, to parametrize an agglomerated cork we need at least two different deformation modes to capture the correct material behavior [24].

In this paper, we will use the uniaxial and the equibiaxial compression tests, schematically represented in Figure 2, to parametrize the extended Mooney–Rivlin model proposed before. For the parametrization procedure adopted we need the experimental data, the analytical solutions for both deformation modes and a good minimization function.

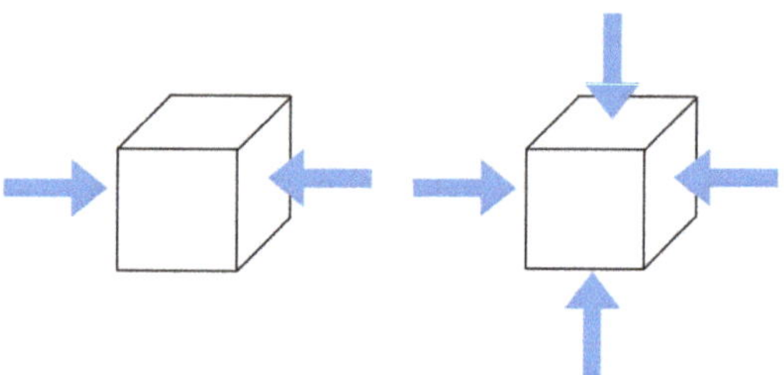

Figure 2. Uniaxial and equibiaxial compression tests schemes.

4.1. Uniaxial Compression

The uniaxial extension is given by

$$x = \lambda X, \quad y = Y, \quad z = Z, \tag{35}$$

where λ is the stretch along the loading. For agglomerated cork, in uniaxial extension, the transversal stretches have the same value and they are near 1, which is why we are not considering them. Therefore, the gradient deformation, the left Cauchy–Green strain tensor and its inverse are given, respectively, by

$$\mathbf{F} = \begin{bmatrix} \lambda & 0 & 0 \\ 0 & 1 & 0 \\ 0 & 0 & 1 \end{bmatrix}, \quad \mathbf{B} = \begin{bmatrix} \lambda^2 & 0 & 0 \\ 0 & 1 & 0 \\ 0 & 0 & 1 \end{bmatrix}, \quad \mathbf{B}^{-1} = \begin{bmatrix} \frac{1}{\lambda^2} & 0 & 0 \\ 0 & 1 & 0 \\ 0 & 0 & 1 \end{bmatrix},$$

and from that, the first, second and third invariants of **B** are

$$\mathrm{I}_{\mathbf{B}} = 2 + \lambda^2, \qquad \mathrm{II}_{\mathbf{B}} = 2\lambda^2 + 1, \qquad \mathrm{III}_{\mathbf{B}} = \lambda^2. \tag{36}$$

Consequently, the Cauchy stress tensor, $T(\lambda)$, can be obtained by inserting (36) into (23):

$$\begin{aligned} \mathbf{T}(\lambda) = &-p\mathbf{I} + [s_1 + s_3(2+\lambda^2) + s_5(2\lambda^2+1)](\mathbf{B} - \mathbf{I}) \\ &- [s_2 + s_4(2+\lambda^2) + s_6(2\lambda^2+1)](\mathbf{B}^{-1} - \mathbf{I}). \end{aligned} \tag{37}$$

Since there is no stretch in the second and third direction we have $T_{U2}(\lambda) = T_{U3}(\lambda) = 0$. Therefore, $p = 0$ and the principal stress in the first direction is defined by:

$$\begin{aligned} T_{U1}(\lambda) = &[s_1 + s_3(2+\lambda^2) + s_5(2\lambda^2+1)](\lambda^2 - 1) \\ &- [s_2 + s_4(2+\lambda^2) + s_6(2\lambda^2+1)](\lambda^{-2} - 1), \end{aligned} \tag{38}$$

4.2. *Equibiaxial Compression*

The equibiaxial extension is given by

$$x = \lambda X, \quad y = \lambda Y, \quad z = Z, \tag{39}$$

Therefore, the deformation gradient, the left Cauchy–Green strain tensor and its inverse are given, respectively, by

$$\mathbf{F} = \begin{bmatrix} \lambda & 0 & 0 \\ 0 & \lambda & 0 \\ 0 & 0 & 1 \end{bmatrix}, \quad \mathbf{B} = \begin{bmatrix} \lambda^2 & 0 & 0 \\ 0 & \lambda^2 & 0 \\ 0 & 0 & 1 \end{bmatrix}, \quad \mathbf{B}^{-1} = \begin{bmatrix} \frac{1}{\lambda^2} & 0 & 0 \\ 0 & \frac{1}{\lambda^2} & 0 \\ 0 & 0 & 1 \end{bmatrix},$$

and from that, the first, second and third invariants of **B** are

$$\mathrm{I}_{\mathbf{B}} = 1 + 2\lambda^2, \qquad \mathrm{II}_{\mathbf{B}} = 2\lambda^2 + \lambda^4, \qquad \mathrm{III}_{\mathbf{B}} = \lambda^4. \tag{40}$$

Consequently, the Cauchy stress tensor, $T(\lambda)$, can be obtained by inserting (40) into (23):

$$\begin{aligned} \mathbf{T}(\lambda) = &-p\mathbf{I} + [s_1 + s_3(1+2\lambda^2) + s_5(2\lambda^2+\lambda^4)](\mathbf{B} - \mathbf{I}) \\ &- [s_2 + s_4(1+2\lambda^2) + s_6(2\lambda^2+\lambda^4)](\mathbf{B}^{-1} - \mathbf{I}). \end{aligned} \tag{41}$$

Since there is no stretch in the third direction, we have $T_{B3}(\lambda) = 0$. Therefore, $p = 0$ and the principal stress in the first and second directions are defined by:

$$\begin{aligned} T_{B1}(\lambda) = T_{B2}(\lambda) = &[s_1 + s_3(1+2\lambda^2) + s_5(2\lambda^2+\lambda^4)](\lambda^2 - 1) \\ &- [s_2 + s_4(1+2\lambda^2) + s_6(2\lambda^2+\lambda^4)](\lambda^{-2} - 1). \end{aligned} \tag{42}$$

Equations (38) and (42) are the analytical solutions for the uniaxial and equibiaxial compression problems, respectively.

4.3. Experimental Tests

To parameterize our model, we have to collect some experimental data for both tests, uniaxial, and equibiaxial compression. The quasi-static uniaxial compression tests were performed in a universal testing machine, the Shimadzu AGS-X-10 kN. The equibiaxial compression tests were performed in an in-house built biaxial machine developed and properly validated in [25]. Both types of experiments were carried out at room temperature.

For uniaxial compression tests, 60 mm agglomerated cork cubes were manufactured, as shown in Figure 3. For the equibiaxial tests, 30 mm thick octagon-shaped samples were produced. The other dimensions are shown in Figure 4.

Figure 3. Agglomerated cork sample positioned for uniaxial compression testing at quasi-static strain rates.

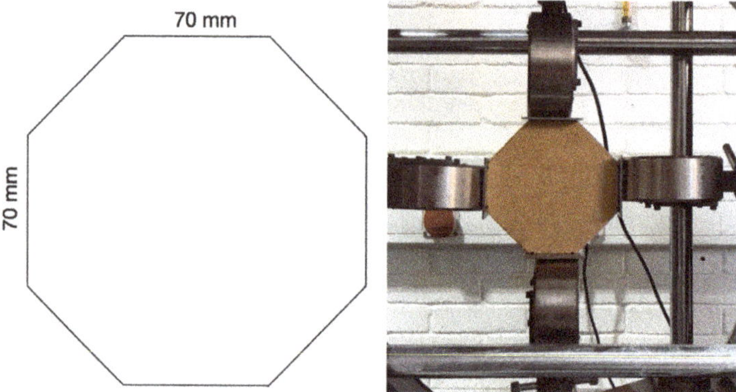

Figure 4. Dimensions of the octagon-shaped sample and the experimental setup for the equibiaxial compression tests.

In the equibiaxial compression tests, we consider the values of stretch and stress at the central part of sample, a square with a 70 mm side. The values for stress and stretch for both tests are presented in Table 1.

Table 1. Values of experimental tests of uniaxial and equibiaxial compression.

Uniaxial Compression		Equibiaxial Compression	
Stretch	**Stress [MPa]**	**Stretch**	**Stress [MPa]**
0.99990	−0.00267	0.99983	−0.00154
0.98716	−0.08793	0.96683	−0.18431
0.97331	−0.15326	0.93350	−0.31465
0.91794	−0.25836	0.90016	−0.38643
0.86256	−0.32187	0.86683	−0.45608
0.80718	−0.38418	0.83350	−0.52095
0.77949	−0.41763	0.81683	−0.54994
0.72411	−0.48701	0.78350	−0.61324
0.69642	−0.52242	0.76683	−0.65323
0.64104	−0.59524	0.73350	−0.73154
0.61335	−0.63417	0.71683	−0.77041
0.55797	−0.72245	0.68350	−0.86132
0.53028	−0.77436	0.66683	−0.92065
0.47490	−0.90147	0.63350	−1.06085
0.44721	−0.98206	0.61683	−1.14073
0.39183	−1.19875	0.58350	−1.38921
0.36414	−1.34949	0.56683	−1.59243
0.33645	−1.54421		
0.30876	−1.80366		
0.25339	−2.65663		

4.4. Curve Fitting

To parametrize the material we use the Wolfram Mathematica software through the function Nminimize. This function implements four methods that don't need derivatives for minimization: Nelder–Mead, Simulated Annealing, Differential Evolution and Random Search. The results presented by the four were very similar with a slight advantage for the Nelder–Mead, used in this work.

This algorithm tries to find a maximum or minimum of an objective function by doing a direct search. This search uses simplexes as input data for the objective function and evaluates how close the minimum of this function is. One of its features is that the derivatives of the function may not be known.

As we have to consider both, uni and equibiaxial tests, our objective function has to take both experiments into account. Let us define T_U^{Exp}, Λ_U^{Exp}, T_B^{Exp}, Λ_B^{Exp} as stress and stretch in uniaxial test and stress and stretch in equibiaxial test, respectively. Therefore, our objective function is given by

$$\Theta(s_1,\ldots,s_6) = Q_1(s_1,\ldots,s_6) + Q_2(s_1,\ldots,s_6), \tag{43}$$

with,

$$\begin{aligned} Q_1(s_1,\ldots,s_6) &= \sum_{r=1}^{R}\left[\frac{T_{U_r}^{Exp} - T_{U1}(\Lambda_{U_r}^{Exp})}{T_{U_r}^{Exp}}\right]^2, \\ Q_2(s_1,\ldots,s_6) &= \sum_{s=1}^{S}\left[\frac{T_{B_s}^{Exp} - T_{B1}(\Lambda_{B_s}^{Exp})}{T_{B_s}^{Exp}}\right]^2, \end{aligned} \tag{44}$$

where R, S are the number of experimental data for each case and $T_{U1}(\Lambda_{U_r}^{Exp})$ and $T_{B1}(\Lambda_{B_s}^{Exp})$ are the uniaxial and biaxial stress calculated, respectively, by (38) and (42), considering experimental deformations $\Lambda_{U_r}^{Exp}$ and $\Lambda_{B_s}^{Exp}$. Q_1 and Q_2 are the normalized error function for each of the tests and both depend on the parameters $s_1, \ldots, s_6$. This objective function was based on that used in the [24].

In Table 2, we present the obtained values for each parameter. The experimental and analytical curves, using the obtained parameters, are presented in Figure 5.

Table 2. Obtained material parameters.

Parameter	Value
s_1	−25.860100
s_2	2.7356100
s_3	19.317400
s_4	0.0407647
s_5	−8.3524100
s_6	−2.5714300

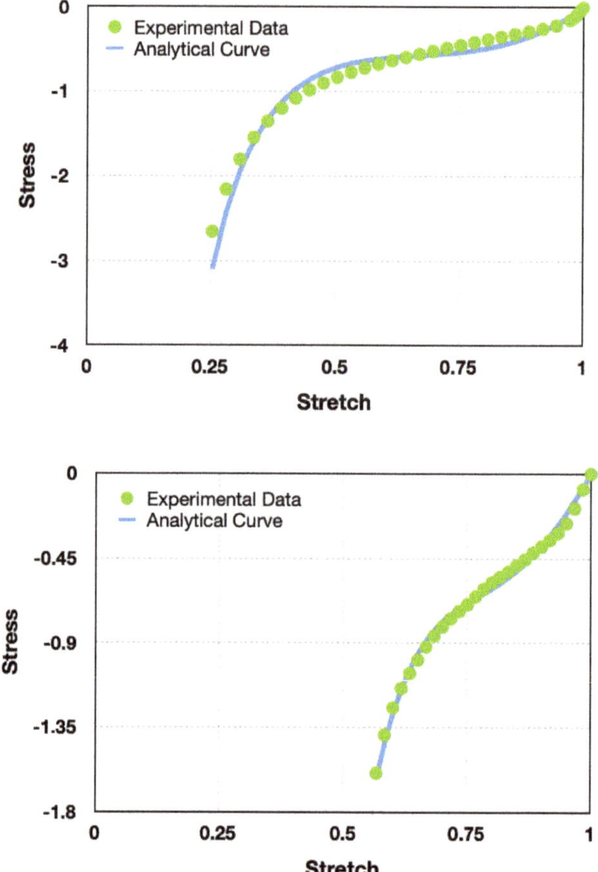

Figure 5. Comparison between analytical and experimental results for uniaxial and equibiaxial tests, respectively.

5. Linearized Partial Differential Equations

Let $\Omega = \{x \in \kappa_t(\mathcal{B})\} \subset \mathbb{R}^3$ be the region occupied by the body at the present configuration κ_t, and let $\partial\Omega = \Gamma_1 \cup \Gamma_2$ be the disjoint unions of its boundary. Let $\boldsymbol{n}(x,t)$ be the exterior unit normal to $\partial\Omega$ at the present time.

At time $\tau > t$, we shall consider an initial boundary value problem in Lagrangian formulation, with the present state at time t as the reference configuration, given by

$$\rho(t)\,\ddot{\boldsymbol{u}}_t(\tau) - \operatorname{div}_x T_t(\tau) = \rho(t)\boldsymbol{b}(\tau), \quad \text{in } \Omega, \tag{45}$$

$$T_t(\tau)\,\boldsymbol{n}(t) = \boldsymbol{f}(\tau), \quad \text{on } \Gamma_1, \tag{46}$$

$$\boldsymbol{u}_t(\tau) = \boldsymbol{d}(\tau), \quad \text{on } \Gamma_2, \tag{47}$$

$$\boldsymbol{u}_t(t) = 0, \quad \dot{\boldsymbol{u}}_t(t) = \boldsymbol{v}(t), \quad \text{in } \Omega. \tag{48}$$

The body is subjected to the surface traction $\boldsymbol{f}(x,\tau)$, the boundary displacement $\boldsymbol{d}(x,\tau)$ on the respective parts of $\partial\Omega$ at time $\tau > t$, and the initial velocity $\boldsymbol{v}(x,t)$ in Ω at the present time t. Note that unlike the explicit time dependence in the above expressions, the spatial dependence is implicitly understood and is not explicitly indicated for simplicity.

In these relations, for simplicity, div_x stands for the divergence operator relative to the coordinate $x \in \kappa_t(\mathcal{B})$, which is the same as the operator $\operatorname{div}_{\kappa_t}$ for the reference configuration κ_t in this case.

Together with the constitutive Equation (23), the mechanical problem is to be solved for the relative displacement vector $\boldsymbol{u}_t(\tau)$. Since the constitutive function $\mathbf{T}$ in (23) is generally nonlinear for finite deformations, the partial differential equation of this problem is genuinely nonlinear. However, in the relative Lagrangian formulation, for small enough incremental time $\Delta t = \tau - t$, we can easily linearize the constitutive equations relative to the present state at time t, so that the boundary value problem becomes linear.

By regarding the present state as the reference state, it is assumed that the state variables are all known functions at the present time t. Those include the deformation gradient $F(t)$ and the stress $T(t)$.

By use of the linearization (14), the partial differential equation of the problem become [26]

$$\rho(t)\,\ddot{\boldsymbol{u}}_t(\tau) - \operatorname{div}_x\Big(\mathcal{K}(t)[\nabla_x \boldsymbol{u}_t(\tau)]\Big) = \operatorname{div}_x T(t) + \rho(t)\boldsymbol{b}(\tau), \tag{49}$$

where the relevant material function $\mathcal{K}$ is defined in (15) and (25).

The Equation (49) is a linear partial differential equation for the relative displacement vector $\boldsymbol{u}_t(x,\tau)$ to be solved with the corresponding initial boundary conditions in the problems (45), for which the state variables of the body at time t are known and the external supplies $\boldsymbol{b}(\tau)$ is given so that the right-hand side of the Equation (49) is known quantity.

In this linearization, we do not assume the deformation is small, rather only the relative displacement gradient with respect to the present state is assumed to be small. This is the idea of "small-on-large", the same as the well-known problem of small deformations superposed on finite deformation in the literature [27]. Therefore, the overall deformation may be of finite values, in contrast to the usual theory of linear elasticity which linearizes the problem with respect to the fixed reference configuration assuming a small deformation gradient.

6. Variational Formulation

Consider

$$V = \{\boldsymbol{v} \in \left(H^1(\Omega)\right)^n; \boldsymbol{v} = 0, \quad \text{on} \quad \Gamma_2\}.$$

Lets $\boldsymbol{u}, \boldsymbol{v} \in V$. By multiplying Equation (49) for $\boldsymbol{v}$, integrating it by parts over Ω and using the definition of the inner product of second-order tensors (if $\boldsymbol{A}$ and $\boldsymbol{B}$ are second-order tensors, $\boldsymbol{A}.\boldsymbol{B} = \mathrm{tr}(\boldsymbol{A}\boldsymbol{B}^T)$), we have, $\forall \boldsymbol{v} \in V$,

$$\begin{aligned} &\int_\Omega \rho(t)\, \ddot{\boldsymbol{u}}_t(\tau) \cdot \boldsymbol{v} d\Omega + \int_\Omega tr\Big[\Big(\mathcal{K}(t)[\nabla_x \boldsymbol{u}_t(\tau)]\Big)\nabla_x \boldsymbol{v}^T\Big] d\Omega \\ &\quad - \int_{\partial\Omega=\Gamma_1} \boldsymbol{v} \cdot \Big(\mathcal{K}(t)[\nabla_x \boldsymbol{u}_t(\tau)]\Big)\boldsymbol{n}_\kappa d\Gamma = \int_\Omega \mathrm{div}_x\, \boldsymbol{T}_e(t) \cdot \boldsymbol{v} d\Omega \\ &\quad + \int_\Omega \rho(t)\, \boldsymbol{b}(\tau) \cdot \boldsymbol{v} d\Omega. \end{aligned} \tag{50}$$

By the boundary condition,

$$\begin{aligned} - \int_{\Gamma_1} \boldsymbol{v} \cdot \Big(\mathcal{K}(t)[\nabla_x \boldsymbol{u}_t(\tau)]\Big)\boldsymbol{n}_\kappa d\Gamma &= - \int_{\Gamma_1} \boldsymbol{v} \cdot \boldsymbol{f}(\tau) d\Gamma \\ &\quad - \int_{\Gamma_1} \boldsymbol{v} \cdot \boldsymbol{T}_e(t) d\Gamma, \qquad \forall \boldsymbol{v} \in V. \end{aligned} \tag{51}$$

Replacing (51) in (50) and using, again, integration by parts, we have the weak formulation,

$$\begin{aligned} &\int_\Omega \rho(t)\, \ddot{\boldsymbol{u}}_t(\tau) \cdot \boldsymbol{v} d\Omega + \int_\Omega tr\Big[\Big(\mathcal{K}(t)[\nabla_x \boldsymbol{u}_t(\tau)]\Big)\nabla_x \boldsymbol{v}^T\Big] d\Omega = \\ &\int_\Omega \rho(t)\, \boldsymbol{b}(\tau) \cdot \boldsymbol{v} d\Omega - \int_\Omega \boldsymbol{T}_e(t) \cdot \nabla_x \boldsymbol{v} d\Omega \quad + \int_{\Gamma_1} \boldsymbol{v} \cdot \boldsymbol{f}(\tau) d\Gamma, \quad \forall \boldsymbol{v} \in V, \end{aligned} \tag{52}$$

or we can rewrite this in terms of the bilinear forms

$$\begin{aligned} \Big(\rho(t)\, \ddot{\boldsymbol{u}}_t(\tau), \boldsymbol{v}\Big) + a\Big(\mathcal{K}(t)[\nabla_x \boldsymbol{u}_t(\tau)], \boldsymbol{v}\Big) &= \Big(\rho(t)\, \boldsymbol{b}(\tau), \boldsymbol{v}\Big) - \\ \Big(\boldsymbol{T}_e(t), \nabla_x \boldsymbol{v}\Big) + \Big(\boldsymbol{v}, \boldsymbol{f}(\tau)\Big)_{\Gamma_1}&, \qquad \forall \boldsymbol{v} \in V. \end{aligned} \tag{53}$$

Let us consider V^h a finite subspace of V. By restricting the formulation (53) to the space V^h we have

$$\begin{aligned} \Big(\rho(t)\, \ddot{\boldsymbol{u}}_t^h(\tau), \boldsymbol{v}^h\Big) + a\Big(\mathcal{K}(t)[\nabla_x \boldsymbol{u}_t^h(\tau)], \boldsymbol{v}^h\Big) &= \Big(\rho(t)\, \boldsymbol{b}(\tau), \boldsymbol{v}^h\Big) - \\ \Big(\boldsymbol{T}_e(t), \nabla_x \boldsymbol{v}^h\Big) + \Big(\boldsymbol{v}^h, \boldsymbol{f}(\tau)\Big)_{\Gamma_1}&, \qquad \forall \boldsymbol{v}^h \in V^h. \end{aligned} \tag{54}$$

Now consider $\{\varphi_1, \varphi_2, \cdots, \varphi_n\}$ a basis of the subspace V^h, that is, all elements $u^h \in V^h$ can be expressed as

$$u^h = \sum_{i=1}^n b_i \varphi_i. \tag{55}$$

Replacing in (54) u_h by (55) and taking $v^h = \varphi_j$, $1 \leq j \leq n$, we have

$$\mathcal{A}\ddot{\boldsymbol{b}} + \mathcal{L}\boldsymbol{b} = \mathcal{N}, \tag{56}$$

where, for $1 \leq i, j \leq n$

$$\begin{aligned} \mathcal{A}_{ij} &= \Big(\rho(t)\, \varphi_i, \varphi_j\Big), \\ \mathcal{L}_{ij} &= a\Big(\mathcal{K}(t)[\nabla_x \varphi_i], \varphi_j\Big), \\ \mathcal{N}_j &= \Big(\rho(t)\, \boldsymbol{b}(\tau), \varphi_j\Big) - \Big(\boldsymbol{T}_e(t), \nabla_x \varphi_j\Big) + \Big(\varphi_j, \boldsymbol{f}(\tau)\Big)_{\Gamma_1}. \end{aligned} \tag{57}$$

7. Numerical Validation

For the numerical simulation, as the two cases studied are examples of large deformations, the SLA method that was presented in Section 2 was used. We consider the material model developed in Section 3 with the parameters defined in Section 4, Table 2. In both cases, we are interested in the deformation after each loading step. Thereby, we are considering quasi-static, time-independent problems.

Regarding the implementation, all the finite element codes were developed by the author. The solver of the linear algebra problem that naturally arises in the finite element context was also implemented by the author. This decision was taken, principally, to solve the linear system with a good algorithm that takes into account the finite element matrix. The Gaussian quadrature was the integration method adopted. All the code is developed in an object-oriented programming language, C++.

For uniaxial compression, we consider a mesh of 50 mm $\times$ 100 mm, with 10 elements in the x-direction and 10 elements in the y-direction. The boundary condition applied considered that a displacement was prescribed on the upper surface and the lateral surface is free, while the base is free horizontally but not vertically. It applied a prescribed displacement equal to 7 mm in each load step.

In the case of the equibiaxial test, we consider a mesh of 200 mm $\times$ 200 mm, with 10 elements in the x-direction and 10 elements in the y-direction. The symmetry of the problem on the x and y axes was considered, and with only a quarter of the model was modeled. Still considering the symmetry, equal displacements were prescribed on both x and y surfaces on the top and the right side, while the left was free to move vertically and the bottom was free to move horizontally. In this case, the prescribed displacement in each load step is equal to 10 mm both in x and y directions.

For both tests, we use 100 steps of SLA and Q4 element, a bilinear quadrilateral element which combines two sets of Lagrange polynomials, i.e., linear isoparametric quadrilateral elements with four nodes. Figure 6 shows the finite element mesh and the boundary conditions for both cases.

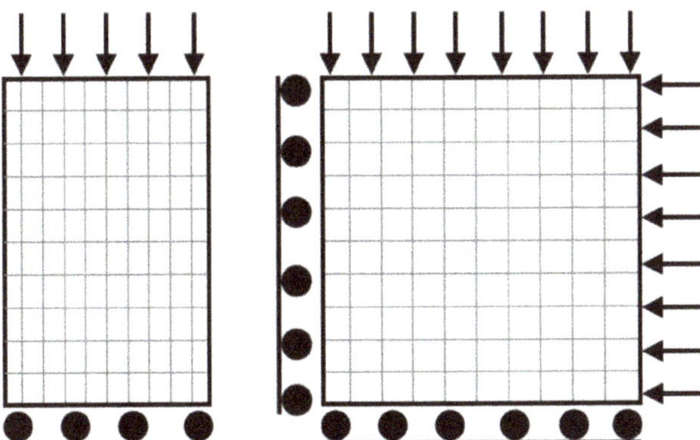

Figure 6. Finite element mesh and boundary condition for both tests, uniaxial (**left**) and equibiaxial (**right**).

In Figure 7, we can see the results obtained for the two cases studied. From left to right, we can see the comparison between the numerical and analytical results of the equibiaxial and uniaxial tests. As expected, the equibiaxial test has a smaller plateau area and densification occurs earlier than in the uniaxial test.

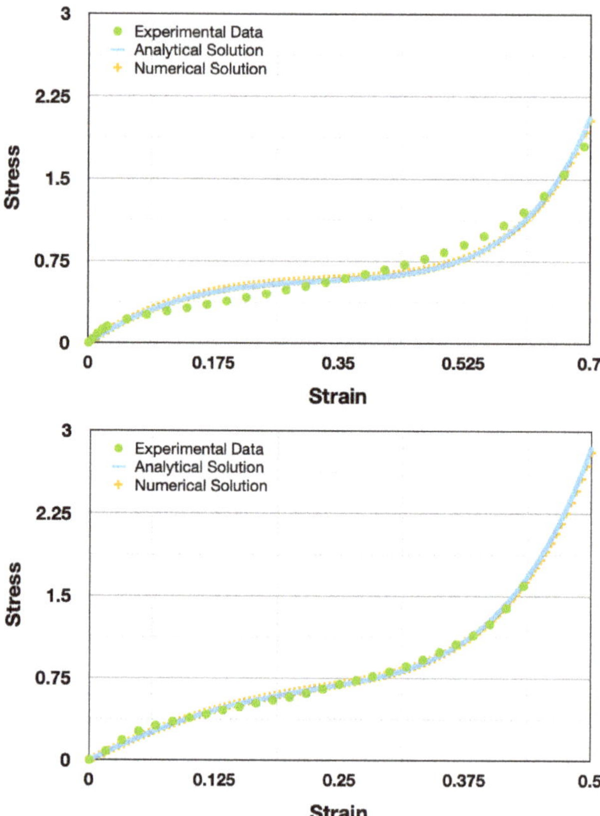

Figure 7. Comparison between numerical, analytical and experimental solutions for uniaxial and equibiaxial compression tests, respectively.

These two simulations show that the agglomerated cork model presented represents with good accuracy the real behavior of the material, and therefore, it is a good option to be studied from now on.

8. Conclusions and Future Work

In this paper, we present a new material model for agglomerated cork. This model is based on the Mooney–Rivlin hyperelastic model and, as the cork agglomerate has a Poisson ratio near zero, has six material parameters. Analytical solutions for uniaxial and equibiaxial tests were developed and used for parameterization. To parameterize, the Nelder–Mead algorithm was used.

We also describe and use the successive linear approximation method (SLA) for the numerical simulation. The description of the SLA was quite general, and can be used for any material.

Our results show that the presented model is a good model and was validated through numerical simulation. One of the advantages of our model is the mathematical simplicity in relation to the classical model [16,17], being a direct extension of a Mooney–Rivlin model.

For future work, it will be interesting to try to relate the parameters s_1, s_2, s_3, s_4, s_5 and s_6 with physical parameters of the material, as was performed directly with the β and the Poisson ratio. Another interesting approach to test the efficiency of the SLA will be the study of a contact-impact problem.

Author Contributions: Conceptualization, G.T.d.A.P., R.J.A.d.S., I.-S.L., M.G.T. and F.A.O.F.; methodology, G.T.d.A.P., R.J.A.d.S., I.-S.L., M.G.T. and F.A.O.F.; formal analysis, G.T.d.A.P., I.-S.L. and M.G.T.; investigation, G.T.d.A.P. and M.G.T.; resources, R.J.A.d.S., M.G.T. and F.A.O.F.; data curation, G.T.d.A.P.; writing—original draft preparation, G.T.d.A.P.; writing—review and editing, G.T.d.A.P., R.J.A.d.S., I.-S.L., M.G.T. and F.A.O.F.; visualization, I.-S.L. and F.A.O.F.; supervision, R.J.A.d.S., I.-S.L. and M.G.T.; project administration, M.G.T. All authors have read and agreed to the published version of the manuscript.

Funding: The author GTAP would like to thank Coordenação de Aperfeiçoamento de Pessoal de Nível Superior (CAPES) for their support (grant number 88881.188598/2018-01). The authors FAOF and RJAS acknowledge the support given by Fundação para a Ciência e a Tecnologia (FCT), through the projects UIDB/00481/2020 and UIDP/00481/2020. In addition, CENTRO-01-0145-FEDER-022083—Centro Portugal Regional Operational Programme (Centro2020) under the PORTUGAL 2020 Partnership Agreement through the European Regional Development Fund.

Data Availability Statement: All data is reported within the article.

Conflicts of Interest: The authors declare no conflict of interest.

References

1. Dart, S.L.; Eugene, G. Elastic properties of cork. I. stress relaxation of compressed cork. *J. Appl. Phys.* **1946**, *17*, 314–318. [CrossRef]
2. Anjos, O.; Pereira, H.; Rosa, M.E. Effect of quality, porosity and density on the compression properties of cork. *Holz Als-Roh-Und Werkst.* **2008**, *66*, 295–301. [CrossRef]
3. Pereira, H.; Graça, J.; Baptista, C. The effect of growth-rate on the structure and compressive properties of cork. *IAWA Bull.* **1992**, *13*, 389–396. [CrossRef]
4. Fortes, M.A.; Nogueira, M.T. The Poisson effect in cork. *Mater. Sci. Eng. A* **1989**, *122*, 227–232 [CrossRef]
5. Delucia, M.; Catapano, A.; Montemurro, M.; Pailhes, J. Pre-stress state in cork agglomerates: Simulation of the compression moulding process. *Int. J. Mater. Form.* **2021**, *14*, 485–498. [CrossRef]
6. Delucia, M.; Catapano, A.; Montemurro, M.; Pailhès, J. Determination of the effective thermoelastic properties of cork-based agglomerates. *J. Reinf. Plast. Compos.* **2019**, *38*, 760–776. [CrossRef]
7. Rosa, M.E.; Osorio, J.; Gree, V. Torsion of cork under compression. *Mater. Sci. Forum* **2004**, *455–456*, 235–238 [CrossRef]
8. Rosa, M.E.; Fortes, M.A. Deformation and fracture of cork in tension. *J. Mater. Sci.* **1991**, *26*, 341–348. [CrossRef]
9. Castro, O.; Silva, J.M.; Devezas, T.; Silva, A.; Gil, L. Cork agglomerates as an ideal core material in lightweight structures. *Mater. Des.* **2010**, *31*, 425–432. [CrossRef]
10. Sarasini, F.; Tirillò, J.; Lampani, L.; Sasso, M.; Mancini, E.; Burgstaller, C.; Calzolari, A. Static and dynamic characterization of agglomerated cork and related sandwich structures. *Compos. Struct.* **2019**, *212*, 439–451. [CrossRef]
11. Ivañez, I.; Sánchez-Saez, S.; Garcia-Castillo, S.K.; Barbero, E.; Amaro, A.; Reis, P.N.B. High-velocity impact behaviour of damaged sandwich plates with agglomerated cork core. *Compos. Struct.* **2020**, *248*, 112520. [CrossRef]
12. Lakreb, N.; Bezzazi, B.; Pereira, H. Mechanical strength properties of innovative sandwich panels with expanded cork agglomerates. *Eur. J. Wood Wood Prod.* **2015**, *73*, 46–473. [CrossRef]
13. Sergi, C.; Tirillò, J.; Sarasini, F.; Barbero Pozuelo, E.; Sanchez-Saez, S.; Burgstaller, C. The Potential of Agglomerated Cork for Sandwich Structures: A Systematic Investigation of Physical, Thermal, and Mechanical Properties. *Polymers* **2019**, *11*, 2118. [CrossRef] [PubMed]
14. Pernas-Sánchez, J.; Artero-Guerrero, J.A.; Varas, D.; Teixeira-Dias, F. Cork Core Sandwich Plates for Blast Protection. *Appl. Sci.* **2020**, *10*, 5180. [CrossRef]
15. Gomez, A.; Barbero, E.; Sanchez-Saez, S. Modelling of carbon/epoxy sandwich panels with agglomerated cork core subjected to impact loads. *Int. J. Impact Eng.* **2022**, *159*, 104047. [CrossRef]
16. Ogden, R.W. Large deformation isotropic elasticity – on the correlation of theory and experiment for incompressible rubberlike solids. *Proc. R. Soc. Lond. Ser. A Math. Phys. Sci.* **1972**, *326*, 565–584. [CrossRef]
17. Hill, R. Aspects of invariance in solid mechanics. *Adv. Appl. Mech.* **1978**, *18*, 1–75.
18. Delucia, M.; Catapano, A.; Montemurro, M.; Pailhés, J. A stochastic approach for predicting the temperature-dependent elastic properties of cork-based composites. *Mech. Mater.* **2020**, *145*, 103399. [CrossRef]
19. Liu, I.-S.; Cipolatti, R.; Rincon, M.A. Sucessive Linear Approximation for Finite Elasticity. *Comput. Appl. Math.* **2010**, *3*, 465–478.
20. Truesdell, C.; Noll, W. *The Non-Linear Field Theories of Mechanics*, 3rd ed.; Springer: Berlin, Germany, 2004.
21. Liu, I.-S. Successive Linear Approximation for Boundary Value Problems of Nonlinear Elasticity in Relative-Descriptional Formulation. *Int. J. Eng. Sci.* **2011**, *49*, 635–645. [CrossRef]
22. Liu, I.-S.; Cipolatti, R.; Rincon, M.A.; Palermo, L.A. Successive Linear Approximation for Large Deformation—Instability of Salt Migration. *J. Elast.* **2014**, *114*, 19–39. [CrossRef]
23. Teixeira, M.G.; Liu, I.-S.; Rincon, M.A.; Cipolatti, R.A.; Palermo, L.A. The Influence of temperature on the formation of salt domes. *Int. J. Model. Simul. Pet. Ind.* **2014**, *8*, 35–41.

24. Kossa, A.; Berezvai, S. Novel strategy for the hyperelastic parameter fitting procedure of polymer foam materials. *Polym. Test.* **2016**, *53*, 149–155. [CrossRef]
25. Pereira, A.B.; Fernandes, F.A.O.; de Morais, A.B.; Maio, J. Biaxial Testing Machine: Development and Evaluation. *Machines* **2020**, *8*, 40. [CrossRef]
26. Teixeira, M.G.; Pereira, G.T.A.; Liu, I.-S. Relative lagrangian formulation in thermo-viscoelastic solid bodies. *Contin. Mech. Thermodunamics* **2021**, *33*, 2263–2277. [CrossRef]
27. Green, A.E.; Rivlin, R.S.; Shield, R.T. General theory of small elastic deformations superposed on finite deformations. *Arch. Ration. Mech. Anal* **1952**, *211*, 128–154.

MDPI
St. Alban-Anlage 66
4052 Basel
Switzerland
Tel. +41 61 683 77 34
Fax +41 61 302 89 18
www.mdpi.com

Mathematical and Computational Applications Editorial Office
E-mail: mca@mdpi.com
www.mdpi.com/journal/mca

www.ingramcontent.com/pod-product-compliance
Lightning Source LLC
LaVergne TN
LVHW070059170726
843515LV00007B/1574